Nanoporous Catalysts
for Biomass Conversion

Wiley Series in Renewable Resources

Series Editor:

Christian V. Stevens, Faculty of Bioscience Engineering, Ghent University, Belgium

Titles in the Series:

Sustainability Assessment of Renewables-Based Products: Methods and Case Studies
Jo Dewulf, Steven De Meester, Rodrigo A. F. Alvarenga

Cellulose Nanocrystals: Properties, Production and Applications
Wadood Hamad

Fuels, Chemicals and Materials from the Oceans and Aquatic Sources
Francesca M. Kerton, Ning Yan

Bio-Based Solvents
François Jérôme and Rafael Luque

Forthcoming Titles:

Biorefinery of Inorganics: Recovering Mineral Nutrients from Biomass and Organic Waste
Erik Meers, Gerard Velthof

The Chemical Biology of Plant Biostimulants
Danny Geelen

Biobased Packaging: Material, Environmental and Economic Aspects
Mohd Sapuan Salit, Muhammed Lamin Sanyang

Thermochemical Processing of Biomass: Conversion into Fuels, Chemicals and Power, 2nd Edition
Robert C. Brown

Nanoporous Catalysts for Biomass Conversion

Edited by

FENG-SHOU XIAO AND LIANG WANG
Zhejiang University, Hangzhou, China

Registered Office(s)
John Wiley & Sons, Inc., 111 River Street, Hoboken, NJ 07030, USA
John Wiley & Sons Ltd, The Atrium, Southern Gate, Chichester, West Sussex, PO19 8SQ, UK

Editorial Office
9600 Garsington Road, Oxford, OX4 2DQ, UK

For details of our global editorial offices, customer services, and more information about Wiley products visit us at www.wiley.com.

Wiley also publishes its books in a variety of electronic formats and by print-on-demand. Some content that appears in standard print versions of this book may not be available in other formats.

Library of Congress Cataloging-in-Publication Data

Names: Xiao, Feng-Shou, 1963- editor. | Wang, Liang, 1986- editor.
Title: Nanoporous catalysts for biomass conversion / edited by Feng-Shou
    Xiao, Zhejiang University, Hangzhou, China, Liang Wang, Zhejiang
    University, Hangzhou, China.
Other titles: Catalysts for biomass conversion
Description: First edition. | Hoboken, NJ : Wiley, 2017. | Includes
    bibliographical references and index. |
Identifiers: LCCN 2017013229 (print) | LCCN 2017013683 (ebook) | ISBN
    9781119128090 (pdf) | ISBN 9781119128106 (epub) | ISBN 9781119128083
    (cloth)
Subjects: LCSH: Biomass conversion. | Porous materials. | Catalysts. |
    Nanopores. | Catalysis.
Classification: LCC TP248.B55 (ebook) | LCC TP248.B55 N36 2017 (print) | DDC
    620.1/16–dc23
LC record available at https://lccn.loc.gov/2017013229

Cover design by Wiley
Cover image: © georgeclerk/Gettyimages; (Top) © LEONELLO CALVETTI/Gettyimages; (Bottom Left) Ingram Publishing/Alamy Stock Photo

Set in 10/12pt TimesLTStd by SPi Global, Chennai, India
Printed and bound in Malaysia by Vivar Printing Sdn Bhd
10 9 8 7 6 5 4 3 2 1

Contents

List of Contributors

Yong Cao   Department of Chemistry, Shanghai Key Laboratory of Molecular Catalysis and Innovative Materials, Fudan University, China

Sheng Dai   Department of Chemistry, The University of Tennessee, USA

Chi-Linh Do-Thanh   Department of Chemistry, The University of Tennessee, USA

Atsushi Fukuoka   Institute for Catalysis, Hokkaido University, Japan

Emiel J.M. Hensen   Laboratory of Inorganic Materials Chemistry, Schuit Institute of Catalysis, Eindhoven University of Technology, The Netherlands

Xiaoming Huang   Laboratory of Inorganic Materials Chemistry, Schuit Institute of Catalysis, Eindhoven University of Technology, The Netherlands

Hirokazu Kobayashi   Institute for Catalysis, Hokkaido University, Japan

Jiechen Kong   Shanghai Key Laboratory of Green Chemistry and Chemical Processes, School of Chemistry and Molecular Engineering, East China Normal University, China

Tamás I. Korányi   Laboratory of Inorganic Materials Chemistry, Schuit Institute of Catalysis, Eindhoven University of Technology, The Netherlands

Changzhi Li   State Key Laboratory of Catalysis, Dalian Institute of Chemical Physics, Chinese Academy of Sciences, China

Guangyi Li   State Key Laboratory of Catalysis, Dalian Institute of Chemical Physics, Chinese Academy of Sciences, China

Ning Li   State Key Laboratory of Catalysis, Dalian Institute of Chemical Physics, Chinese Academy of Sciences, China

Shu-Shuang Li   Department of Chemistry, Shanghai Key Laboratory of Molecular Catalysis and Innovative Materials, Fudan University, China

Xin-Hao Li   School of Chemistry and Chemical Engineering, Shanghai Jiao Tong University, China

Yao Lin   Department of Chemistry and Institute of Materials Science, University of Connecticut, USA

Fei Liu   State Key Laboratory of Catalysis, Dalian Institute of Chemical Physics, Chinese Academy of Sciences, China

Fujian Liu   Department of Chemistry, Shaoxing University, China

Yong-Mei Liu   Department of Chemistry, Shanghai Key Laboratory of Molecular Catalysis and Innovative Materials, Fudan University, China

Zhicheng Luo   Shanghai Key Laboratory of Green Chemistry and Chemical Processes, School of Chemistry and Molecular Engineering, East China Normal University, China

Xiangju Meng   Key Laboratory of Applied Chemistry of Zhejiang Province, Department of Chemistry, Zhejiang University, China

Jifeng Pang   State Key Laboratory of Catalysis, Dalian Institute of Chemical Physics, Chinese Academy of Sciences, China

Abhijit Shrotri   Institute for Catalysis, Hokkaido University, Japan

Hui Su   School of Chemistry and Chemical Engineering, Shanghai Jiao Tong University, China

Lei Tao   Department of Chemistry, Shanghai Key Laboratory of Molecular Catalysis and Innovative Materials, Fudan University, China

Noritatsu Tsubaki   Department of Applied Chemistry, School of Engineering, University of Toyama, Japan

Aiqin Wang   State Key Laboratory of Catalysis, Dalian Institute of Chemical Physics, Chinese Academy of Sciences, China

Hong-Hui Wang   School of Chemistry and Chemical Engineering, Shanghai Jiao Tong University, China

Liang Wang   Key Laboratory of Applied Chemistry of Zhejiang Province, Department of Chemistry, Zhejiang University, China

Yanqin Wang   Shanghai Key Laboratory of Functional Materials Chemistry, Research Institute of Industrial Catalysis, School of Chemistry and Molecular Engineering, East China University of Science and Technology, China

Liubi Wu   Shanghai Key Laboratory of Green Chemistry and Chemical Processes, School of Chemistry and Molecular Engineering, East China Normal University, China

Qineng Xia   Shanghai Key Laboratory of Functional Materials Chemistry, Research Institute of Industrial Catalysis, School of Chemistry and Molecular Engineering, East China University of Science and Technology, China

Feng-Shou Xiao   Key Laboratory of Applied Chemistry of Zhejiang Province, Department of Chemistry, Zhejiang University, China

Chuang Xing   School of Biological and Chemical Engineering, Zhejiang University of Science and Technology, China

Jinming Xu   State Key Laboratory of Catalysis, Dalian Institute of Chemical Physics, Chinese Academy of Sciences, China

Shaodan Xu   Key Laboratory of Applied Chemistry of Zhejiang Province, Department of Chemistry, Zhejiang University, China

Guohui Yang   Department of Applied Chemistry, School of Engineering, University of Toyama, Japan

Ruiqin Yang   School of Biological and Chemical Engineering, Zhejiang University of Science and Technology, China

Tao Zhang   State Key Laboratory of Catalysis, Dalian Institute of Chemical Physics, Chinese Academy of Sciences, China

Chen Zhao   Shanghai Key Laboratory of Green Chemistry and Chemical Processes, School of Chemistry and Molecular Engineering, East China Normal University, China

Tian-Jian Zhao   School of Chemistry and Chemical Engineering, Shanghai Jiao Tong University, China

Xiaochen Zhao   State Key Laboratory of Catalysis, Dalian Institute of Chemical Physics, Chinese Academy of Sciences, China

Mingyuan Zheng   State Key Laboratory of Catalysis, Dalian Institute of Chemical Physics, Chinese Academy of Sciences, China

Xiang Zhu   Department of Chemistry, The University of Tennessee, USA

Series Preface

Renewable resources, their use and modification are involved in a multitude of important processes with a major influence on our everyday lives. Applications can be found in the energy sector, paints and coatings, and the chemical, pharmaceutical, and textile industry, to name but a few.

The area interconnects several scientific disciplines (agriculture, biochemistry, chemistry, technology, environmental sciences, forestry ...), which makes it very difficult to have an expert view on the complicated interaction. Therefore, the idea to create a series of scientific books that will focus on specific topics concerning renewable resources, has been very opportune and can help to clarify some of the underlying connections in this area.

In a very fast changing world, trends are not only characteristic for fashion and political standpoints; science is also not free from hypes and buzzwords. The use of renewable resources is again more important nowadays; however, it is not part of a hype or a fashion. As the lively discussions among scientists continue about how many years we will still be able to use fossil fuels – opinions ranging from 50 to 500 years – they do agree that the reserve is limited and that it is essential not only to search for new energy carriers but also for new material sources.

In this respect, renewable resources are a crucial area in the search for alternatives for fossil-based raw materials and energy. In the field of energy supply, biomass and renewables-based resources will be part of the solution alongside other alternatives such as solar energy, wind energy, hydraulic power, hydrogen technology, and nuclear energy. In the field of material sciences, the impact of renewable resources will probably be even bigger. Integral utilization of crops and the use of waste streams in certain industries will grow in importance, leading to a more sustainable way of producing materials. Although our society was much more (almost exclusively) based on renewable resources centuries ago, this disappeared in the Western world in the nineteenth century. Now it is time to focus again on this field of research. However, it should not mean a "retour à la nature," but it should be a multidisciplinary effort on a highly technological level to perform research towards new opportunities, to develop new crops and products from renewable resources. This will be essential to guarantee a level of comfort for a growing number of people living on our planet. It is "the" challenge for the coming generations of scientists to develop more sustainable ways to create prosperity and to fight poverty and hunger in the world. A global approach is certainly favoured.

This challenge can only be dealt with if scientists are attracted to this area and are recognized for their efforts in this interdisciplinary field. It is, therefore, also essential that consumers recognize the fate of renewable resources in a number of products.

Furthermore, scientists do need to communicate and discuss the relevance of their work. The use and modification of renewable resources may not follow the path of the genetic engineering concept in view of consumer acceptance in Europe. Related to this aspect, the series will certainly help to increase the visibility of the importance of renewable resources. Being convinced of the value of the renewables approach for the industrial world, as well as for developing countries, I was myself delighted to collaborate on this series of books focusing on different aspects of renewable resources. I hope that readers become aware of the complexity, the interaction and interconnections, and the challenges of this field and that they will help to communicate on the importance of renewable resources.

I certainly want to thank the people of Wiley's Chichester office, especially David Hughes, Jenny Cossham and Lyn Roberts, in seeing the need for such a series of books on renewable resources, for initiating and supporting it, and for helping to carry the project to the end.

Last, but not least, I want to thank my family, especially my wife Hilde and children Paulien and Pieter-Jan, for their patience and for giving me the time to work on the series when other activities seemed to be more inviting.

Christian V. Stevens,
Faculty of Bioscience Engineering
Ghent University, Belgium
Series Editor 'Renewable Resources'
June 2005

Acknowledgements

We would like to thank the National Natural Science Foundation of China for the constant encouragement and financial support (NO. 91634201, 21333009, 21403192, 91645105, and U1462202) to our investigation in *Nanoporous Catalyst Synthesis and Biomass Conversion*.

We are also grateful to Shagun Chaudhary and Emma Strickland, from Wiley, whose great patience was much appreciated in 'polishing' the text of the book.

1

Nanoporous Organic Frameworks for Biomass Conversion

Xiang Zhu, Chi-Linh Do-Thanh, and Sheng Dai

Department of Chemistry, The University of Tennessee, USA

1.1 Introduction

Porosity, a profound concept that helps to understand Nature and create novel fascinating architectures, is inherent to natural processes, as seen in hollow bamboo, hexagonal honeycomb, and the alveoli in the lungs (Figure 1.1) [1,2]. These advanced natural porous frameworks and their promising applications have widely inspired scientists with the idea of mimicking them in artificial structures down to the micro- and nanoscale range [1,2]. The rational design and synthesis of advanced nanoporous materials, which play a crucial role in established processes such as catalysis and gas storage and separations and catalysis [3–12], have long been an important science subject and attracted tremendous attention. During the past two decades, the linking of molecular scaffolds by covalent bonds to create crystalline extended structures has afforded a broad family of novel nanoporous crystalline structures [13] such as like metal–organic frameworks (MOFs) [14] and covalent organic frameworks (COFs) [15]. The key advance in this regard has been the versatility of covalent chemistry and organic synthesis techniques, which give rise to a wide variety of target applications for these extended organic frameworks, for example, the use of MOFs and COFs in the context of biomass conversion. In addition to crystalline frameworks, nanoporous organic resins have long been extensively studied as heterogeneous catalysts for the conversion of biomass because of their commercial synthesis [16].

Upgrading biomass into fuel and fine chemicals has been considered a promising renewable and sustainable solution to replacing petroleum feedstocks, owing to the

Nanoporous Catalysts for Biomass Conversion, First Edition. Edited by Feng-Shou Xiao and Liang Wang.
© 2018 John Wiley & Sons Ltd. Published 2018 by John Wiley & Sons Ltd.

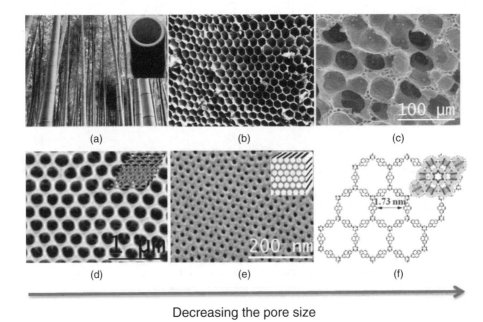

(a)   (b)   (c)

(d)   (e)   (f)

Decreasing the pore size

Figure 1.1 *Illustration of porosity existing in Nature and synthesized frameworks with a decreasing pore size. (a) Bamboo; (b) honeycomb; (c) scanning electron microscopy (SEM) image of alveolar tissue in mouse lung; (d) SEM image of an ordered macroporous polymer; (e) SEM image of an ordered mesoporous polymer from self-assembly of block copolymers; (f) structural representation of the COF structure. (See color plate section for the color representation of this figure.)*

rich family of biomass raw materials, which mainly includes cellulose, hemicellulose, and lignin [17,18]. For example, the carbohydrates, present in the cellulosic and hemicellulosic parts of biomass, can be converted into renewable platform chemicals such as 5-hydroxymethylfurfural (HMF), via acid-catalyzed dehydration for the production of a wide variety of fuels and chemical intermediates [19]. Despite great progress, including unprecedented yields and selectivities, having been made in biomass conversion using conventional homogeneous catalysts, the cycling abilities have long been the main drawbacks that inhibit their large-scale applications. As a result, heterogeneous nanoporous solid catalysts hold great promise in these diverse reactions [16]. High porosities of nanoporous catalysts may help to access reactants, mass transfer, and functionalization of task-specific active sites, such that the product selectivities can be easily controlled. To this end, nanoporous materials with high surface areas, tunable pore sizes and controllable surface functionalities have been extensively prepared and studied. Significantly, nanoporous crystalline organic frameworks, with well-defined spatial arrangements where their properties are influenced by the intricacies of the pores and ordered patterns onto which functional groups can be covalently attached to produce chemical complexity, exhibit distinct advantages over other porous catalysts. For instance, post-synthetic modification (PSM) techniques [20] provide a means of designing the intrinsic pore environment without losing their long-range order to improve the biomass conversion performance. The inherent 'organic effect' enables the architectures to function with task-specific moieties such as the acidic sulfonic acid ($-SO_3H$) group. The desired microenvironment can also be generated by rationally modifying the organic building units or metal nodes. In addition, the attractive large porosity allows the frameworks to become robust solid supports to immobilize active units such as polyoxometalates and polymers [21]. In essence, nanoporous crystalline organic frameworks including MOFs and COFs have demonstrated strong potential as heterogeneous catalysts for biomass conversion [21]. The ability to reticulate task-specific functions into frameworks not only allows catalysis to be performed in a high-yield manner but also provides a means of facile control of product selectivity.

HMF, as a major scaffold for the preparation of furanic polyamides, polyesters, and polyurethane analogs, exhibits great promise in fuel and solvent applications [19,22]. The efficient synthesis of HMF from biomass raw materials has recently attracted major research efforts [23–29]. Via a two-step acyclic mechanism, HMF can be prepared from the dehydration of C-6 sugars such as glucose and fructose. First, glucose undergoes an isomerization to form fructose in the presence of either base catalysts or Lewis acid catalysts by means of an intramolecular hydride shift [30]. Subsequently, Brønsted acid-catalyzed dehydration of the resultant fructose affords the successful formation of HMF with the loss of three molecules of H_2O (Scheme 1.1) [31]. The development of novel nanoporous acidic catalysts for the catalytic dehydration of sugars to HMF is of great interest, and is highly desirable. Hence, design strategies for the construction of nanoporous crystalline organic frameworks that are capable of the efficient transformation of sugars to HMF are discussed in this chapter, and some nanoporous organic resins for the conversion of raw biomass

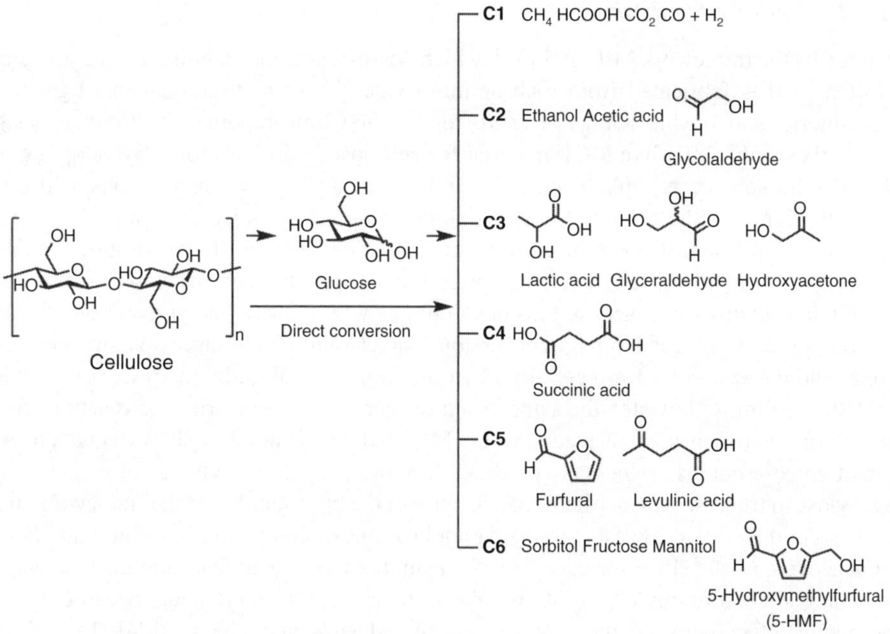

Scheme 1.1 *Possible valuable chemicals based on carbohydrate feedstock.*

materials are highlighted. By examining the common principles that govern catalysis for dehydration reactions, a systematic framework can be described that clarifies trends in developing nanoporous organic frameworks as new heterogeneous catalysts while highlighting any key gaps that need to be addressed.

1.2   Nanoporous Crystalline Organic Frameworks

1.2.1   Metal–Organic Frameworks

The Brønsted acidity of nanoporous catalysts is very essential for the dehydration of carbohydrates towards the formation of HMF [32–34]. One significant advantage of metal–organic frameworks (MOFs) is their highly designable framework, which gives rise to a versatility of surface features within porous backbones. Whereas, a wide variety of functional groups has been incorporated into MOF frameworks, exploring the Brønsted acidity of MOFs [14], the introduction of sulfonic acid groups in the framework remains a challenge and less explored, mainly because of the weakened framework stability. In this regard, several different synthetic techniques have been developed and adopted to introduce sulfonic acid ($-SO_3H$) groups for MOF-catalyzed dehydration processes: (i) de-novo synthesis using organic linkers with $-SO_3H$ moiety; (ii) pore wall engineering by the covalently postsynthetic modification [20] (PSM) route; and (iii) modification of the pore microenvironment through the introduction of additional active sites. These novel MOF materials featuring strong Brønsted acidity show great promise as solid nanoporous acid catalysts in biomass conversion.

1.2.1.1   De-Novo Synthesis

Inspired by the framework MIL-101 [35], which possesses strong stability in aqueous acidic solutions and is fabricated from a chromium oxide cluster and terephthalate ligands in hydrofluoric acid media, Kitagawa *et al.* for the first time reported the rational design and synthesis of a MIL-like MOF material for cellulose hydrolysis [36]. By, adopting the MIL-101 framework as a platform, these authors created a novel nanoporous acid catalyst with highly acidic $-SO_3H$ functions along the pore walls by the innovative use of 2-sulfoterephthalate instead of the unsubstituted terephthalate in MIL-101 (Figure 1.2). The resultant Cr-based MOF MIL-101-SO_3H was shown to exhibit a clean catalytic activity for the cellulose hydrolysis reaction, thus opening a new window on the preparation of novel nanoporous catalysts for biomass conversion. On account of the unsatisfactory yields of mono- and disaccharides from cellulose hydrolysis being caused by the poor solubility of crystalline cellulose in water, the same group further studied isomerization reactions from glucose to fructose in aqueous media, where MIL-101-SO_3H not only shows a high conversion of glucose but also selectively produces fructose [37]. A catalytic one-pot conversion of amylose to fructose was also achieved because of the high stability of the framework in an acidic solution, which suggests promising applications of compound in the biomass field.

On account of the HMF formation mechanism, the Lewis acid featuring metal center – for example, chromium (II) – allows for a high-yield isomerization because of the coordinate effect between the Lewis acidic metal sites and glucose [24]. In addition

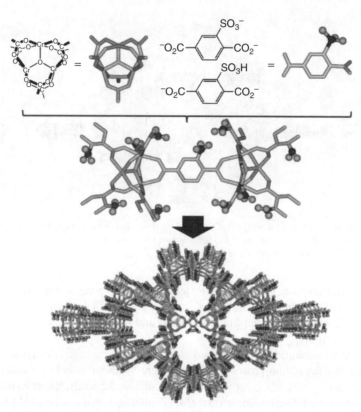

Figure 1.2 *Schematic representation of the structure of MIL-101-SO₃H. (See color plate section for the color representation of this figure.)*

to strong Brønsted acidity caused by the $-SO_3H$ moieties, MIL-101-SO_3H also bears Cr(III) sites within the structure, which is similar to $CrCl_2$ and may act as active sites for the isomerization of glucose to fructose, whereas the fructose dehydration can be initiated with the aid of $-SO_3H$ groups. Bao *et al.* carried out an integrated process using nanoporous MIL-101-SO_3H as the catalyst and biomass-derived solvent (γ-valerolactone; GVL) for the conversion of glucose into HMF (Figure 1.3) [38]. The batch heterogeneous reaction was shown to give a HMF yield of 44.9% and a selectivity of 45.8%. The glucose isomerization in GVL with 10 wt% water was found to follow second-order kinetics, with an apparent activation energy of 100.9 kJ mol^{-1} according to the reaction kinetics study. Clearly, the bifunctional MIL-101-SO_3H framework can serve as a potential platform for the dehydration reaction of biomass-derived carbohydrate to generate platform chemicals.

Solvents play another crucial role in the green biomass conversion processes, and the development of water-based heterogeneous systems for dehydration is important for the industrial reaction of fructose conversion to HMF [19]. Janiak *et al.* adopted MIL-101Cr (MIL-SO_3H) as the heterogeneous catalyst and achieved a 29% conversion of glucose to HMF in a THF : H_2O (39 : 1, v : v) mixture [39]. Recently, Du *et al.* reported a 99.9%

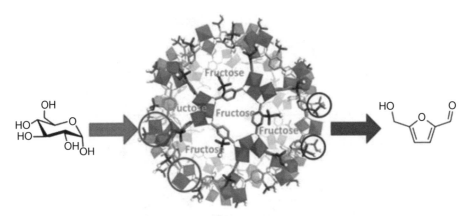

Figure 1.3   *Bifunctional catalyst MIL-101(Cr)-SO₃H used for glucose conversion to HMF. Reproduced with permission from Ref. [38]. Copyright 2016, American Institute of Chemical Engineers. (See color plate section for the color representation of this figure.)*

glucose conversion and an excellent HMF yield of 80.7% in water using a new bifunctional PCP(Cr)-SO₃HCr(III) material [40]. These authors showed that the sulfonic acid group in the framework was the essential function center, and more Lewis acid sites resulted in a better catalyst activity.

Despite the aforementioned nanoporous Cr-based MOF materials displaying promising applications in dehydration processes, the stringent synthetic conditions and toxic inorganic reagents of Cr MOFs greatly limit their real use. Recently, the research group of Zhao reported a modulated hydrothermal (MHT) approach that can be used to synthesize a series of highly stable Brønsted acidic NUS-6(Zr) and NUS-6(Hf) MOFs in a green and scalable way, although the linker 2-sulfotherephthalate was previously reported to make unstable UiO-type frameworks. The hafnium (Hf)-based material NUS-6(Hf) exhibited a superior performance for the dehydration of fructose to HMF (Figure 1.4) [41], outperforming all other presently known MOFs or heterogeneous catalysts with a yield of 98% under the same reaction conditions. Nevertheless, although such a high transformation yield can be achieved in organic dimethyl sulfoxide (DMSO), the attempts at dehydration with the same MOF material in aqueous media resulted in only negligible amounts of HMF (ca. 5%). Therefore, it is very valuable and significant to rationally design and develop stable Brønsted acidic MOF materials for high-performance dehydrations to prepare HMF in green aqueous media, even though this is an enormous challenge.

1.2.1.2   *Postsynthetic Modification*

In addition to direct synthesis using building linkers to modify MOF materials *in situ*, the PSM route is becoming an important technique to introduce various functions inside the MOFs for diverse applications, such as heterogeneous catalysis and gas storage and separation [20]. Therefore, PSM provides another means of introducing Brønsted acid groups to MOF backbones for the dehydration of biomasses to HMF. The presence of organic scaffolds in MOFs allows for a convenient employment of a variety of organic transformations. Furthermore, the acid strength on the surface can be precisely controlled

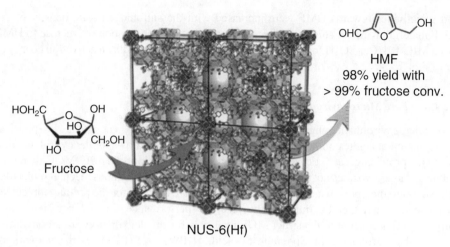

Figure 1.4 *NUS-6(Hf) used as a heterogeneous catalyst for fructose conversion to HMF. Reproduced by permission of Ref. [41]. Copyright 2012 American Chemical Society. (See color plate section for the color representation of this figure.)*

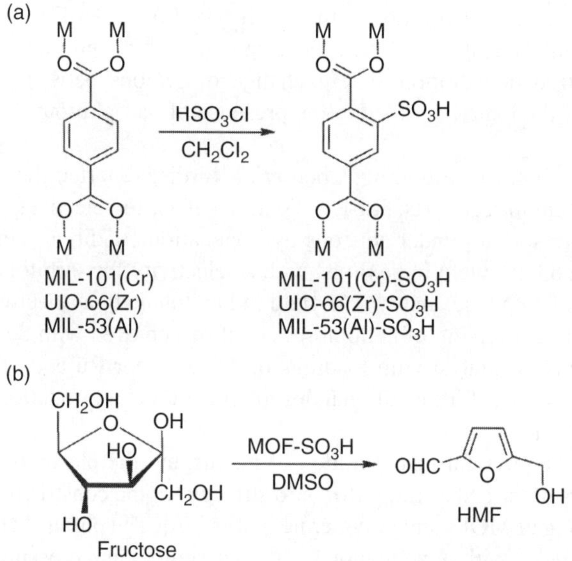

(a)

MIL-101(Cr)
UIO-66(Zr)
MIL-53(Al)

HSO_3Cl
CH_2Cl_2

MIL-101(Cr)-SO_3H
UIO-66(Zr)-SO_3H
MIL-53(Al)-SO_3H

SO_3H

(b)

Fructose

MOF-SO_3H
DMSO

HMF

Figure 1.5 *Synthetic routes to MOF-SO_3H and the conversion of fructose into HMF.*

by the PSM through varying the grafting rate of the reaction. As shown in Figure 1.5, Chen *et al.* reported the synthesis of a family of MOF frameworks functionalized with the –SO_3H group through the PSM of the organic linkers, using chlorosulfonic acid [42]. The resultant framework MIL-101(Cr) [MIL-101(Cr)-SO_3H] exhibited a full fructose conversion with a HMF yield of 90% in DMSO. The –SO_3H groups was found to have a significant effect on fructose-to-HMF transformation [42]. Both the conversions of fructose and selectivities towards HMF were increased with the sulfonic acid-site density of the MOF material. Kinetics studies further suggested that the dehydration of fructose to HMF using MIL-101(Cr)-SO$_3$H followed pseudo-first-order kinetics with an activation energy of 55 kJ mol^{-1}.

1.2.1.3 Pore Microenvironment Modification

Pore microenvironment engineering inside MOF frameworks via either the physical impregnation of acidic compounds or *in-situ* polymerization, allows for the hybrid material to obtain promising catalytic activities towards the formation of HMF. MOF materials featuring large porosities not only give rise to easy access of voluminous reactant molecules diffusing into the pores, but also provide an appropriate channel for the product molecules to move out, and thereby improve the catalytic performance. Li and coworkers first introduced this concept to construct MOF materials for the dehydration of carbohydrates to HMF by introducing phosphotungstic acid, H$_3$PW$_{12}$O$_{40}$ (PTA), to the network of MIL-101 [43]. PTA and other transition metal-substituted polyoxometalates have been widely encapsulated into MIL-101 for other catalysis processes, such as the oxidation of alkanes and alkenes and esterification of n-butanol with acetic acid in the liquid phase. Using ionic liquids as reaction media, PTA(3.0)/MIL-101 showed the best catalytic performance in fructose dehydration with a HMF yield of 79% [43]. The immobilization of PTA significantly changed the pore environment, which promoted the initiation of the proton-catalyzed dehydration. The exchange of protons between the PTA and the organic cations of the ionic liquid led to the presence of partial protons in the ionic liquid medium [43].

Inspired by this innovative finding, Chen *et al.* further reported the synthesis of PTA/MOF hybrid ruthenium catalysts, Ru-PTA/MIL-100(Cr), to selectively convert cellulose and cellobiose into sorbitol under aqueous hydrogenation conditions (Figure 1.6) [44]. It was shown that a 63.2% yield in hexitols, with a selectivity for sorbitol of 57.9% at complete conversion of cellulose, and 97.1% yield in hexitols with a selectivity for sorbitol of 95.1% at complete conversion of cellobiose, could be achieved with the creative use of a Ru-PTA/MIL-100(Cr) catalyst with loadings of 3.2 wt% for Ru and 16.7 wt% for PTA. This opened new avenues for the rational design of acid/metal bifunctional MOF catalysts for biomass conversion.

In addition to the physical incorporation of metals and the phosphotungstic acid, Hatton and coworkers described a straightforward strategy for the construction of novel hybrid materials consisting of MOFs and polymer networks [45]. The porous MIL-101(Cr) framework was impregnated with organic monomers (maleimide) and crosslinkers for an *in-situ* free-radical (co)polymerization (P), and thereby the functionalization (F) of the pore environment with poly(N-bromomaleimide) (Figure 1.7). Although the porosity of the composite material was decreased because of the presence of polymers, the pore of the MOF material was still large enough to enable ready access of the reactants to the active sites inside the framework. As a result, the resultant MOF-brominated polymer hybrids exhibited potential applications for the industrially important reaction of fructose conversion to HMF, suggesting an innovative and effective method for controlling the pore environment for target applications and thus promoting the development of MOF materials for biomass conversion.

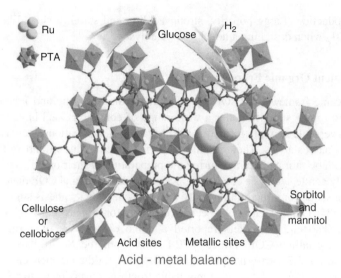

Figure 1.6 *Ru-PTA/MIL-100(Cr) used as a heterogeneous catalyst for biomass conversion. Reproduced with permission from Ref. [44]. Copyright 2013, Wiley.*

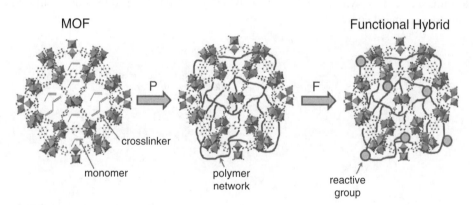

Figure 1.7 *Scheme of fabrication of functional hybrids of MOFs and polymer networks. Reprinted with permission from Ref. [45]. Copyright 2014, American Chemical Society.* (See color plate section for the color representation of this figure.)

Even though MOF compounds show significant advantages for biomass conversions over other heterogeneous catalysts such as zeolites and metal oxides, only a handful of studies have been reported dealing with the conversion of biomass using MOF catalysts. Herbst and Janiak recently highlighted the promising potential in the field of MOF-catalyzed biomass conversion [21]. These authors discussed the synthetic conditions of MOF materials, mechanisms of biomass reactions, product yields, and the corresponding selectivity in detail. Meanwhile, in-depth reviews by Jiang and Yaghi which focused on Brønsted acidic MOFs were recently published [14,46]. Clearly, there exist great opportunities and potential to address investigations into MOF catalysis in the context of biomass-based fine chemical production. Large porosity, strong acidity, and stability should be considered simultaneously when designing a novel MOF catalyst.

1.2.2   Covalent Organic Frameworks

Covalent organic frameworks (COFs) possess molecular ordering and inherent porosity in addition to robust stability under a wide range of conditions, and have recently been identified as versatile new platforms for a wide range of applications such as gas storage or heterogeneous catalysis. Their promising performance is due in part to the extent to which they can be rationally tuned during the preparation of their organic constituents, as comparatively gentle synthesis conditions are required for the final COF material [47–52]. Although the scope for structural diversity and high-yielding synthesis suggests potential advantages [53–61], few COFs have been investigated for biomass conversion [62].

Recently, Zhao and coworkers reported a de-novo synthesis of a sulfonated two-dimensional crystalline COF termed TFP-DABA from the Schiff base condensation reaction between 1,3,5-triformylphloroglucinol and 2,5-diaminobenzenesulfonic acid (Scheme 1.2) [63]. For the first time, remarkable fructose conversion yields (97% for HMF and 65% for 2,5-diformylfuran), good selectivity and recyclability were obtained using COF TFP-DABA as the heterogeneous catalyst. This innovative study paves a solid way towards the de-novo synthesis of novel catalytically active COFs and their applications in biobased chemical conversion.

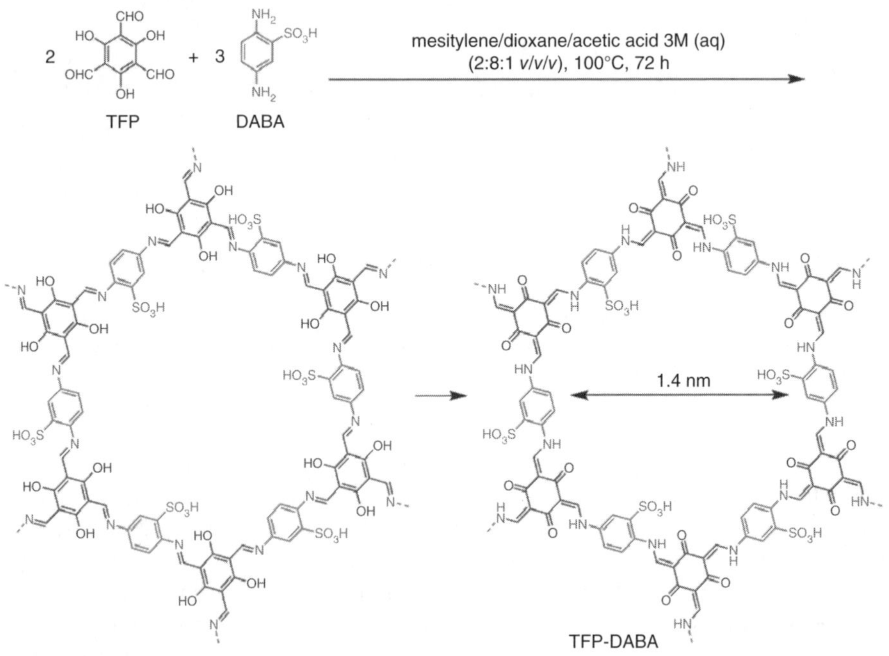

Scheme 1.2   *Scheme of fabrication of COF-SO_3H for the conversion of fructose into HMF.*

1.3 Nanoporous Organic Sulfonated Resins

Nanoporous organic sulfonated resins, such as Amberlyst® and Nafion®, have attracted tremendous attention and show significant promise for the conversion of biomass because of their facile synthesis and high catalytic activities [16]. The intrinsic inherent porosity allows for an easy installation of the Brønsted sulfonated group ($-SO_3H$) on the pore surface, thereby affording an enhanced interaction between the catalyst framework and the reactants. A high-performance transformation of sugars to HMF is therefore achieved.

1.3.1 Amberlyst Resins

A facile polymerization of styrene and further sulfonation results in the formation of one famous nanoporous resin catalyst, Amberlyst® 15 (Figure 1.8), that enables a wide variety of acid-catalyzed liquid-phase biomass conversions, for example the dehydration of fructose to HMF. Shimizu *et al.* reported a 100% yield of HMF in a water-separation reactor, using Amberlyst® 15 powder in a size of 0.15–0.053 mm [64]. The removal of adsorbed water by the small-sized resin particles resulted in an excellent yield, which was supported by the near-infrared spectroscopic characterization. It should be noted that the intrinsic nanoporous structure of Amberlyst® 15 features a relatively low stability that needs to be enhanced in future investigations.

1.3.2 Nafion Resins

Compared with Amberlyst® 15, Nafion® perfluorosulfonic acid resins show better stability in the nanoporous structure and bear strong acid strength that is comparable to that of pure sulfuric acid. However, unswollen Nafion® resin suffers from a serious drawback due to its very low surface area that may impede application in the conversion of sugars to HMF (Figure 1.9) [16]. Commercialized Nafion® NR50, with a pellet diameter of 0.6–1.5 mm, was used as a catalyst for the preparation of HMF from fructose, and a moderate HMF yield (45–68%) was obtained in the presence of ionic liquid 1-butyl-3-methyl imidazolium chloride ([BMIM][Cl]) [65]. The ionic liquid and sulfonic ion-exchange resin could be recycled for seven successive trials.

On account of these promising results and low surface areas of Nafion® resins, Shen and coworkers developed a novel Nafion®-modified mesocellular silica foam (MCF) hybrid material for the catalytic dehydration of D-fructose to HMF [66]. By using a physical

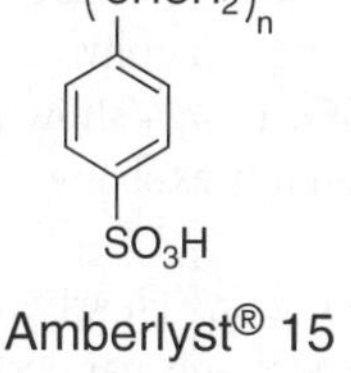

Amberlyst® 15

Figure 1.8 *Chemical structure of Amberlyst® 15.*

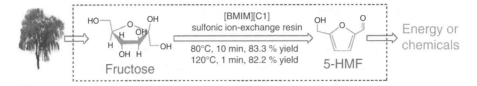

Figure 1.9   *Nafion® NR50 used for the conversion of fructose into HMF.*

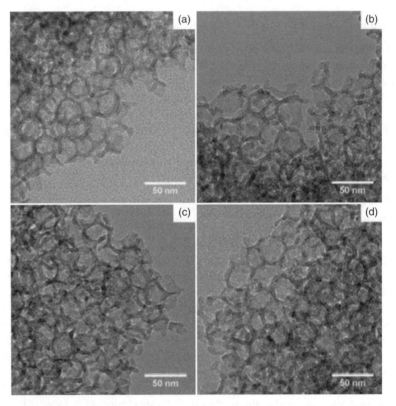

Figure 1.10   *Transmission electron microscopy (TEM) images of (a) MCF, (b) Nafion®(15)/MCF, (c) Nafion®(30)/MCF, and (d) Nafion®(45)/MCF.*

impregnation approach, the Nafion® resin was highly dispersed in the ultra-large pores of the MCFs (Figure 1.10), affording a high conversion performance with an 89.3% HMF yield and 95.0% selectivity in DMSO. It was shown that the Nafion®(15)/MCF catalyst could be easily regenerated through an ion-exchange method, and a high yield was retained after five cycles.

As discussed above, the sulfonated resins clearly hold promise as heterogeneous acidic catalysts for biomass conversion, where the nanoporous structure plays crucial roles in the acid concentration and/or strength of the resins. However, the fact must be noted that the sulfonated resin catalysts still have a relatively low thermal stability due to collapse of the porous architectures, which greatly hinders their wide application in biomass conversion.

1.4 Conclusions and Perspective

This chapter has provided a brief overview of recent studies which have centered on the rational design, synthesis, and catalytic applications of nanoporous organic frameworks as acidic heterogeneous catalysts for the conversion of raw biomass materials to the renewable platform chemical HMF. Crystalline frameworks including MOFs and COFs and amorphous sulfonated resin-based networks were discussed in detail. Clearly, these novel catalysts display competitive activity towards the acid-catalyzed synthesis of HMF. The intrinsic nanoporous properties play a crucial role in the generation and dispersion of the acidic active sites, access of reactants, mass transfer, and control of product selectivity. Although in recent times extensive fundamental research and experiments have been carried out to optimize the conversion of biomass to HMF with the innovative use of these novel nanoporous organic frameworks, their commercial application for the production of HMF remains a challenge from an economic point of view. The following aspects should be considered simultaneously when designing new organic frameworks such as MOFs to improve the catalysis performance in the future: (i) better porosities consisting of hierarchical meso/macropores for faster mass transport of reactants and runoff of the products; (ii) better thermal and chemical stabilities that enable an efficient dehydration; and (iii) modulation of the degree of acidity and hydrophobicity on the surface via pore wall engineering, thereby providing a means for the controlled adsorption of substrates/intermediates. To summarize, the innovative utilization of novel nanoporous organic frameworks for the conversion of biomass to HMF as a bio-renewable energy source will lead to a sustainable and bright future.

References

1. Wu, D., Xu, F., Sun, B., Fu, R., He, H., Matyjaszewski, K. (2012) Design and preparation of porous polymers. *Chem. Rev.*, **112** (7), 3959–4015.
2. Das, S., Heasman, P., Ben, T., Qiu, S. (2016) Porous organic materials: strategic design and structure–function correlation. *Chem. Rev.* **117** (3), 1515–1563.
3. Corma, A., García, H., Llabrés i Xamena, F.X. (2010) Engineering metal organic frameworks for heterogeneous catalysis. *Chem. Rev.*, **110** (8), 4606–4655.
4. Phan, A., Doonan, C.J., Uribe-Romo, F.J., Knobler, C.B., O'Keeffe, M., Yaghi, O.M. (2010) Synthesis, structure, and carbon dioxide capture properties of zeolitic imidazolate frameworks. *Acc. Chem. Res.*, **43** (1), 58–67.
5. Zhu, X., Tian, C., Veith, G.M., Abney, C.W., Dehaudt, J., Dai, S. (2016) In situ doping strategy for the preparation of conjugated triazine frameworks displaying efficient CO_2 capture performance. *J. Am. Chem. Soc.*, **138** (36), 11497–11500.
6. Kreno, L.E., Leong, K., Farha, O.K., Allendorf, M., Van Duyne, R.P., Hupp, J.T. (2012) Metal–organic framework materials as chemical sensors. *Chem. Rev.*, **112** (2), 1105–1125.
7. Yoon, M., Srirambalaji, R., Kim, K. (2012) Homochiral metal-organic frameworks for asymmetric heterogeneous catalysis. *Chem. Rev.*, **112** (2), 1196–1231.
8. Sumida, K., Rogow, D.L., Mason, J.A., McDonald, T.M., Bloch, E.D., Herm, Z.R., Bae, T.-H., Long, J.R. (2012) Carbon dioxide capture in metal-organic frameworks. *Chem. Rev.*, **112** (2), 724–781.

9. Li, J.-R., Sculley, J., Zhou, H.-C. (2012) Metal-Organic Frameworks for Separations. *Chem. Rev.*, **112** (2), 869–932.

10. Horcajada, P., Gref, R., Baati, T., Allan, P.K., Maurin, G., Couvreur, P., Férey, G., Morris, R.E., Serre, C. (2012) Metal-organic frameworks in biomedicine. *Chem. Rev.*, **112** (2), 1232–1268.

11. Liang, C., Li, Z., Dai, S. (2008) Mesoporous carbon materials: synthesis and modification. *Angew. Chem. Int. Ed.*, **47** (20), 3696–3717.

12. Zhai, Y., Dou, Y., Zhao, D., Fulvio, P.F., Mayes, R.T., Dai, S. (2011) Carbon materials for chemical capacitive energy storage. *Adv. Mater.*, **23** (42), 4828–4850.

13. Jiang, J., Zhao, Y., Yaghi, O.M. (2016) Covalent chemistry beyond molecules. *J. Am. Chem. Soc.*, **138** (10), 3255–3265.

14. Jiang, J., Yaghi, O.M. (2015) Brønsted acidity in metal–organic frameworks. *Chem. Rev.*, **115** (14), 6966–6997.

15. Waller, P. J., Gándara, F., Yaghi, O.M. (2015) Chemistry of covalent organic frameworks. *Acc. Chem. Res.*, **48** (12), 3053–3063.

16. Wang, L., Xiao, F.-S. (2015) Nanoporous catalysts for biomass conversion. *Green Chem.*, **17** (1), 24–39.

17. Besson, M., Gallezot, P., Pinel, C. (2014) Conversion of biomass into chemicals over metal catalysts. *Chem. Rev.*, **114** (3), 1827–1870.

18. Mäki-Arvela, P., Simakova, I.L., Salmi, T., Murzin, D.Y. (2014) Production of lactic acid/lactates from biomass and their catalytic transformations to commodities. *Chem. Rev.*, **114** (3), 1909–1971.

19. Zakrzewska, M.E., Bogel-Łukasik, E., Bogel-Łukasik, R. (2011) Ionic liquid-mediated formation of 5-hydroxymethylfurfural – A promising biomass-derived building block. *Chem. Rev.*, **111** (2), 397–417.

20. Cohen, S.M. (2012) Postsynthetic methods for the functionalization of metal–organic frameworks. *Chem. Rev.*, **112** (2), 970–1000.

21. Herbst, A., Janiak, C. (2017) MOF catalysts in biomass upgrading towards value-added fine chemicals. *CrystEngCommun.* DOI:10.1039/C6CE01782G.

22. Tian, C., Zhu, X., Chai, S.-H., Wu, Z., Binder, A., Brown, S., Li, L., Luo, H., Guo, Y., Dai, S. (2014) Three-phase catalytic system of H_2O, ionic liquid, and $VOPO_4$-SiO_2 solid acid for conversion of fructose to 5-hydroxymethylfurfural. *ChemSusChem.*, **7** (6), 1703–1709.

23. Román-Leshkov, Y., Chheda, J.N., Dumesic, J.A. (2006) Phase modifiers promote efficient production of hydroxymethylfurfural from fructose. *Science*, **312** (5782), 1933–1937.

24. Zhao, H., Holladay, J.E., Brown, H., Zhang, Z.C. (2007) Metal chlorides in ionic liquid solvents convert sugars to 5-hydroxymethylfurfural. *Science*, **316** (5831), 1597–1600.

25. Richter, F.H., Pupovac, K., Palkovits, R., Schüth, F. (2013) Set of acidic resin catalysts to correlate structure and reactivity in fructose conversion to 5-hydroxymethylfurfural. *ACS Catal.*, **3** (2), 123–127.

26. Tucker, M.H., Crisci, A.J., Wigington, B.N., Phadke, N., Alamillo, R., Zhang, J., Scott, S.L., Dumesic, J.A. (2012) Acid-functionalized SBA-15-type periodic mesoporous organosilicas and their use in the continuous production of 5-hydroxymethylfurfural. *ACS Catal.*, **2** (9), 1865–1876.

27. Pagán-Torres, Y.J., Wang, T., Gallo, J.M.R., Shanks, B.H., Dumesic, J.A. (2012) Production of 5-hydroxymethylfurfural from glucose using a combination of Lewis and Brønsted acid catalysts in water in a biphasic reactor with an alkylphenol solvent. *ACS Catal.*, **2** (6), 930–934.

28. Mellmer, M.A., Sener, C., Gallo, J.M.R., Luterbacher, J.S., Alonso, D.M., Dumesic, J.A. (2014) Solvent effects in acid-catalyzed biomass conversion reactions. *Angew. Chem. Int. Ed.*, **53** (44), 11872–11875.

29. Alamillo, R., Crisci, A.J., Gallo, J.M.R., Scott, S.L., Dumesic, J.A. (2013) A tailored microenvironment for catalytic biomass conversion in inorganic-organic nanoreactors. *Angew. Chem. Int. Ed.*, **52** (39), 10349–10351.

30. Moliner, M., Román-Leshkov, Y., Davis, M.E. (2010) Tin-containing zeolites are highly active catalysts for the isomerization of glucose in water. *Proc. Natl Acad. Sci. USA*, **107** (14), 6164–6168.

31. Horvath, I.T., Mehdi, H., Fabos, V., Boda, L., Mika, L.T. (2008) [gamma]-Valerolactone – a sustainable liquid for energy and carbon-based chemicals. *Green Chem.*, **10** (2), 238–242.

32. Sievers, C., Musin, I., Marzialetti, T., Valenzuela Olarte, M.B., Agrawal, P.K., Jones, C.W. (2009) Acid-catalyzed conversion of sugars and furfurals in an ionic-liquid phase. *ChemSusChem.*, **2** (7), 665–671.

33. Rosatella, A.A., Simeonov, S.P., Frade, R.F.M., Afonso, C.A.M. (2011) 5-Hydroxymethylfurfural (HMF) as a building block platform: Biological properties, synthesis and synthetic applications. *Green Chem.*, **13** (4), 754–793.

34. Yang, G., Pidko, E.A., Hensen, E.J.M. (2012) Mechanism of Brønsted acid-catalyzed conversion of carbohydrates. *J. Catal.*, **295**, 122–132.

35. Hong, D.-Y., Hwang, Y.K., Serre, C., Férey, G., Chang, J.-S. (2009) Porous chromium terephthalate MIL-101 with coordinatively unsaturated sites: surface functionalization, encapsulation, sorption and catalysis. *Adv. Funct. Mater.*, **19** (10), 1537–1552.

36. Akiyama, G., Matsuda, R., Sato, H., Takata, M., Kitagawa, S. (2011) Cellulose hydrolysis by a new porous coordination polymer decorated with sulfonic acid functional groups. *Adv. Mater.*, **23** (29), 3294–3297.

37. Akiyama, G., Matsuda, R., Sato, H., Kitagawa, S. (2014) Catalytic glucose isomerization by porous coordination polymers with open metal sites. *Chem. Asian J.*, **9** (10), 2772–2777.

38. Su, Y., Chang, G., Zhang, Z., Xing, H., Su, B., Yang, Q., Ren, Q., Yang, Y., Bao, Z. (2016) Catalytic dehydration of glucose to 5-hydroxymethylfurfural with a bifunctional metal-organic framework. *AIChE J.*, **62** (12), 4403–4417.

39. Herbst, A., Janiak, C. (2016) Selective glucose conversion to 5-hydroxymethylfurfural (5-HMF) instead of levulinic acid with MIL-101Cr MOF-derivatives. *New J. Chem.*, **40** (9), 7958–7967.

40. Chen, D., Liang, F., Feng, D., Xian, M., Zhang, H., Liu, H., Du, F. (2016) An efficient route from reproducible glucose to 5-hydroxymethylfurfural catalyzed by porous coordination polymer heterogeneous catalysts. *Chem. Eng. J.*, **300**, 177–184.

41. Hu, Z., Peng, Y., Gao, Y., Qian, Y., Ying, S., Yuan, D., Horike, S., Ogiwara, N., Babarao, R., Wang, Y., Yan, N., Zhao, D. (2016) Direct synthesis of hierarchically porous metal-organic frameworks with high stability and strong Brønsted acidity: The decisive role of hafnium in efficient and selective fructose dehydration. *Chem. Mater.*, **28** (8), 2659–2667.

42. Chen, J., Li, K., Chen, L., Liu, R., Huang, X., Ye, D. (2014) Conversion of fructose into 5-hydroxymethylfurfural catalyzed by recyclable sulfonic acid-functionalized metal-organic frameworks. *Green Chem.*, **16** (5), 2490–2499.

43. Zhang, Y., Degirmenci, V., Li, C., Hensen, E.J.M. (2011) Phosphotungstic acid encapsulated in metal-organic framework as catalysts for carbohydrate dehydration to 5-hydroxymethylfurfural. *ChemSusChem.*, **4** (1), 59–64.

44. Chen, J., Wang, S., Huang, J., Chen, L., Ma, L., Huang, X. (2013) Conversion of cellulose and cellobiose into sorbitol catalyzed by ruthenium supported on a polyoxometalate/metal-organic framework hybrid. *ChemSusChem.*, **6** (8), 1545–1555.

45. Bromberg, L., Su, X., Hatton, T.A. (2014) Functional networks of organic and coordination polymers: catalysis of fructose conversion. *Chem. Mater.*, **26** (21), 6257–6264.

46. Jiang, J., Gándara, F., Zhang, Y.-B., Na, K., Yaghi, O.M., Klemperer, W.G. (2014) Superacidity in sulfated metal-organic framework-808. *J. Am. Chem. Soc.*, **136** (37), 12844–12847.

47. Ding, S.-Y., Wang, W. (2013) Covalent organic frameworks (COFs): from design to applications. *Chem. Soc. Rev.*, **42** (2), 548–568.

48. Feng, X., Ding, X., Jiang, D. (2012) Covalent organic frameworks. *Chem. Soc. Rev.*, **41** (18), 6010–6022.

49. Díaz, U., Corma, A. (2016) Ordered covalent organic frameworks, COFs and PAFs. From preparation to application. *Coord. Chem. Rev.*, **311**, 85–124.

50. Vyas, V.S., Lau, V.W.-h., Lotsch, B.V. (2016) Soft photocatalysis: Organic polymers for solar fuel production. *Chem. Mater.*, **28** (15), 5191–5204.

51. DeBlase, C.R., Dichtel, W.R. (2016) Moving beyond boron: the emergence of new linkage chemistries in covalent organic frameworks. *Macromolecules*, **49** (15), 5297–5305.

52. Zeng, Y., Zou, R., Zhao, Y. (2016) Covalent organic frameworks for CO_2 capture. *Adv. Mater.*, **28** (15), 2855–2873.

53. Nguyen, H.L., Gándara, F., Furukawa, H., Doan, T.L.H., Cordova, K.E., Yaghi, O.M. (2016) A titanium-organic framework as an exemplar of combining the chemistry of metal- and covalent-organic frameworks. *J. Am. Chem. Soc.*, **138** (13), 4330–4333.

54. Pang, Z.-F., Xu, S.-Q., Zhou, T.-Y., Liang, R.-R., Zhan, T.-G., Zhao, X. (2016) Construction of covalent organic frameworks bearing three different kinds of pores through the heterostructural mixed linker strategy. *J. Am. Chem. Soc.*, **138** (14), 4710–4713.

55. Ma, H., Liu, B., Li, B., Zhang, L., Li, Y.-G., Tan, H.-Q., Zang, H.-Y., Zhu, G. (2016) Cationic covalent organic frameworks: a simple platform of anionic exchange for porosity tuning and proton conduction. *J. Am. Chem. Soc.*, **138** (18), 5897–5903.

56. Dalapati, S., Jin, E., Addicoat, M., Heine, T., Jiang, D. (2016) Highly emissive covalent organic frameworks. *J. Am. Chem. Soc.*, **138** (18), 5797–5800.

57. Xu, H.-S., Ding, S.-Y., An, W.-K., Wu, H., Wang, W. (2016) Constructing crystalline covalent organic frameworks from chiral building blocks. *J. Am. Chem. Soc.*, **138** (36), 11489–11492.

58. Lohse, M.S., Stassin, T., Naudin, G., Wuttke, S., Ameloot, R., De Vos, D., Medina, D.D., Bein, T. (2016) Sequential pore wall modification in a covalent organic framework for application in lactic acid adsorption. *Chem. Mater.*, **28** (2), 626–631.

59. Ascherl, L., Sick, T., Margraf, J.T., Lapidus, S.H., Calik, M., Hettstedt, C., Karaghiosoff, K., Döblinger, M., Clark, T., Chapman, K.W., Auras, F., Bein, T. (2016) Molecular docking sites designed for the generation of highly crystalline covalent organic frameworks. *Nat. Chem.*, **8** (4), 310–316.

60. Huang, N., Zhai, L., Coupry, D.E., Addicoat, M.A., Okushita, K., Nishimura, K., Heine, T., Jiang, D. (2016) Multiple-component covalent organic frameworks. *Nat. Commun.*, **7**, 12325.

61. Vyas, V.S., Vishwakarma, M., Moudrakovski, I., Haase, F., Savasci, G., Ochsenfeld, C., Spatz, J.P., Lotsch, B.V. (2016) Exploiting noncovalent interactions in an imine-based covalent organic framework for quercetin delivery. *Adv. Mater.*, **28** (39), 8749–8754.

62. Yu, S.-B., Lyu, H., Tian, J., Wang, H., Zhang, D.-W., Liu, Y., Li, Z.-T. (2016) A polycationic covalent organic framework: a robust adsorbent for anionic dye pollutants. *Polym. Chem.*, **7** (20), 3392–3397.

63. Peng, Y., Hu, Z., Gao, Y., Yuan, D., Kang, Z., Qian, Y., Yan, N., Zhao, D. (2015) Synthesis of a sulfonated two-dimensional covalent organic framework as an efficient solid acid catalyst for biobased chemical conversion. *ChemSusChem.*, **8** (19), 3208–3212.

64. Shimizu, K.-i., Uozumi, R., Satsuma, A. (2009) Enhanced production of hydroxymethylfurfural from fructose with solid acid catalysts by simple water removal methods. *Catal. Commun.*, **10** (14), 1849–1853.

65. Qi, X., Watanabe, M., Aida, T.M., Smith, J.R.L. (2009) Efficient process for conversion of fructose to 5-hydroxymethylfurfural with ionic liquids. *Green Chem.*, **11** (9), 1327–1331.

66. Huang, Z., Pan, W., Zhou, H., Qin, F., Xu, H., Shen, W. (2013) Nafion-resin-modified mesocellular silica foam catalyst for 5-hydroxymethylfurfural production from D-fructose. *ChemSusChem.*, **6** (6), 1063–1069.

2

Activated Carbon and Ordered Mesoporous Carbon-Based Catalysts for Biomass Conversion

Xiaochen Zhao, Jifeng Pang, Guangyi Li, Fei Liu, Jinming Xu, Mingyuan Zheng, Ning Li, Changzhi Li, Aiqin Wang, and Tao Zhang

State Key Laboratory of Catalysis,
Dalian Institute of Chemical Physics, Chinese Academy of Sciences, China

2.1 Introduction

Carbon materials represent versatile supports or catalysts for biomass conversion due to their special physical and chemical properties. Carbon materials have diversified porosities and large surface areas, with rather stable properties under acidic, basic or hydrothermal conditions, which renders them some of the most promising candidates for biomass conversion. Among different biomass resources (e.g., lignocelluloses, crops, aquatic cultures, and biowastes), inedible lignocelluloses are now considered as the central feedstock for the synthesis of chemicals and fuels. Lignocellulose comprises approximately 40–50 wt % cellulose to frame the plants structure, approximately 16–33 wt% hemicelluloses to bind the cellulose fibers, and approximately 15–30 wt% lignin to harden the structure [1]. Although differences exist between the chemical structures of these components, the activation and cleavage of C–O bonds represents a crucial step in unlocking the biomass potential and releasing the valuable products [2]. Owing to sustainable requirements, most of these processes take place in green solvents (e.g., waters and alcohols) and require relatively harsh conditions (i.e., $373K < T < 523K$ and $P^*H_2O(T) < P < 100$ bar) [3].

Nanoporous Catalysts for Biomass Conversion, First Edition. Edited by Feng-Shou Xiao and Liang Wang.
© 2018 John Wiley & Sons Ltd. Published 2018 by John Wiley & Sons Ltd.

Unfortunately, most oxides suffer from partial dissolution in the condensed phase and are unlikely to offer satisfactory long-term performance, and consequently, a new class of carbon-based catalysts, bearing high hydrothermal stability, represent an interesting advance in this field. Moreover, various carbon materials can be obtained and functionalized from biomass which, in contrast, may reduce the 'carbon-footprint' of the biomass transformation process. Moreover, advances in nanocarbon technology have enabled a controllable tailoring of the pore structure and surface functionalities to some extent, and thus have provided the possibility to clarify the role that carbon plays and also allows the rational design of new catalysts for superior biomass conversion performance [4].

In this chapter, brief information is first provided relating to activated carbon and mesoporous carbon, such as their preparation methods and chemical and physical properties. Details of the conversion of cellulose over carbon materials are then highlighted. Hemicellulose and cellulose are polymers of xylose and/or D-glucose that are interconnected with glycosidic bonds, whereas hemicelluloses are more reactive due to their lack of structure uniformity; hence, the conversion of hemicellulose has been included in the topic of cellulose conversion (Section 2.3). In contrast to cellulose, lignin is a complex three-dimensional amorphous polymer that bears methoxylated phenylpropane units, with the different monomers being linked via a variety of different C–O bonds (e.g., β-O-4, α-O-4, 4-O-5) and C–C bonds (e.g., 5-5, β-5, β-1, β-β). Since the structure and composition of lignin depends heavily on the origin of the biomass, the decision was taken to focus on the transformation of raw lignin and its model compounds in Section 2.4, which covers lignin conversion. Finally, in addition to these top-down methods, the use of C5 or C6 sugars (the products of catalytic conversion of lignocelluloses) to fabricate biofuels are summarized to provide an alternative bottom-up viewpoint on biomass conversion over carbon-based catalysts.

2.2   Activated Carbon and Mesoporous Carbon

2.2.1   Preparation of Activated Carbon and Mesoporous Carbon

Activated carbons are usually synthesized from natural sources such as coal and biomass, or synthetic polymers such as phenolic resin, polyacrylonitrile and furfural resin [5]. Certain degrees of porosity are created in activated carbons during the carbonization of an organic precursor, but these are not sufficiently developed for most applications and hence some activation processes are needed in a prerequisite step. This can usually be achieved in two ways involving the creation of further porosity. One method is to employ a physical activation process [6] whereby, after carbonization, the carbonized materials are exposed to oxidizing atmospheres (such as oxygen, carbon dioxide, etc.) at 873–1473K. The second approach is via a chemical activation process [7] where, prior to carbonization, the raw materials are impregnated with certain chemicals (e.g., zinc chloride, phosphoric acid and potassium hydroxide) and are then carbonized and activated simultaneously at 723–1173K. Chemical activation is preferred over physical activation owing to the lower temperatures and shorter time needed for activating carbon materials. Since the carbonization and activation temperatures are usually below 1273K, the graphitization level of activated carbon is relatively low.

During the carbonization process, carbon atoms can be removed from porous carbons, and large amounts of micropores with dimensions <2 nm are formed. Hence, when activated carbons are used as catalysts or catalyst supports for large sizes of reactants, diffusion limitations and accessibility to active sites become an important issue. To overcome these limitations, major efforts have been made towards developing synthetic methods for mesoporous carbons [8].

Mesoporous carbons can be prepared via several methods, including the carbonization of aerogels or xerogels, the catalytic activation of carbon precursors with metal species, carbonization of polymer blends, and a high degree of activation by physical or chemical methods. A more precise controlled synthesis on ordered mesoporous structures can be achieved by employing a hard or soft templating strategy. In a hard templating strategy [9], the as-synthesized mesoporous oxides (e.g., silica and alumina) serve as molds, and this is followed by the introduction of carbon precursors. After carbonization and etching of the hard silicate templates using HF or NaOH solutions, ordered mesoporous carbons with an inverse structure are obtained. As hard material templating is typically time-consuming and expensive to carry out on a large scale, a soft-template method based on supramolecular assemblies of surfactants and carbon precursors is highly desirable as an alternative option [10]. However, the soft templating strategy requires more carefully controlled synthetic conditions, such as the interactions between surfactants templates and carbon precursors, mixing ratios, solvents, and temperatures.

2.2.2 Properties of Carbon in Catalysis

Activated carbon and mesoporous carbon have long been used as either catalysts or catalyst supports due to their high surface areas and unique physicochemical properties. Their performances in catalysis are determined not only by their structural parameters but also by their chemical compositions and surface functional groups. Compared to other catalysts or catalyst supports, carbons are easily functionalized, and hence their acid–base, oxidation–reduction characteristics, as well as their specific adsorption properties, can be conveniently adjusted.

Activated carbon is widely accepted as comprising three-dimensional pores (macro-, meso-, and micro-pores), with micro-sized pores being dominant (>0.2 cm^3 g^{-1}) and contributing to most of the internal surface area (>400 m^2 g^{-1}). The presence of abundant micropores endows the activated carbon with a higher surface area than conventional oxide supports such as silica and alumina. As a consequence, activated carbons have been used in simultaneous SO_2/NO_x removal for decades [11], where the active carbons act simultaneously as adsorbents and as catalysts in the temperature range 383–443K. Activated carbons are also used in the synthesis of phosgene and thionyl chloride [12], in the oxidative dehydrogenation of hydrocarbons [13], the dehydration of alcohols [14], catalytic wet air oxidation [15], and the dehydrochlorination of alkyl chlorides to the corresponding alkenes [16]. In the case of biomass conversion, however, abundant micropores may limit the diffusion and accessibility of large-sized biomass-derived reactants to active sites [17]. Whilst activation treatments, as mentioned above, can broaden the pore size to some extent, a more effective means of manipulating the porosity of carbon materials is to employ 'templates' to fabricate mesoporous carbon and hierarchical structured carbon (see Section 2.2.1). In principle, a highly interconnected hierarchical structure will benefit the reactant diffusion and its accessibility to active sites.

Besides porosities, the surface chemistry of carbon also plays important roles in catalysis. Particularly in the conversion of lignocelluloses, solid–solid interactions take place at the initial stage, and this calls for a relatively hydrophilic catalyst surface to facilitate mass transfer between the feedstock and the catalyst [18]. When formed spontaneously by exposure to the atmosphere, the carbon surface usually contains a certain number of oxygenated functional groups, which confers on the carbon surface a hydrophilic character and hence benefits the catalytic biomass transformation. Moreover, amorphous carbon may also play the role of a solid Brønsted acid [19], taking advantage of the abundant carboxyl, anhydride, hydroxyl, lactone, and lactol functionalities. Besides oxygen functionalities, heteroatom (N, P, S, etc.) -modified carbons have also been confirmed active for the hydrodeoxygenation (HDO) reaction of biomass (see Chapter 6). On the other hand, the presence of these functionalities confers the carbon surface with hydrophilic and chelating characters, which are beneficial to the adsorption of metal precursors in the impregnation process, and may thus enhance the dispersion of metal sites and their subsequent catalytic activity.

Moreover, as the carbon supports can be easily burned off, the recovery and recycling of metals – particularly of noble metals – is simplified on carbon supports, which provides the carbon support with a unique advantage in terms of its practical applications.

2.2.3   Functionalization of Carbon Materials

Carbon can be functionalized either in gaseous or liquid phase over a temperature range from 473 to 1293K [20]. In biomass conversion, one of the most widely used applications of carbon is as solid acid to catalyze the hydrolysis reaction, and therefore the sulfonation method to increase acid density and strength is highly desirable. Sulfuric acid is the most widely used source to functionalize carbon, but while higher temperatures will create more functionalities, the original carbon structure will be sacrificed. Sulfuric acid, together with other oxidizing agents, can even enhance the sulfonation effect and provide more effective solid acids [21].

Besides the function of solid acids, porous carbons also serve as outstanding supports to survive fierce, condensed-phase reaction conditions. However, their interactions with metal precursors are much weaker than with oxide supports, which in turn makes the effective dispersion of metal on carbon supports rather challenging. In order to overcome this disadvantage, an oxidation method is widely used to functionalize carbon supports, since oxygen functionalities are beneficial for metal loading and dispersion. Generally, gas-phase oxidation (using air and oxygen, etc.) and plasma treatment are employed to create weakly phenolic groups [22]. To obtain more acidic functionalities, however, liquid-phase oxidation (using nitric acid, potassium permanganate, sodium peroxydisulfide, etc.) is required and carboxylic groups are the dominant functionalities on the modified carbon surface [23]. Hydrogen peroxide-treated carbon bears increasing phenolic groups exceptionally well, owing to the moderate oxidation capabilities of hydrogen peroxide [24].

Similar to oxygen, nitrogen-containing functionalities can also improve the dispersion of metal species on carbon surfaces. In general, ammonia, melamine, carbonzole, nitrogen-enriched polymers and so forth are used to obtain N-containing functionalities [25]. The types and amounts of these functionalities depend heavily on the treatment temperature: low-temperature treatment (<800K) endows the carbon surface with slightly acidic functionalities, while high-temperature treatment modifies the carbon surface to its basic form, owing to the presence of pyridine and pyrrole functionalities [25d]. In some cases, the N function was approved to modify the electronic structures of carbon, which can then catalyze certain types of reaction, especially in the field of electrochemistry.

Finally, it is important to discuss washing as an important treatment, though it is not a typical functionalization method. As most activated carbons are generated from different sources of biomass, the presence of impurities being included in the resulting carbon materials (e.g., ashes and mineral components) during the activation process is inevitable. Washing the activated carbon with water or dilute acid may significantly reduce the impurity content [26], and this is usually necessary before any further modification can be made. In some cases, washing is also required after the functionalization (e.g., sulfonation, oxidation) to remove excess reagents and physically adsorbed entities.

2.3 Cellulose Conversion

Cellulose, one of the most abundant renewable feedstock in lignocellulosic biomass, is produced from CO_2 and water via photosynthesis. Cellulose is a non-food biomass with a high production, which makes it an ideal feedstock for the synthesis of chemicals, fuels, and materials [27]. Cellulose is a water-insoluble polymer composed of glucose units linked by β-1,4-glycosidic bonds, and has strong inter- and intra-molecular hydrogen bonds with highly crystal structures (see Scheme 2.1). Hence, it remains a major challenge to unlock the rigid structure of cellulose to produce target materials with high selectivity [28]. The latest developments of cellulose transformation into some important chemicals, such as glucose, hexitols, and glycols, using carbon materials as supports or catalysts, are introduced in the following sections (see Scheme 2.2).

2.3.1 Cellulose Hydrolysis

Glucose is expected to be an important renewable platform molecule that can be used for the synthesis of a variety of chemicals and fuels, besides being used for foods and in the

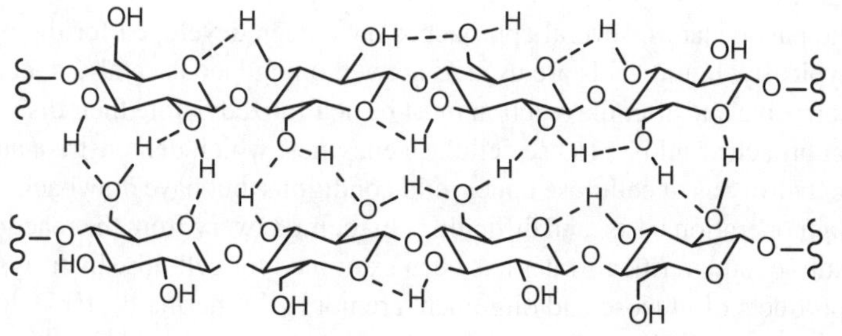

Scheme 2.1 *Structure of cellulose. (See color plate section for the color representation of this figure.)*

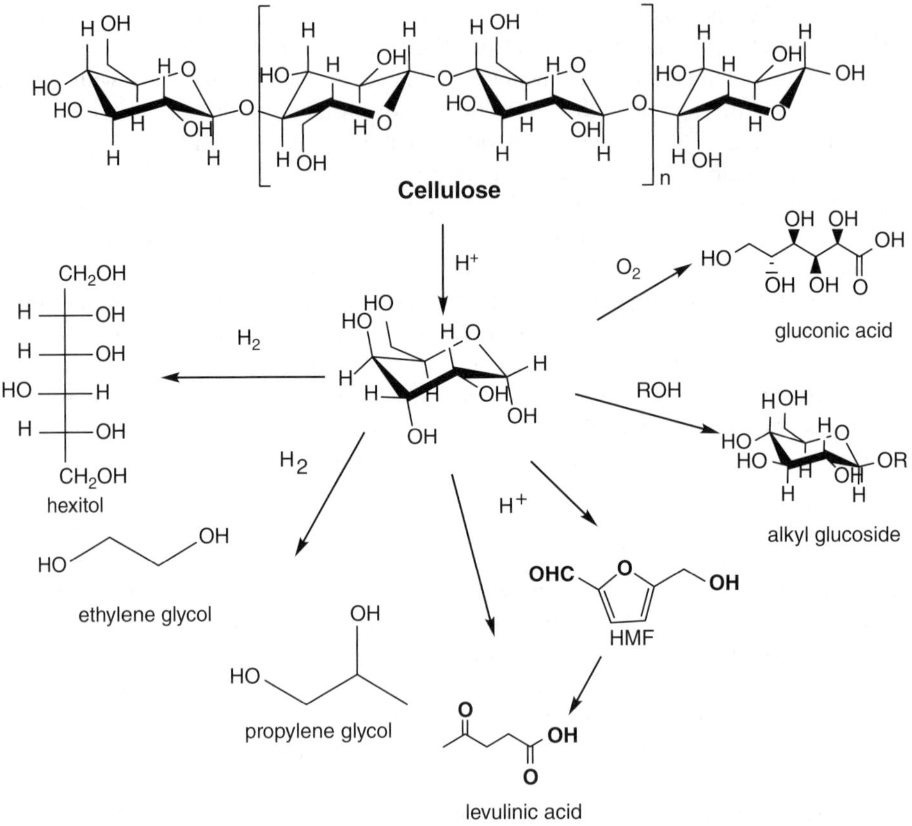

Scheme 2.2 *Conversion of cellulose to chemicals over carbon-based catalysts.* (See color plate section for the color representation of this figure.)

medical industry [29]. Because of the large amounts of cellulose present in lignocellulosic biomass, the hydrolysis of cellulose to glucose is regarded as a key process in the beneficial use of lignocellulosic biomass. While the effective hydrolysis of cellulose to glucose, with a low environmental impact, remains a strong challenge, much effort has been dedicated to this target [30].

   During the past decades, several approaches have been developed for the hydrolysis of cellulose. Typically, liquid acids are used as active homogeneous catalysts, but these acids are highly corrosive and after the reaction must be neutralized before their disposal [31]. An alternative approach would be to use cellulase enzymes, which demonstrate advantages of the selective hydrolysis of cellulose under mild conditions, but have drawbacks of low reaction rates, high operation costs, and difficulties in their recovery from the reaction mixtures. The use of sub- or supercritical water has been examined for cellulose hydrolysis. Unfortunately, the products of glucose and oligomers are not hydrothermally stable, but are rather prone to decompose or polymerize into byproducts and humins. Hence, the highest yield of glucose available is still less than 40% [32].

Solid-acid catalysts represent an attractive type of catalyst for cellulose hydrolysis because they have the advantages of easy product separation, recyclability, less damage to the reactor, and minimal impact on the environment. A variety of solid-acid catalysts have been used for cellulose hydrolysis, including mixed oxides, polymer-based acids, carbon-based catalysts, heteropoly acids, and zeolites [30a,33]. Among solid-acid catalysts, carbon-based catalysts have shown superior activity due to their tunable surface functional groups and versatile pore structures. The typical results of cellulose hydrolysis are shown in Table 2.1.

In 2005, Hara *et al.* prepared a robust solid catalyst by sulfonating incompletely carbonized natural organic materials obtained from sugar, starch, and cellulose. The carbon material was composed of small polycyclic aromatic carbon sheets in a three-dimensional sp^3-bonded structure, bearing a high density of active sites [21c,34]. A large amount of hydrophilic molecules was contained in the bulk and surface of the carbon material, providing a ready access for reactants to the $-SO_3H$ groups in aqueous solution. The new material demonstrated a high catalytic performance for acid-catalyzed reactions such as esterification and hydration reactions [35]. These materials were further employed for cellulose hydrolysis and showed high activity. Following a 3 h reaction, the cellulose conversion reached 68%, with glucose and soluble β-1,4 glucan yields of 4% and 64%, respectively [21b]. Compared to other solid catalysts, the enhanced catalytic performance of carbon materials cannot be explained solely by conventional factors such as surface area or acid strength [36]. Subsequently, the reason for this outstanding hydrolytic performance was further investigated with cellobiose adsorption experiments, the results of which showed that the abundant $-SO_3H$, $-OH$, and $-COOH$ groups on the carbon material were active sites for cellulose hydrolysis (Figure 2.1). The three key factors that endowed the carbon materials with superior performance for this reaction were the ability to adsorb β-1,4 glucan, the large effective surface area in water, and the tolerability of the $-SO_3H$ groups towards hydration. The apparent activation energy for cellulose hydrolysis over the carbon catalyst was estimated at $110 \, kJ \, mol^{-1}$, which was less than required by sulfuric acid ($170 \, kJ \, mol^{-1}$) under optimal conditions. In order to increase the accessibility of cellulose, ionic liquids were used to dissolve the cellulose in readiness for hydrolysis [37]. For example, by utilizing the advantage of cellulose insolubility, Liu *et al.* conducted cellulose hydrolyses over sugar-derived carbon catalysts in ionic liquids. With fourfold additions of water (150 μl) in [BMIM]Cl within the first 30 min, glucose and the total product yields were maximized to 57% and 80%, respectively, after reaction at 393 K for 4 h [37a]. Qi *et al.* also employed [BMIM]Cl for cellulose hydrolysis over sulfonated catalysts which were derived from glucose. The catalyst containing $-SO_3H$, $-COOH$ and phenolic $-OH$ groups exhibited a high catalytic activity for cellulose hydrolysis, and gave 72.7% reducing sugar yield at 383 K for 240 min [37b].

A direct sulfonating raw biomass method was developed by Fu *et al.* for cellulose hydrolysis [38]. The biomass carbon sulfonic acid (BC-SO_3H) catalysts were prepared from natural bamboo, cotton and starch and subsequently treated with 80% sulfuric acid and oleum. The yield of glucose reached 19.8% at 363 K for 60 min under microwave irradiation conditions, and the turnover number (TON) at 1.33–1.73 was much higher than that of a dilute H_2SO_4 solution (TON = 0.02) for cellulose hydrolysis. The high activity was attributed to the introduction of microwave irradiation, which activated cellulose

Table 2.1 Hydrolysis of cellulose and lignocellulosic biomass.

Entry	Precursor	Catalyst	Surface area (m² g⁻¹)	Acid density (mmol g⁻¹)	Reaction conditions	Main products	Yield (%)	Reference
1	Cellulose	carbon-SO_3H	2	-SO_3H: 1.9	Cellulose 373K, 3 h	Glucose β-1,4-glucan	4 64	[21b]
2	Sugar	Sugar based carbon-SO_3H	–	–	Cellulose, [BMIM]Cl, 393K, 4 h, 30 min	Glucose	57	[37a]
3	Glucose	Glucose derived carbon-SO_3H	0.5	Total: 2.15	Cellulose, [BMIM]Cl, 383K, 240 min	Reducing sugar	72.7	[37b]
4	Raw biomass	BC-SO_3H	2–5	-SO_3H: 2.0	Cellulose, 363K, 60 min, microwave	Glucose	19.8	[38]
5	Corncob	CSA-SO_3H	2	Total: 1.76	Corncob, 403K, 1 h, microwave	Glucose	34.6	[40]
6	AC	AC-SO_3H	806	-SO_3H: 0.44	BM-cellulose, 423K, 16 h	Glucose	40.5	[41]
7	AC	AC-N-SO_3H-250	762	-SO_3H: 0.44	BM-cellulose, 423K, 24 h	Glucose	62.6	[42]
8	CNT	CNT-SO_3H	–	–	crop stalks, 423K, 2 h	Sugar	30	[43]
9	CMK-3	10 wt% Ru/CMK-3	1100	–	BM-cellulose 230 °C, 24 h	Glucose	34.2	[45]
10	CMK-3	CMK-3-SO_3H	412	-SO_3H: 0.63	BM-cellulose, 423K, 24 h	Glucose	74.5	[42]
11	–	$Si_{33}C_{66}$-823-SO_3H	–	Total: 0.37	BM-cellulose, 423K, 24 h	Glucose	50.4	[46]

BM-cellulose: Ball-milled cellulose.

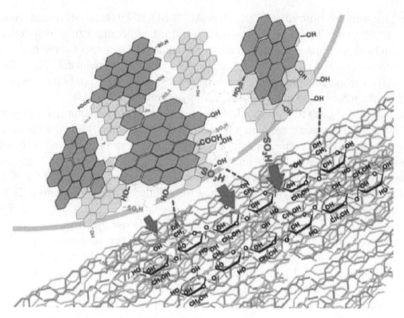

Figure 2.1 *Reaction mechanism for the hydrolysis of cellulose by carbon materials bearing $-SO_3H$, $-COOH$, and phenolic $-OH$ groups. Reprinted with permission from Ref. [36].*

molecules and strengthened the particles' collisions, and consequently resulted in a greatly accelerated rate of hydrolysis [39]. Meanwhile, the catalyst had a high density of $-SO_3H$ groups and a strong affinity to the β-1,4-glycosidic bonds of cellulose, which also promoted the activation of cellulose. Jiang and Mu *et al.* used the hydrolyzed residues of corncob to prepare carbonaceous solid-acid (CSA) catalysts and used these to hydrolyze corncobs. Under microwave irradiation conditions, the cellulose and hemicellulose in corncobs were converted into corresponding sugars, with yields of xylose and arabinose as high as 78.1% at 403K [40]. The as-prepared CSA catalysts showed excellent catalytic hydrolysis performance for the lignocellulosic biomass, especially of the conversion of hemicellulose in the biomass. In this way, all parts of the corncob were utilized and two products were obtained; that is, sugars were obtained from cellulose and hemicellulose hydrolysis, while CSA catalysts were obtained from the unconverted solid residues.

Changing the surface structure of commercial carbon materials represents another strategy for preparing solid-acid catalysts. Onda *et al.* sulfonated active carbon with concentrated H_2SO_4 at 423K for 16 h and obtained sulfonated activated-carbon (AC-SO_3H) [41]. Under mildly hydrothermal conditions, the glucose yield reached 40.5% at 423K, with 95% selectivity of glucose and water-soluble organics. These results suggested that the high catalytic activity should be attributed to the high hydrophilicity of graphene planes and strong acidic functional groups of the AC-SO_3H catalyst. The sulfonated conditions greatly affect the surface structure of catalysts and its consequent activity for cellulose hydrolysis. When Zhang *et al.* prepared the AC-N-SO_3H catalyst from activated-carbon (AC) at different sulfonation temperatures [42], they found that the acid density of the sulfonated carbon increased as the sulfonation temperature increased. The specific surface areas of the samples were maximized at 523K. Over AC-N-SO$_3$H-250 catalysts, the glucose yield reached 62.6%, with 74.3% cellulose conversion for hydrolyzing ball-milled cellulose at 423K. Liu *et al.* sulfonated magnetic carbon nanotube (CNT) arrays for the hydrolysis of polysaccharides in crop stalks, and obtained a sugar yield of approximately 30%. The catalyst displayed an excellent magnetic property and a good magnetic stability, which proved to be convenient for its separation and recycling after the reaction [43].

Mesoporous carbon (MC) materials have a well-ordered pore structure, a large pore volume, high specific surface areas. and tunable mesopore diameters, all of which were associated with attractive properties in catalysis [44]. Fukuoka *et al.* developed a Ru/CMK-3 catalyst for ball-milled cellulose hydrolysis, and the glucose yield reached 27.6% at 503K with rapid heating and cooling processes [45]. It was interesting to find that CMK-3 itself exhibited considerable catalytic performance for cellulose hydrolysis, with a 21% yield of glucose being achieved under the same reaction conditions as Ru/CMK-3 catalysts. In another study, the MC of CMK-3 was sulfonated for cellulose hydrolysis, and provided a 94.4% cellulose conversion and 74.5% yield of glucose, which was surprisingly higher than that of other microporous carbon materials [42]. Characterization of the material revealed that the sulfonated CMK-3 had a high acid density and a reasonably large surface area, both of which accelerated the hydrolysis of cellulose into glucose. In addition, the mesoporous structure also facilitated the transportation of large molecules (such as glucose, cellobiose, cellotriose) in comparison with microporous carbons. Sels *et al.* prepared silica/carbon nanocomposites from a sucrose/silica/F127 solution, and then treated them with sulfuric acid to obtain sulfonated silica/carbon nanocomposites [46]. The nanocomposite of Si$_{33}$C$_{66}$-823-SO$_3$H provided a 50% glucose yield at 423K for 24 h. These authors proposed that the presence of accessible strong Brønsted acid sites and the hybrid surface structures facilitated the adsorption of glucan.

Besides –SO$_3$H groups, –Cl, phenols and lactone groups were also introduced to carbon materials, and showed promotional effects for cellulose hydrolysis. Qiu *et al.* synthesized a carbon catalyst by the sulfonation and carbonization of microcrystalline cellulose, which was pretreated with dilute hydrochloric acid. This catalyst contained both –Cl and –SO$_3$H groups, both of which provided a glucose selectivity of 95.8% at a moderate temperature of 428K under hydrothermal conditions [47]. The authors proposed that the active electronic states of carbon catalyst make the –Cl groups more liable to form hydrogen bonds with cellulose, while the –SO$_3$H groups with stronger acidity easily break the glycosidic bonds of cellulose to produce glucose during the cellulose hydrolysis process. Lou *et al.* developed a sucralose-derived solid-acid catalyst (SUCRA-SO$_3$H) which contained –Cl and –SO$_3$H groups. Over this catalyst, the glucose yield reached 55%, with a selectivity of 98% in converting the ionic-liquid-pretreated cellulose within 24 h at 393K [48]. The –Cl groups showed a positive effect on cellulose hydrolysis. In fact, the apparent activation energy for the hydrolysis of cellobiose with SUCRA-SO$_3$H was found to be 94 kJ mol^{-1}, which was much lower than that of the sucrose-derived catalyst (SUCRO-SO$_3$H) without chlorine groups (114 kJ mol^{-1}). When Zhang *et al.* used graphite oxide (an extremely hydrophilic material with a variety of oxygen functionalities) for cellulose hydrolysis, the results showed that the high catalytic activity over graphite oxide can be attributed to the synergy of its layered, soft structure, and the abundance of its hydroxyl/carboxyl functionalities [49]. Sievers *et al.* investigated the influence of structural and chemical modifications of active carbon after H$_2$O$_2$ and H$_2$SO$_4$ treatments on cellulose hydrolysis [50], and found that treatment with H_2O_2 enlarged the pore size and imparted functional groups such as phenols, lactones, and carboxylic acids, whereas H_2SO_4 treatment targeted the edges of carbon sheets primarily. Even though most of acid sites were weak sites, the material still showed a high activity for cellulose hydrolysis due to the synergistic effect between defect sites and functional groups.

As discussed above, great progress has been made in the depolymerization of cellulose to generate glucose, which in turn promotes the utilization of biomass. On the other hand, most of cellulose was pretreated prior to hydrolysis in order to unlock its rigid structure. Moreover, glucose is not very stable under hydrothermal conditions, and is prone to decompose to humins. Thus, the hydrolysis of cellulose should be conducted under mild conditions with rather long reaction times. These drawbacks greatly reduce the reaction efficiency and limit the applications of cellulose hydrolysis over solid-acid catalysts.

2.3.2   Conversion of Cellulose to Hexitols

Sorbitol is a high-value-added chemical, which is widely used in the food and pharmaceutical industries. For instance, it is the important precursor for producing vitamin C [51], and is also a platform chemical for the synthesis of various value-added chemicals and fuels such as isosorbide, 1,4-sorbitan, glycols, lactic acid, L-sorbose, and alkane fuels. Currently, sorbitol is mainly produced through the hydrogenation of glucose. With the rapid development of biomass conversion, various catalysts have been exploited for the direct catalytic conversion of cellulose to sorbitol and mannitol [52].

In a one-pot reaction process, elevated temperatures were always employed to accelerate the cellulose conversion. Most oxide supports or catalysts are unstable under the fierce reaction conditions, and this hinders their applications. Carbon materials are inert under most reaction conditions, and can effectively disperse the metal sites. More important, they are generally hydrothermally stable and show positive effects on biomass hydrolysis, as noted above. Hence, they are widely used as catalyst supports for the synthesis of hexitols (Scheme 2.3).

In 2006, Fukuoka *et al.* reported the conversion of cellulose to sorbitol over a Pt/Al_2O_3 catalyst. The yield of sorbitol reached 25% with 6% mannitol in the presence of H_2 at 463K for 24 h [54]. In a following study, the alumina support was found to be prone to hydrolysis under hydrothermal conditions [55]. More stable catalysts of activated carbon-supported ruthenium (Ru/C) were employed by Liu *et al.* [53]. Over Ru/C, a 39.3% yield of hexitols was achieved for the conversion of microcrystalline cellulose at 518K under an H_2 pressure of 6 MPa. Wang *et al.* treated CNTs with concentrated HNO_3 to gain a better support for loading Ru nanoparticles, and the as-prepared 1.0 wt% Ru/CNT catalyst showed a high activity for cellulose conversion [56]. A sorbitol yield of 69% was obtained at 458K under an H_2 pressure of 5 MPa for the conversion of cellulose after pretreatment by 85% H_3PO_4. The characterization results showed that both the acidic functional groups and the higher concentration of adsorbed hydrogen species on the CNT surfaces played key roles in sorbitol formation. Zhang and coworkers exploited an activated carbon-supported Ni_2P catalyst for cellulose conversion [57] and, with good matching of the bifunctional sites, promising performances were obtained over 16%Ni_2P/AC catalysts, though the catalysts quickly deactivated owing to P leaching and Ni sintering. Sels *et al.* designed a carbon nanofiber (CNF)-supported Ni catalyst with Ni nanoparticles on the tips of CNFs for the

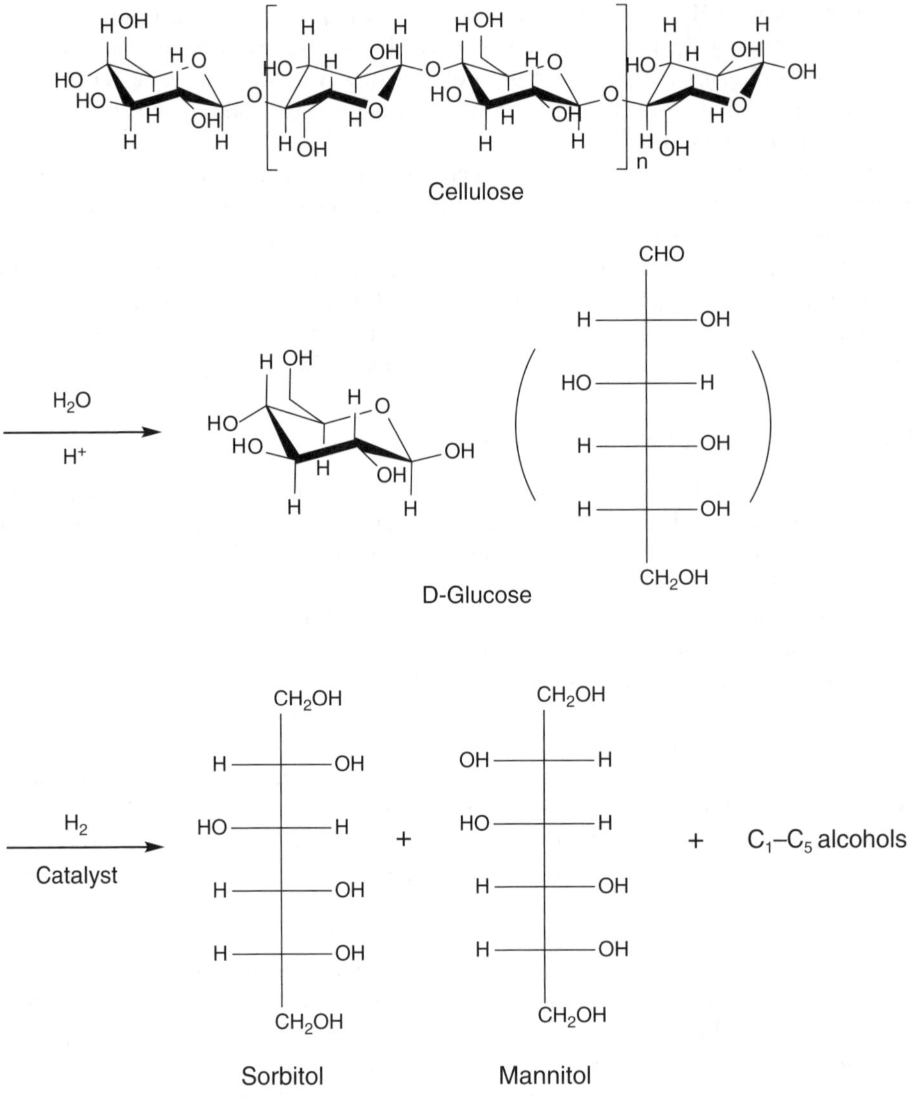

Scheme 2.3   *Catalytic conversion of cellulose into polyols.*

conversion of cellulose to hexitols [58]. The location of the Ni particles was proposed to be a key factor for a high hexitol yield, as Ni particles attached to the tips of the CNFs might be more easily accessed by cellulose; this led to a hexitols yield of 27.6% in the conversion of microcrystalline cellulose at 503K for 4 h. When Fukuoka *et al.* used carbon black BP2000 as a support in their catalysts [59], the hexitols yield reached 63.5% at 463K for 24 h over a Pt(N)/BP2000 catalyst. The Cl-free nature of the catalyst prevented side-reactions of C–C cleavage and dehydration, while the water-tolerant carbon support allowed recycling with good durability.

The conversion of cellulose to hexitols is a consecutive reaction, in which the rates of cellulose hydrolysis and sugar hydrogenation reactions should be balanced for a high hexitols yield. According to the reaction mechanism, Zhang *et al.* developed bifunctional catalysts of Ru/(AC-SO$_3$H) to convert Jerusalem artichoke tubers to hexitols. The hexitol yields reached 92.6% over the 3%Ru/(AC-SO$_3$H) catalyst after 5 h of reaction at 373K, and the sulfonation process was found to have introduced an abundance of –OH, –COOH and –SO$_3$H groups onto the carbon surface, which provided acidity and anchoring sites for the Ru nanoparticles [60]. When Tsubaki *et al.* hydrogenolyzed cellulose to sorbitol over Pt/reduced graphene oxide, the yield of sorbitol was as high as 91.5% from cellobiose and 58.9% from cellulose. Remarkably, the catalyst induced hydrogen spillover and created in situ acid sites to promote cellulose hydrolysis. The resultant glucose rapidly proceeded via hydrogenation to sorbitol over nearby active Pt sites. The synergetic effect between surface acid sites and Pt nanoparticles led to the superior performance in cellulose conversion [61].

Mesoporous carbon (MC) materials offer great advantages over conventional AC owing to their well-controlled pore structures, which are favorable to the transportation of large molecules. Zhang *et al.* took advantage of MC's properties for cellulose conversion when, over a 20%Ni/MC catalyst, the cellulose conversion reached 84.5% and the hexitol yield was up to 42.1%. However, the activity decreased sharply in the following runs due to the aggregation of Ni particles. Noble metals were introduced to improve the catalysts' stability. Interestingly, bimetallic catalysts showed a better result for producing hexitols; over a 1%Rh-5%Ni/MC catalyst the hexitols yield was improved to 59.8%. Characterization results showed that the mesoporous structure of the catalyst facilitated reactant transportation [62]. Meanwhile, the MC-supported bimetallic catalysts promoted hydrogenation activity. By efficiently coupling the hydrolysis reaction and the 'in situ' hydrogenation of glucose, the hexitols yield was maximized. Zhang *et al.* successfully synthesized carbon–alumina nanocomposites with highly ordered mesoporous structures via the co-assembly of resol, alumina sol, and triblock copolymer F127. This showed a high activity for converting cellulose to hexitols after being loaded with Pt nanoparticles, over which the hexitols yield reached 47.5% with 99% cellulose conversion [63]. The high performance was attributed to the tunable hydrophobicity–hydrophilicity properties, adjustable metal–support interaction, and good hydrothermal stability.

Besides hydrothermal reactions, some strategies have also been exploited for converting cellulose to hexitols under mild conditions. For instance, Fukuoka *et al.* took advantage of the hydrogen-transfer properties of carbon-supported Ru catalysts for cellulose conversion. In this case, the hexitols yield reached 47% with 2-propanol as the hydrogen resource at 463K for 18 h [64]. Although it is the highly dispersed cationic Ru species that active for the transfer hydrogenation reaction, the support is crucial for product selectivity. Carbon materials, such as AC, CMK-3 and BP2000, exhibited superior activities to oxide supports, owing to their special functionalities on their surface. Fukuoka *et al.* also used the Ru-supported AC catalysts for the hydrogenation of cellulose oligomers with transfer hydrogenation of 2-propanol [65]. A hexitols yield of 85 % was obtained in less than 1 h in a batch reactor for the conversion of milled, acidulated microcrystalline cellulose. In another study, Schüth *et al.* depolymerized cellulose via a ball-milling method to produce hexitols with Ru/C as catalysts; here, the yield of hexitols was up to 94% at 423K in an overall process time of 4 h [66]. The catalyst was rather stable and maintained its activity throughout six runs.

Based on the viewpoint of application, silver grass (a type of raw biomass) was converted to sugar alcohols on a Pt/carbon catalyst by Fukuoka *et al.* The sorbitol and xylitol yields over Pt/BP200 catalysts were 2.8 wt% and 7.3 wt%, respectively, and these were increased to 13 wt% and 14 wt% after removing the lignin and inorganic salts by alkali-explosion and neutralization steps [67]. The resulting alcohols could be further converted to other platform chemicals as the catalyst's support was tolerant to both acidic and basic conditions.

2.3.3   Conversion of Cellulose to Glycols

Glycols of ethylene glycol (EG) and 1,2-polyethylene glycol (1,2-PG) are important chemicals for preparing materials and value-added products. The diol structure of the compounds renders them candidate precursors in the polyesters industry, which accounts for 58% of the application market [68]. In 2014, the global demand for EG and 1,2-PG reached 25 and 3 million tons, respectively, with an estimated increase of 5% year on year. At present, EG and 1,2-PG are produced from ethylene and propylene, which are obtained from steam cracking of ethane and propane, or from the MTO/MTP processes [69]. During the past decade, promising results have been obtained in the synthesis of renewable glycols, that is, the conversion of (hemi)cellulose to glycols.

Owing to insertion of the C atom into the lattice of the parent metals, carbides of the metals in Groups 4, 5 and 6 show catalytic performances similar to those of the Pt-group metals in a variety of reactions involving hydrogen [70]. Zhang *et al.* were the first to develop the one-pot conversion process of cellulose to EG over carbide catalysts. Initially, the EG yield reached 27% over a W_2C/AC catalyst, and was then improved to 61%, with 100% cellulose conversion with the promotion of Ni [71]. Over AC support, the reduction temperature for the carbide catalyst was 973K, which was much lower than that of an oxide-supported catalyst with methane as the carbon resource. More importantly, the carbide catalyst also showed a higher activity than oxide-supported catalysts for cellulose conversion [72]. In a subsequent study, the same group investigated the performance of tungsten phosphide loaded onto AC for this reaction. Similar to the performance of tungsten carbide, the EG yield was maximized with the doping of Ni on tungsten phosphide [73]. Later, a series of bimetal, binary catalysts was developed for the synthesis of EG, which improved the EG yield up to 76 wt% [74]. Zhang *et al.* employed binary catalysts of Ru/AC-tungstic acid for cellulose conversion, and obtained a 50% EG yield. The temperature-controlled-phase-transfer properties of tungstic acid and good hydrothermal stability of Ru/AC were found to be crucial for the reusability of catalysts, which could be run for 20–30 reaction [74b]. Liu *et al.* combined the Ru catalyst with WO_3 for the controllable synthesis of 1,2-PG and EG. The competitive reactions of sugar hydrogenation and degradation were adjusted, and the selectivity of 1,2-PG and EG was improved to 40.9% and 22.7%, respectively, with 21.2% cellulose conversion over $50\%WO_3/Al_2O_3 + C_{act}$ catalysts [74c]. The addition of carbon significantly changed the selectivity of the glycols, which turned from EG to 1,2-PG. It was suggested that the AC would accelerate the isomerization of glucose into fructose by its surface basicity.

The use of traditional AC often results in a low dispersion and poor accessibility of the active sites, owing to its microporous structure. In particular, for the high loading of tungsten carbide the particles are aggregated to more than 50 nm. In contrast, MC has remarkable advantages of good accessibility of the pores, which benefits molecular diffusion. Moreover, recent advances in the synthesis of MCs has provided many possibilities for manipulating the catalytic properties of carbon-supported catalysts. Zhang *et al.* prepared a MC support bearing three-dimensional pores for the conversion of cellulose to EG. For this, the tungsten carbide was highly dispersed on the carbon support, and the average diameter of tungsten carbide was <10 nm. Over the WxC/MC catalyst, a 72.9% EG yield was obtained even without the addition of Ni, and this was further improved to 74.4% with the addition of 2% Ni [75]. These authors proposed that the high catalytic performance over MC-supported tungsten carbide was attributed to the high dispersion of tungsten carbide and improvements in transport between the reactant and product molecules.

Aiming at practical applications, different types of raw biomass, including corn stalk, poplar wood and *Miscanthus*, were used as feedstocks for EG synthesis over these carbon-supported catalysts [76]. Zhang *et al.* found that the lignin in the corn stalk inhibited the cellulosic biomass conversion and decreased the EG yield [76a]. After successive pretreatments with ammonia and H_2O_2, both the cellulose and hemicellulose in the corn stalk were effectively converted to EG and 1,2-PG, with a global yield of 48% over Ni-W_2C/AC catalysts. For woody biomass such as poplar wood, the sugars and lignin could be converted to EG (or 1,2-PG) and phenols, respectively, even without any pretreatment [76b]. When Claus *et al.* investigated the conversion of cellulose and woody biomass over Ru/W/AC catalysts, the EG yield was about 30% at 5% feedstock concentration at 493K for 3 h [76f]. Zhang *et al.* investigated the conversion of Jerusalem artichoke tuber using Ni–W_2C/AC catalysts, and obtained a 1,2-PG yield of 38.5% at 518K, 6 MPa H_2, and 80 min.

The reaction mechanism of this process was investigated on the basis of catalyst characterization and intermediates capture. It is believed that the reaction was a cascade reaction involving cellulose hydrolysis, retro-aldol condensation and hydrogenation reactions (Scheme 2.4) [77]. Cellulose was hydrolyzed to sugars by the acids arising from catalysts and hot water, after which the C–C bonds of the sugars were selectively cleaved by tungsten species following a retro-aldol condensation pathway. Finally, the EG precursor, glycolaldehyde, was hydrogenated to EG by hydrogenation catalysts such as Ru and Ni catalysts. In hot water and a H_2 atmosphere, the tungstic acid was transformed to soluble H_xWO_3, which was deemed a genuine active species for the catalytic retro-aldol condensation of sugars. After reaction, the soluble H_xWO_3 precipitated in a form of insoluble tungstic acid or oxide when the reaction solution was exposed to air [74b,77a].

The reaction kinetics for the conversion of cellulose to EG was investigated by using glucose as a starting material over AMT and Ru/AC catalysts. Zhang *et al.* disclosed that the pseudo first-order reaction of glycolaldehyde formation versus the pseudo second-order reaction of the side reactions of glycolaldehyde accounted for the sensitivity of EG formation. The major discrepancy in the activation energies between glucose hydrogenation and retro-aldol condensation of glucose resulted in an EG selectivity that was sensitive to the reaction temperatures [78]. Subsequently, it was disclosed that the retro-aldol condensation of glucose to form erythrose and glycolaldehyde is a first-order reaction with an apparent activation energy of 141.3 kJ mol^{-1}, while the retro-aldol condensation of erythrose to form two moles of glycolaldehyde and glycolaldehyde that further convert to side products are 1.7th- and 2.5th-order reactions, with apparent activation energies of 79.9 and 52.7 kJ mol^{-1}, respectively. The presence of tungstate species significantly retarded the

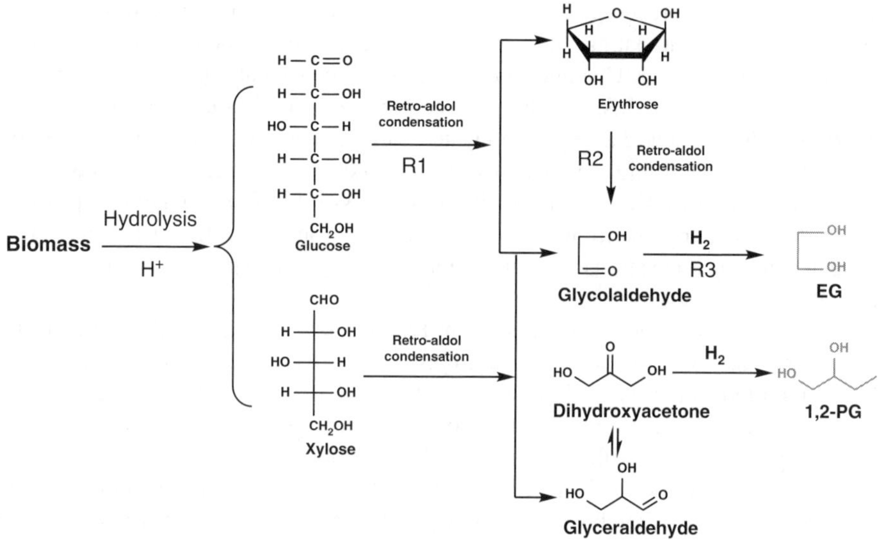

Scheme 2.4   *The reaction pathway for conversion of cellulose and hemicellulose to EG.*

aldose hydrogenation due to the competitive adsorption of aldoses and tungstate species, which in turn facilitated cleavage of the C–C bonds of the aldoses and increased the EG yield [79].

Besides carbon-supported catalysts, many oxide-supported catalysts have been developed for the conversion of cellulose to glycols [80]. For example, the Ni-W/SBA-15 catalyst showed considerable activity for cellulose conversion, with EG yields up to 76.1% [80a]. Nevertheless, the carbon-supported catalysts showed much better stability than did the oxide catalysts for cellulose conversion under hydrothermal conditions.

2.3.4   Conversion of Cellulose to Other Important Chemicals

Other valuable chemicals such as methyl glucosides, 5-hydroxymethylfurfural (5-HMF) and gluconic acid can also be produced from cellulose. Among these reactions, carbon-supported catalysts with highly dispersed active sites showed a promotion effect for product selectivity [81].

Methyl glucosides are important materials used for preparing surfactants, detergents, and cosmetics, which are stable even under harsh conditions owing to the protection of a hemiacetal group [27a]. Consequently, the methanolysis of cellulose to methyl glucosides has been studied extensively. Wang *et al.* reported that cellulose can be transformed into methyl glucosides in methanol with a yield of 62% in the presence of sulfonated carbon catalysts [82]. These carbon materials with –SO_3H groups are highly dispersed in methanol, which may be the reason for their superior catalytic performances. Bhaskar *et al.* developed an amorphous carbon-based catalyst by sulfonating bio-char obtained from biomass [83]. In this case, the cellulose was completely converted over the carbon catalyst, giving an approximately 90% yield of α, β-methyl glucoside in the reaction at 548K for 15 min.

HMF is a versatile platform chemical for the synthesis of biofuels, high-value-added chemicals, and plastics. Its production from cellulose has attracted considerable attention in the biorefinery process, and many progressive studies have been conducted on the conversion of monomeric and polymeric carbohydrates such as glucose, fructose, and cellulose, to HMF over carbon-based catalysts [84]. For instance, Prati *et al.* employed phosphorylated mesoporous carbon as catalyst in the dehydration of fructose to HMF, and achieved 70% selectivity [85], while Shotipruk *et al.* used a sulfonated carbon-based catalyst for the production of HMF and furfural from cassava waste. In the latter case a 12.1% yield of HMF and 2% yield of furfural were obtained with acetone/DMSO and water as the reaction medium at 523K in 1 min [86]. Although the HMF and furfural yields were rather low, the activity of the carbon-based acid catalyst was comparable to that of the homogeneous acid catalyst such as sulfuric acid. When Hu *et al.* prepared low-cost, carbon-based, solid-acid catalysts from cellulose for the conversion of sugars to HMF, a 46.4% yield of HMF was obtained from glucose in the presence of catalysts and [BMIM]Cl at 433K for 15 min. These authors confirmed that the catalysts and [BMIM]Cl system were suitable for converting most carbohydrates such as fructose, sucrose, maltose, cellobiose, starch and cellulose into HMF [87].

Gluconic acid is widely used in the food industry and in pharmaceuticals, but its selective synthesis from biomass remains a challenge [88]. Biella *et al.* investigated the oxidation of glucose to gluconic acid performed at both controlled and free pH values in an aqueous solution in the presence of a gold-on-carbon catalyst, using dioxygen as the oxidant [89]. Under mild conditions of 323–373K, no isomerization of glucose to fructose was observed, while Au/C with a particle size smaller than 10 nm showed a higher activity than did Pd–Bi/C and Pt–Pd–Bi/C catalysts. Wang *et al.* applied nitric acid-pretreated CNTs as a support for Au loading, and prepared the Au/CNT catalyst for the conversion of cellobiose to gluconic acid [90]. The gluconic acid yield reached 70% at 418K under 1.0 MPa O_2 pressure, without bases addition. Onda *et al.* used a bifunctional sulfonated AC-supported platinum catalyst for gluconic acid production from polysaccharides, such as starch and cellobiose. In this case, the gluconic acid yield reached about 60% in the reaction at 393K under 0.1 MPa of air without pH control [91].

In summary, the feedstock of cellulose is rather reluctant to be converted due to its high crystallinity, hydrogen bonding, and insolubility in most solvents. Various strategies have been developed to unlock the rigid structure of cellulose for the production of important chemicals, among which the conversion of cellulose at elevated temperature is very promising. By taking advantage of their good hydrothermal stability and tunable surface structure and porosity, carbon materials represent a type of ideal candidate for cellulose conversion. Hence, carbon materials will surely find wide applications in cellulose conversion with the rapid developments of biomass conversion and carbon material preparation methods.

2.4   Lignin Conversion

The carbon content in lignin (up to 50 wt%) is significantly higher than that in cellulose and hemicellulose. Therefore, lignin is a good candidate for the preparation of new carbon materials. In terms of its composition, lignin is mainly an amorphous tridimensional polymer of three primary fragments, namely, *p*-hydroxyphenyl units (H), guaiacyl units (G)

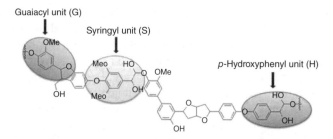

Figure 2.2   *Schematic representative structures of lignin.*

and syringyl units (S), which are linked by ether C–O or C–C bonds (Figure 2.2). Nowadays, all of the industrial methods used to produce aromatic chemicals rely heavily on petrochemicals as the starting materials. As lignin is by far the most abundant renewable substance that is composed of aromatic moieties, it is a sustainable feedstock for aromatic chemicals [92].

During the past ten years, much effort has been devoted to the rational design of catalysts for the conversion of lignin [93]. Different types of phenolic compounds have been obtained via catalytic hydrogenolysis [94], hydrodeoxygenation [95], hydrogenation [96] and other methods [97]. The use of carbon materials as catalysts for lignin conversion is scare, however, and therefore attention in the following section will mainly be focused on the application of carbon-based catalysts in lignin conversion.

2.4.1   Hydrogenolysis (Hydrocracking)

Xu and coworkers [98] reported the hydrogenolysis of lignosulfonate into phenols over heterogeneous catalysts. These authors first compared the activities of precious metal catalysts and transition metal catalysts for the conversion of lignosulfonate to simple aromatics. It was shown that Ni-based catalysts had the best catalytic activity, especially in the case of Ni/AC, which offered a high conversion of 60% in methanol as solvent. Meanwhile, 75–95% of the liquid products were alkane-substituted guaiacols, which implied that the highly selective reductive cleavage of aliphatic carbon–oxygen bonds was catalyzed by Ni/AC. The same authors also investigated the hydrogenolysis of lignin model compounds in the presence of Pd/AC and Ni/AC. In this case, Ni/AC was found to be active not only for the cleavage of aryl-alkyl C–O bond of lignin model compounds without disturbing the arenes, but also for breaking aliphatic C–O bonds of lignosulfonate in methanol as solvent. The products over Pd/C consisted mainly of dimers and cyclohexane, which was remarkably different from that over Ni/C (100% aromatic products). This result confirmed that Pd/C can be used to facilitate the hydrogenolysis of β-O-4 bonds concurrent with the hydrogenation of aromatic rings in lignin (Figure 2.3).

An attempt has been made to improve the catalytic activity of Ni/C catalysts by the addition of a second metal. Zhang and coworkers [76b] reported that bimetallic Ni-W$_2$C/C provided a 46.5% yield of phenolic compounds from the hydrogenolysis of lignin components in various woody biomasses. The catalytic activities of different carbon supported

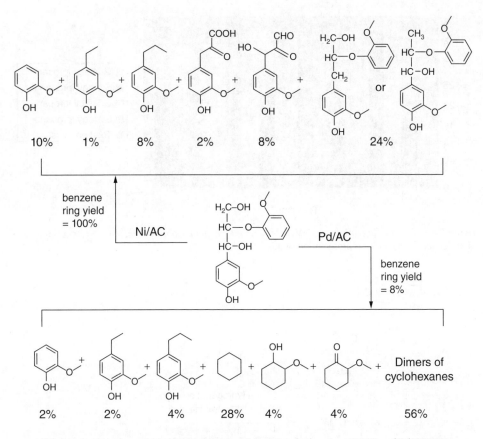

Figure 2.3   *Hydrogenolysis of guaiacylglycerol-β-guaiacyl ether over Ni/AC and Pd/AC [98].*

catalysts were also investigated, results showed that Pd/AC favored hydrogenation with aliphatic hydroxyl functions untouched, whereas Ni-W$_2$C/AC and other catalysts catalyzed further dehydroxylation reactions. In addition, Ni-W$_2$C/AC has excellent recycling ability, without any loss of activity after being reused three times (Figure 2.4).

The hydrogenolysis of lignin into monomeric phenols over carbon-supported Pd, Pt and Ru was carried out via selective cleavage of the aryl-O-aliphatic and aryl-O-aryl linkages, while corn stalk lignin with different ratios of H, G, and S units was hydrogenolyzed to 4-ethylphenolics (4-EP) and 4-ethylguaiacol (4-EG) with 3.1% and 1.4%, respectively (Figure 2.5) [99].

2.4.2   Hydrodeoxygenation (HDO)

It has been well documented that supported metal catalysts have shown potential applications in many HDO reactions. Wheeler and coworkers [100] reported the MoS$_2$/AC-catalyzed hydrodeoxygenation of lignin model compounds, and found that the surface chemistry of the activated carbon greatly affected the dispersion of the MoS$_2$ species.

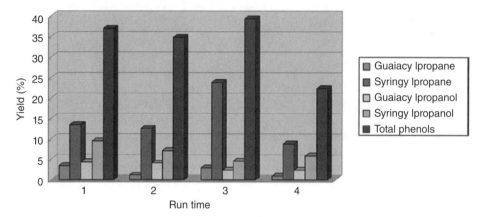

Figure 2.4   *The recycling results of Ni-W$_2$C/AC catalyst for lignin hydrocracking reaction [76b]. (See color plate section for the color representation of this figure.)*

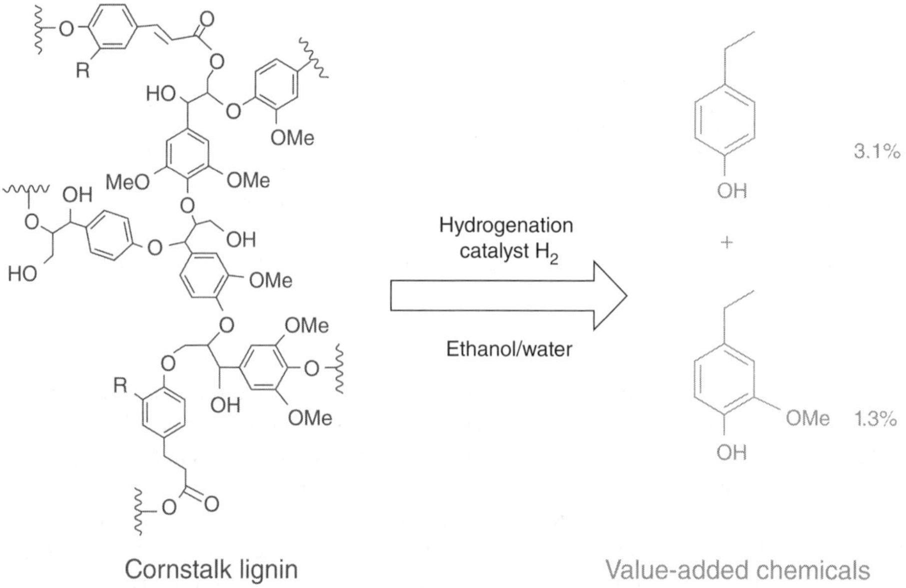

Figure 2.5   *Conversion of lignin over carbon-supported noble metals catalysts [99].*

A high dispersion of catalyst showed a significantly high activity in demethylation/dehydroxylation to produce catechol and phenol (Figure 2.6).

Nitrogen-containing mesoporous carbons have attracted much attention recently. Wang and coworkers [101] synthesized Pd nanoparticles supported on mesoporous N-doped carbon for vanillin HDO reaction. The high performance of the Pd@CN catalyst for HDO (100% conversion, 100% selectivity for 2-methoxy-4-methylphenol) was attributed to the structure of the catalytic N-doped carbon–metal heterojunction, which enhanced the

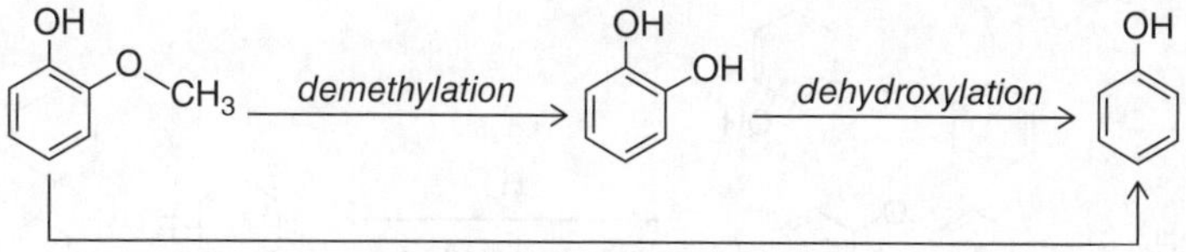

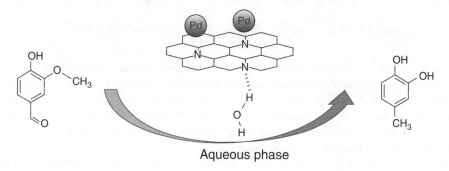

Figure 2.6   *Hydrodeoxygenation of guaiacol with MoS$_2$/AC catalysts.*

Aqueous phase

Figure 2.7   *Aqueous-phase HDO of vanillin with Pd@CN catalyst.*

catalyst dispersion in water and gave better activity under mild conditions in aqueous media (Figure 2.7).

A dual acid/redox catalytic process has been widely used in many heterogeneous reactions. A bifunctional catalyst, containing two types of active site, not only leads to multiple steps occurring in a single reactor but also shortens the work-up procedure. Lercher and coworkers [102] investigated the influence of a bifunctional catalyst on the overall conversion and product selectivity of lignin model compound reactions. In these studies, a combination of carbon-supported noble metals (Pd, Pt, Ru, or Rh) and phosphoric acid was used as catalysts, and the results showed that the bifunctional catalyst could efficiently catalyze the HDO of phenolic monomers to cycloalkanes with a high turnover frequency (TOF) (>1000 h^{-1}). A systematic kinetic study confirmed that the dual catalytic functions were indispensable for the HDO reaction. Furthermore, the general pathway for the conversion of lignin model compounds was proposed as: (i) metal-catalyzed hydrogenation of the benzene ring, followed by (ii) the acid-catalyzed hydration of cyclodiol, and (iii) a subsequent hydrogenation of cycloalkenes, yielding the target cycloalkanes (Figure 2.8).

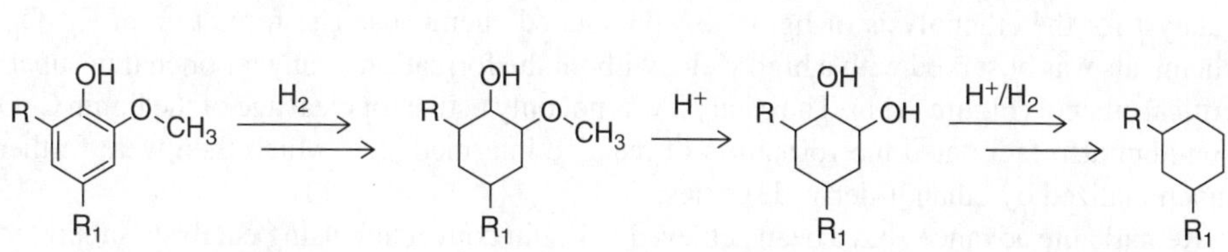

Figure 2.8   *Possible reaction pathways from lignin model compounds to cycloalkanes.*

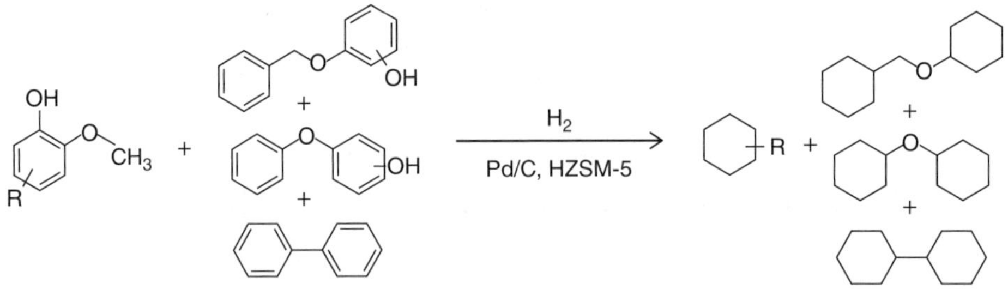

Figure 2.9   *HDO of phenolic monomers/dimers on Pd/C and HZSM-5 catalysts.*

In this reaction, it was necessary to find a compromise between the quantity of Pd/C and mineral acid to precisely control the rate of hydrogenation and dehydration step. In order to better understand the role of a bifunctional catalyst in this reaction, Lercher and coworkers, in a separate study, reported a detailed mechanistic comparison of different acid/base catalysts [103]. The results confirmed that a dual-functional catalyst provided a more efficient and less energy-demanding upgrading process.

Solid-acid catalysts have great potential in the development of new eco-friendly production processes, and have been widely used in various reactions today. Solid acids include heteropolyacids, metal complexes, and zeolites, for example, which have relatively high acid densities that can replace liquid mineral acids with an increasing efficiency. In a previous study, the development of a new bifunctional catalyst using a solid acid was carried out by Lercher and coworkers [104]. The results showed that HZSM-5 catalyzed not only the HDO of phenolic monomers but also the HDO of phenolic dimers (Figure 2.9). The latter one requires a cascade metal-acid-catalyzed cleavage of C$-$O bonds in aryl ethers and integrated hydrogenation and dehydration reactions.

2.4.3   Hydrogenation and Ethanolysis

Activated carbon cloth (ACC) shows high rates of adsorption, high surface areas and the potential for easy *in situ* regeneration, and consequently has recently received increasing attention (Figure 2.10). Saffron and coworkers [105] showed that Ru supported on ACC was an effective cathodic catalyst for the electrocatalytic hydrogenation of phenolic compounds. The catalyst performance was assessed based on original and demineralized ACC. In addition, the catalyst activity and conversion of guaiacol were affected by the reaction temperature (298–353K) and the pH value of the solution.

Li and coworkers [106] found that AC-supported *a*-molybdenum carbide was an efficient catalyst for the ethanolysis of lignin to value-added chemicals. The formation of C_6–C_{10} chemicals was observed with a high yield, without the formation of any tar or char in supercritical ethanol (Figure 2.11). The catalyst was not only active for cleavage of the lignin C$-$O bond but also facilitated the formation of reactive intermediates, which then were further functionalized by ethanol-derived species.

Remarkable advances have been achieved in lignin conversion using catalysts supported on carbon materials. These catalysts are not only highly resistant to acidic media but also

Figure 2.10 *Scanning electron microscopy image of the blank ACC FM100. Scale bar = 500 μm [105].*

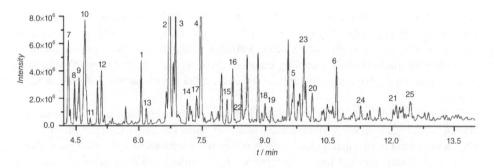

Figure 2.11 *Total-ion chromatogram (TIC) of the liquid products.*

low cost. It can be predicted, therefore, that carbon materials will play a more important role in biomass conversion in the foreseeable future.

2.5 Synthesis of Biofuel (Diesel or Jet Fuel) from Lignocellulose

As an efficient solution to the gradual diminishing of nonrenewable fossil energy, with the consequent environmental issues, an increasing amount of attention is being paid to the development of new energy (e.g., solar energy and biomass resources) as substitutes [107]. Lignocellulose, as the cheapest and most abundant nonedible biomass, is an ideal feedstock for the production of renewable fuels. Compared with fossil fuels, biofuels are renewable and CO_2-neutral.

The cellulose and hemicellulose fractions of lignocellulosic biomass can be hydrolyzed to C_5 or C_6 sugars, which could be further transformed into different platform chemicals or fuels and fuel additives. In the case of fuel production, one key step is elongation of the carbon chain (C–C coupling) to finally produce longer-chain hydrocarbons that closely meet the standards of current transportation fuels derived from oil refining (e.g., C_5–C_{12} for gasoline, C_8–C_{16} for jet fuel, and C_9–C_{22} for diesel). Another step is the defunctional-iztion of the oxygenated derivatives obtained from the aforementioned C–C coupling step. Such defunctionaliztion has two benefits. On the one hand, the oxygen in sugars could be fully removed from hydrogenation to generate gasoline-range alkanes. On the other hand, the sugars could be partially reduced to produce a series of reactive platform chemicals. Versatile C–C bond-formation reactions could then be carried out to produce heavy oxy-genates with longer carbon chains. Finally, these derived oxygenates may be thoroughly hydrodeoxygenated to the desired hydrocarbons. The hydrocarbons as-obtained can also be directly blended with conventional fuels.

Carbon materials are types of versatile support or catalyst for the production of biofuels, due to their unique physical and chemical properties. The primary focus of the following section is to provide an overview of the catalytic strategies employing carbon materials to produce biofuels in the C–C coupling reaction and the following HDO reaction.

2.5.1   C–C Coupling Reactions

Acids (sulfuric acid, molecular sieves, acidic resin, etc.) and bases (alkali hydroxide, alka-line earth oxides, hydrotalcites, etc.) are typical catalysts for C–C coupling reactions (such as aldol condensation, ketonization, oligomerization, alkylation). The synthetic methods of creating biofuels via C–C coupling reactions have recently been reviewed [108], and in the following section, attention will be focused on the use of carbon materials in these reactions.

Carbon materials have been widely used as supports due to their good hydrothermal stability and high specific surface area, which facilitates the better distribution of active sites and promotes stronger interactions between the reactants and the catalyst surface. Faba *et al.* [109] reported that supporting a Mg-Zr mixed oxide on carbon nanofibers (CNFs) and high-surface-area graphites (HSAGs) could improve the activity of the Mg-Zr mixed oxide in the aldol condensation of furfural and acetone. The best result was obtained over the HSAGs-supported catalyst prepared by a coprecipitation method. In the cross-condensation of acetone and furfural at 323K, the carbon yield of C_{13} and C_8 adducts over the carbon-supported catalysts was higher than that over the bulk oxides (84.7% versus 62.0%). The stability of the catalyst could also be increased by tuning the morphology of the carbonaceous support. Zapata *et al.* [110] investigated a biphasic emulsion system for the synthesis of fuel-range molecules by the aldol condensation of furfural and acetone derived from biomass. This showed that the nanohybrids obtained by the fusion of basic oxide nanoparticles and CNTS could simultaneously stabilize water/oil emulsions and catalyze the aldol condensation reaction. Among the investigated catalysts, the MWCNT/MgO nanohybrid was the most active for the aldol-condensation reaction. Under the same conditions, it was also found that higher yields of fuel-range condensation products (C_8–C_{13}) were obtained in the emulsion than those in the single phase.

Pd/C was used for hydrocarbon production via a stepwise dehydrogenation, aldol condensation, and hydrogenation reactions [111]. First, ethanol and butanol could

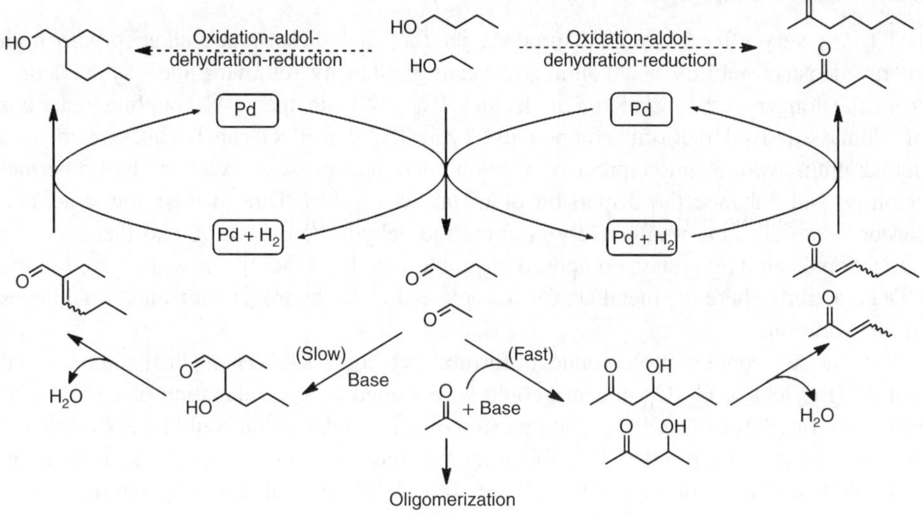

Figure 2.12 *Possible mechanism for the palladium-catalyzed alkylation of acetone [111].*

be transformed into hydrogen and aldehydes via dehydrogenation, and the aldehydes as-generated could then react with acetone under the reaction conditions. Hydrogenation of the aldol products to saturated ketones could be realized, with the hydrogen generated in the dehydrogenation step over Pd/C (see Figure 2.12). Anbarasan *et al.* [111] created this integration of biological and chemical routes to convert acetone–butanol–ethanol (ABE) fermentation products efficiently into ketones. In this way, they could obtain either biomass-based petrol or jet and diesel precursors by selectively controlling the reaction conditions. According to the results obtained, high yields of hydrocarbon fuel blend stocks could be created by catalytic conversion of the compounds in-situ extracted from the ABE fermentation products (extractive fermentation). These investigations provided a novel and prospective route for the economical conversion of biomass into liquid transportation fuels.

Besides catalyst support, carbon materials can also be used as catalysts by surface modification. For instance, sulfonated carbon materials [112] (i.e., AC, MC and CMK-3) have been used as solid-acid catalysts for the hydroxyalkylation/alkylation of 2-methylfuran and carbonyl compounds (e.g., furfural, acetone and butanal) and have exhibited good performances. Among these materials, the sulfated mesoporous carbon materials (CMK-3-SO$_3$H) exhibited the highest activity. The better activity of the sulfated mesoporous carbon materials can be rationalized in two main ways: (i) a higher surface area, which favors the adsorption of reactants over sulfate mesoporous carbon; and (ii) the larger pore sizes of the mesoporous carbon materials are beneficial to the mass transfer, which is important in the low-temperature liquid-phase reactions.

Sulfonated carbon materials offer a great advantage in catalyzing C–C coupling reactions in terms of their low cost and excellent catalytic performances. Therefore, increasing numbers of carbon material catalysts with specifically designed structures will be used in these reactions in the future.

2.5.2   Hydrodeoxygenation (HDO)

HDO is a very effective way to produce jet fuel and diesel-range alkanes with high energy density, stability and volumetric heating value by removing the oxygen atoms from the longer carbon skeleton molecules (C_8–C_{22}) via the C–C coupling reactions of biomass-derived platform compounds. Typically, the HDO can be interpreted as a dehydration–hydrogenation process. Carbon materials possess excellent hydrothermal stability, and enhance the dispersion of metal active sites. Due to their lower acidity, carbon materials cannot effectively promote the dehydration reaction, and therefore the bifunctional catalytic system composed of [noble-metal/C + acid] was widely used in the HDO reaction, where the metal acts as the active sites for hydrogenation and/or acid sites for dehydration.

With the self-condensation product of furfural as feedstock, Fu *et al.* [113] demonstrated that a high yield of C_8–C_{10} alkanes could be obtained by a combination of commercial Pt/C with $TaOPO_4$ or $NbOPO_4$. The first step of self-condensation could be easily carried out with cheap metal powders at ambient temperature in water. The extended application of $TaOPO_4$ for the production of C_{12} alkanes from HMF was subsequently reported [114]. 2,5-Dihydroxymethylfuran (DHMF), which can be obtained by the self-coupling of HMF, was converted into liquid hydrocarbon fuels over a bifunctional catalytic system consisting of Pt/C + $TaOPO_4$. The highest alkane selectivities (27.0% *n*-decane, 22.9% *n*-undecane and 45.6% *n*-dodecane) were reached at 573K for 3 h.

Sutton *et al.* [115] reported that saturated hydrocarbons can be produced by using $La(OTf)_3$ coupled with a Pd/C catalyst in a water/acetic acid reaction medium by the gradual defunctionalization of a series of biomass-derived molecules (Figure 2.13). There was no skeletal rearrangement by isomerization during the HDO reaction to produce transportation fuel-range linear alkanes. The greatest advantage was that the *n*-alkanes could be obtained at high yield and selectivity under mild reaction conditions. The HDO reactions could be carried out by either a stepwise process or a one-pot process. In the acidic medium (50% aqueous acetic acid solution), the acetic acid would rapidly promote hydrolysis of the furan ring by water, rather than hydrogenation of the furan ring over Pd/C, resulting in 2,5,8-nonanetrione as the sole product. Subsequently, as a Lewis acid, the $La(OTf)_3$ favored the HDO reaction via dehydration–hydrogenation pathways. That is, 2,5,8-nonanetrione was hydrogenated with Pd/C to alcohols, and the alcohols were dehydrated with $La(OTf)_3$ and finally saturated over Pd/C to give *n*-nonane as the final product under mild reaction conditions (573K, 3.45 MPa H_2). Another bifunctional catalytic system consisting of Pd/C + $La(OTf)_3$ + acetic acid was applied to convert DHMF into liquid hydrocarbon fuels at 523K and 2 MPa H_2 (16 h reaction) [116]. Alkanes and oxygenates mixtures were produced with a 64% selectivity to *n*-$C_{12}H_{26}$ and an overall C/H/O ratio of 84/11/5.

The hydrogenative ring-opening of furan compounds over Pd/C combined with acidic cocatalysts represents another strategy for effectively reducing the HDO temperature and increasing the selectivity of the target paraffins. The synthesis of furoins from biomass-derived furfural and 2-methylfurfural is realized at high yields in green and renewable solvents using N-heterocyclic carbene organocatalysts [117]. The resulting furoin molecules are used as precursors for fuels, using cascade catalysis (Figure 2.14). Two main products, identified as 1,2-bis(5-methyltetrahydrofuran-2-yl) ethane and

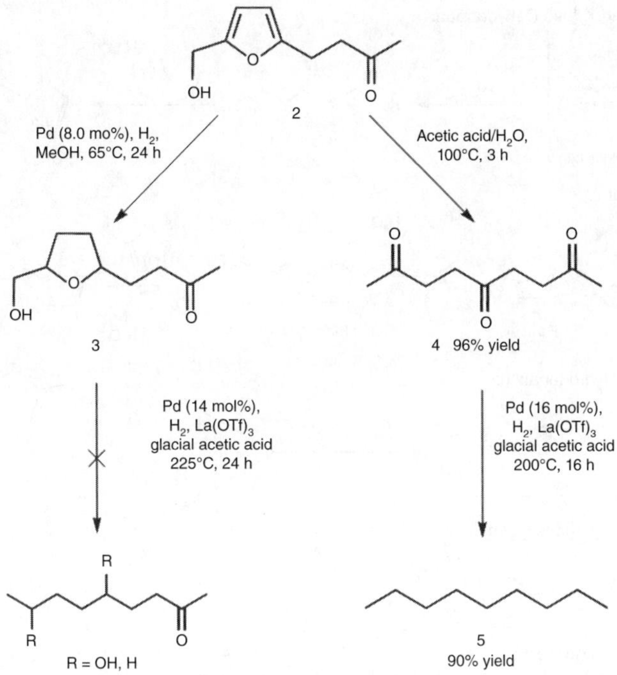

Figure 2.13   *Divergent pathway leads to different products [115].*

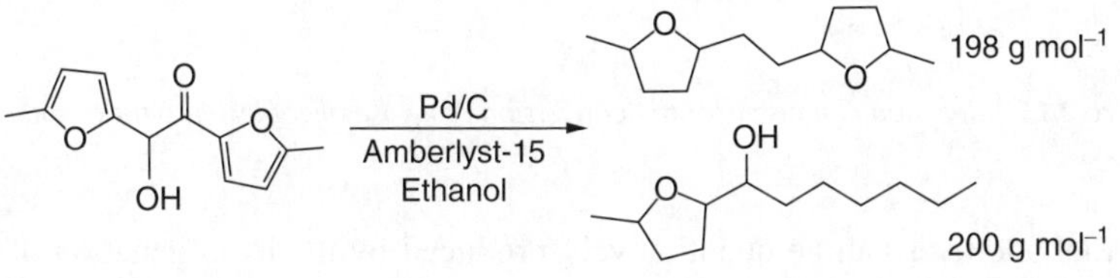

Figure 2.14   *One-pot HDO treatment of Me$_2$-furoin at 393K under 2 MPa H$_2$ [117].*

1-(5methyltetrahydrofuran-2-yl) heptanol, were synthesized over Pd/C with acidic cocatalysts under very mild conditions. These compounds could be used as fuel additives.

Prior to the HDO reaction, hydrogenation of the furan compounds obtained from the C–C coupling reaction could effectively suppress the generation of char during the HDO reaction. Xing *et al.* [118] reported a route for the production of liquid alkane fuels from a hemicellulose extract. This process involved four steps (Figure 2.15): (i) the preparation of furfural via the dehydration of hemicelluloses; (ii) production of the furfural–acetone–furfural (FAF) dimer by aldol condensation; (iii) hydrogenation of FAF to a hydrogenated product (H-FAF) at low temperatures over 5 wt% Ru/C catalyst (110–130 °C, 5.5 MPa H$_2$) in a batch reactor; and (iv) hydrodeoxygenation of H-FAF (in THF) to jet fuel-range alkanes over a Pt/Al$_2$O$_3$-SiO$_2$ catalyst in a fixed-bed reactor. Alkane yields of up to 91% (primarily C$_{13}$ and C$_{12}$ alkanes) were achieved. Finally, 76% of the theoretical yield for the overall process was obtained experimentally.

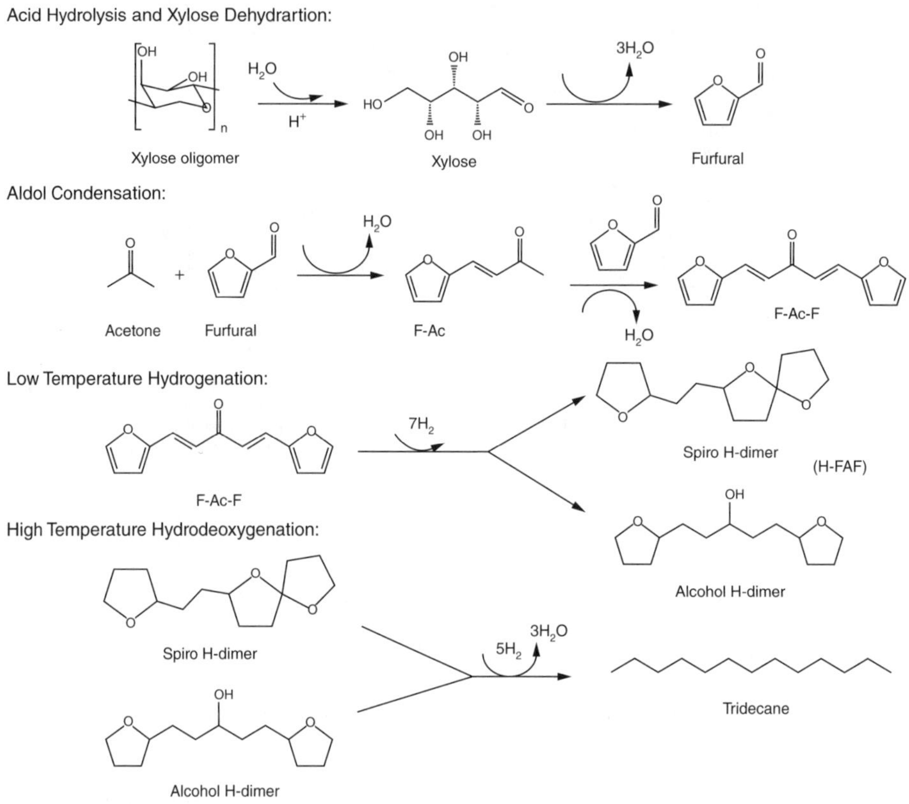

Acid Hydrolysis and Xylose Dehydrartion:

Aldol Condensation:

Low Temperature Hydrogenation:

High Temperature Hydrodeoxygenation:

Figure 2.15   *Reaction chemistry for the conversion of xylose oligomers into tridecane [118].*

Some fuel additives can be quantitatively produced by the hydrogenation of biomass-derived oxygenates over Pd/C and Ru/C catalysts. Chatterjee *et al.* [119] reported that 4-(tetrahydrofuran-2-yl)butan-2-one was solely produced from 4-(furan-2-yl)butan-2-one (temperature = 353 K; P_{H2} = 4 MPa and P_{CO2} = 14 MPa) over the Pd/C catalyst with high selectivity (>99%). Pholjaroen *et al.* [120] reported the synthesis of oxygenate fuel additives by hydrogenation of the aldol condensation product of furfural and methyl isobutyl ketone with Ru/C at low temperature (433K).

Hydrogenolysis represents another way to remove oxygen through hydrogenating ketones to alcohol, and then dehydrating alcohol to olefins. Dumesic *et al.* [121] reported the hydrogenation of 5-nonanone to 5-nonanol over a Ru/C catalyst; a mixture of C_9 olefins was then formed by the dehydration of 5-nonanol over Amberlyst 70, and a 43% yield of nonene was achieved after a holding time of 4 h at 433K. To attain C_{18}–C_{27} olefins of longer carbon chains, the mixture of C_9 olefins was further oligomerized over Nafion SAC 13.

Carbon material-supported noble metal or non-platinum group precious metal catalysts can directly remove the oxygen of C–C coupling reaction products via HDO reaction at elevated temperatures (623K). Corma *et al.* [122] reported that

2,2′-butyldiene-bis(methylfuran) was synthesized from the hydroxyalkylation–alkylation (HAA) of 2-methylfuran (2-MF) and butanal with a *p*-toluene sulfonic acid catalyst under solvent-free conditions at 333 K, with a yield of 78%. Subsequently, HDO of the C_{14} intermediate was carried out in a fixed-bed reactor at 633 K and 5 MPa H_2 over a Pt/C catalyst, and an up to 89% carbon yield of C_9–C_{14} alkanes was obtained. When 20% of the carbon support was replaced by titanium dioxide, the diesel-range alkanes yield was further improved to 92% as the acidic TiO_2 prompted the complete elimination of oxygenated compounds in the aforementioned liquid organic phase (<1%). Li *et al.* [112a] reported a similar route for the synthesis of jet fuel-range alkanes in three steps: (i) HAA of furfural and 2-MF; (ii) hydrogenating the HAA product over Pd/C at a low temperature (433K); and (iii) the hydrogenated product was further hydrodeoxygenated over a Pt/ZrP catalyst at 623K. In this process, the hydrogenated product was pumped directly into the fixed-bed reactor without any solvent, and C_9–C_{14} alkanes were obtained with 94% yield. In subsequent studies, Li *et al.* [112b,123] prepared a series of diesel and jet fuel precursors by the HAA of 2-methylfuran and carbonyl compounds (furfural, butanal, acetone and hydroxyacetone). These C_{13}–C_{15} oxygenate intermediates could be directly hydrodeoxygenated to diesel and jet fuel-range branched alkanes over Pd/C, Pt/C and Ni-W_xC/C at 623K. More than 93% carbon yields of diesel-range alkanes were achieved over the Ni-W_xC/C catalyst, which also has a good stability in the hydrodeoxygenation reaction. Compared with H-β-supported noble catalysts [124], which can realize the HDO reaction at much lower temperatures, a higher yield of 6-propyl-undecane (86.1% Pd/C versus 57.3% Pd/H-β) can be obtained when neutral carbon was used as a support in the HDO reaction. This improvement is due to the acidic supports not only promoting C–O bond cleavage via the dehydration reaction, but also to unexpected C–C cracking.

Yang *et al.* [125] also reported the direct HDO of 1-(furan-2-yl)-5-methylhex-1-en-3-one (the aldol condensation product of methyl isobutyl ketone and furfural) over Pt/C, Ir/C, Pd/C and Pd-Fe/C catalysts under solvent-free conditions. At 643K and 6 MPa H_2, more than 90% carbon yields of jet fuel-range alkanes were achieved. Interestingly, with the Fe-modified Pd/C catalyst, the decarboxylation was conspicuously suppressed and the molar ratio of 2-methyl-decane/2-methyl-nonane greatly increased, from 0.4 to 14.

Recently, Mascal *et al.* [126] reported a two-step process to produce branched C_7–C_{10} gasoline-range alkanes at high yields. The first step involved the formation of an angelica lactone dimer, while in the second step the dimer was quantitatively hydrodeoxygenated over ReO_x/SiO_2 and Pt-ReO_x/C catalysts. A high yield (88%) of C_7–C_{10} alkanes was achieved at 513K, which indicated that the ReO_x-modified Pt/C catalyst could directly produce alkanes via the HDO reaction at low temperature.

During recent years, biomass conversion to biofuels has become a hot topic in the transportation fuel field. This section has summarized the recent advances in upgrading biomass platform compounds to long-chain alkanes using carbon materials as support or solid-acid catalysts. Currently, the routes from biomass to biofuels typically involve several steps. In order to reduce the separation steps and production costs, the development of one-pot synthesis of alkanes from lignocellulose is very important and, indeed, is imperative. The special design of multifunctional materials would play a major role in the development of simplified processes. Carbon materials, possessing characteristics such as high specific surface areas, hierarchical pores, easy functionalization and good thermostability, have great potential and prospects in future biofuels production.

2.6 Summary

In summary, various carbon materials are synthesized from sugars, starch or lignocellulosic biomass. In contrast to oxides, carbon materials are more structurally and chemically stable under hydrothermal or solvent–thermal reaction conditions. On the one hand, considerable studies have focused on the porosity architecture to improve mass transfer and effective catalytic surface areas during biomass conversion. On the other hand, the morphology of metals supported on carbon can be significantly modulated by varying the surface chemistry of the supports with N- and/or O-containing functionalities. After surface modification and/or metal loading, the carbon-based catalysts show considerable activity for biomass conversion, and therefore both physical and chemical properties should be considered when designing catalysts with high activity and selectivity. Currently, carbon-supported catalysts and carbon catalysts have been widely used for cellulose conversion (e.g., hydrolysis, hydrolytic hydrogenation, hydrocracking reactions, esterification, dehydration and oxidation reactions), lignin conversion (e.g., hydrogenolysis, hydrodeoxygenation, hydrogenation and ethanolysis), as well as for biofuel synthesis.

Although great progress has been made with regards to catalytic biomass conversion over carbon/carbon-supported catalysts, several issues remain that require considerable research and development: (i) the role of the carbon surface and carbon hybridization, besides their influence on metallic active sites; (ii) the development of facile methods to fabricate mesoporous/hierarchically structured carbons and to functionalize them with certain groups; and (iii) strategies to transform biomass as a whole, rather than its individual components or model compounds.

References

1. Mäki-Arvela, P., Holmbom, B., Salmi, T., Murzin, D.Y. (2007) Recent progress in synthesis of fine and specialty chemicals from wood and other biomass by heterogeneous catalytic processes. *Catal. Rev.*, **49**, 197–340.
2. Deng, W., Zhang, H., Xue, L., Zhang, Q., Wang, Y. (2015) Selective activation of the C–O bonds in lignocellulosic biomass for the efficient production of chemicals. *Chin. J. Catal.*, **36**, 1440–1460.
3. Matthiesen, J., Hoff, T., Liu, C., Pueschel, C., Rao, R., Tessonnier, J.-P. (2014) Functional carbons and carbon nanohybrids for the catalytic conversion of biomass to renewable chemicals in the condensed phase. *Chin. J. Catal.*, **35**, 842–855.
4. Lam, E., Luong, J.H.T. (2014) Carbon materials as catalyst supports and catalysts in the transformation of biomass to fuels and chemicals. *ACS Catal.*, **4**, 3393–3410.
5. Zhai, Y., Dou, Y., Zhao, D., Fulvio, P.F., Mayes, R.T., Dai, S. (2011) Carbon materials for chemical capacitive energy storage. *Adv. Mater.*, **23**, 4828–4850.
6. Figueiredo, J.L., Poco, J.G.R., Thomaz, O., Aldeia, W., DiGiorgi, V., Sakamoto, R.G. (1996) Evaluation of the efficiency of activation in the production of carbon adsorbents. *Carbon*, **34**, 679–681.
7. Tian, H.Y., Buckley, C.E., Wang, S.B., Zhou, M.F. (2009) Enhanced hydrogen storage capacity in carbon aerogels treated with KOH. *Carbon*, **47**, 2128–2130.

8. Liang, C., Li, Z., Dai, S. (2008) Mesoporous carbon materials: Synthesis and modification. *Angew. Chem. Int. Ed.*, **47**, 3696–3717.

9. Ryoo, R., Joo, S.H., Jun, S. (1999) Synthesis of highly ordered carbon molecular sieves via template-mediated structural transformation. *J. Phys. Chem. B*, **103**, 7743–7746.

10. (a) Meng, Y., Gu, D., Zhang, F.Q., Shi, Y.F., Yang, H.F., Li, Z., Yu, C.Z., Tu, B., Zhao, D.Y. (2005) Ordered mesoporous polymers and homologous carbon frameworks: Amphiphilic surfactant templating and direct transformation. *Angew. Chem. Int. Ed.*, **44**, 7053–7059; (b) Liang, C.D., Hong, K.L., Guiochon, G.A., Mays, J.W., Dai, S. (2004) Synthesis of a large-scale highly ordered porous carbon film by self-assembly of block copolymers. *Angew. Chem. Int. Ed.*, **43**, 5785–5789.

11. (a) Mochida, I., Ogaki, M., Fujitsu, H., Komatsubara, Y., Ida, S. (1983) Catalytic activity of coke activated with sulphuric acid for the reduction of nitric oxide. *Fuel*, **62**, 867–868; (b) Tsuji, K., Shiraishi, I. (1997) Combined desulfurization, denitrification and reduction of air toxics using activated coke: 2. *Process applications and performance of activated coke. Fuel*, **76**, 555–560.

12. Abrams, L., Cicha, W.V., Manzer, L.E., Subramoney, S. (2000) in *Studies in Surface Science and Catalysis*, Vol. **130** (eds F.V.M.S.M. Avelino Corma, G.F. José Luis), Elsevier, pp. 455–460.

13. Silva, I.F., Vital, J., Ramos, A.M., Valente, H., do Rego, A.M.B., Reis, M.J. (1998) Oxydehydrogenation of cyclohexanol over carbon catalysts. *Carbon*, **36**, 1159–1165.

14. Grunewald, G.C., Drago, R.S. (1991) Carbon molecular-sieves as catalysts and catalyst supports. *J. Am. Chem. Soc.*, **113**, 1636–1639.

15. Santos, A., Yustos, P., Rodriguez, S., Garcia-Ochoa, F. (2006) Wet oxidation of phenol, cresols and nitrophenols catalyzed by activated carbon in acid and basic media. *Appl. Catal. B. Environ.*, **65**, 269–281.

16. Xu, J., Zhao, X., Wang, A., Zhang, T. (2014) Synthesis of nitrogen-doped ordered mesoporous carbons for catalytic dehydrochlorination of 1,2-dichloroethane. *Carbon*, **80**, 610–616.

17. Verboekend, D., Perez-Ramirez, J. (2011) Design of hierarchical zeolite catalysts by desilication. *Catal. Sci. Tech.*, **1**, 879–890.

18. Zhao, X., Xu, J., Wang, A., Zhang, T. (2015) Porous carbon in catalytic transformation of cellulose. *Chin. J. Catal.*, **36**, 1419–1427.

19. Yamaguchi, D., Kitano, M., Suganuma, S., Nakajima, K., Kato, H., Hara, M. (2009) Hydrolysis of cellulose by a solid acid catalyst under optimal reaction conditions. *J. Phys. Chem. C*, **113**, 3181–3188.

20. Bandosz, T.J. (2008) in *Carbon Materials for Catalysis*, John Wiley & Sons, Inc., pp. 45–92.

21. (a) Toda, M., Takagaki, A., Okamura, M., Kondo, J.N., Hayashi, S., Domen, K., Hara, M. (2005) Green Chemistry - Biodiesel made with sugar catalyst. *Nature*, **438**, 178–178; (b) Suganuma, S., Nakajima, K., Kitano, M., Yamaguchi, D., Kato, H., Hayashi, S., Hara, M. (2008) Hydrolysis of cellulose by amorphous carbon bearing SO$_3$H, COOH, and OH groups. *J. Am. Chem. Soc.*, **130**, 12787–12793; (c) Okamura, M., Takagaki, A., Toda, M., Kondo, J.N., Domen, K., Tatsumi, T., Hara, M., Hayashi, S. (2006) Acid-catalyzed reactions on flexible polycyclic aromatic carbon in amorphous carbon. *Chem. Mater.*, **18**, 3039–3045.

22. (a) Domingo-Garcá, M., López-Garzón, F.J., Pérez-Mendoza, M. (2000) Modifications produced by O$_2$ plasma treatments on a mesoporous glassy carbon. *Carbon*, **38**, 555–563; (b) Montes-Morán, M.A., Martínez-Alonso, A., Tascón, J.M.D., Paiva, M.C., Bernardo, C.A. (2001) Effects of plasma oxidation on the surface and interfacial properties of carbon fibres/polycarbonate composites. *Carbon*, **39**, 1057–1068; (c) Lee, Y.-J., Kim, H.-H., Hatori, H. (2004) Effects of substitutional B on oxidation of carbon nanotubes in air and oxygen plasma. *Carbon*, **42**, 1053–1056; (d) Park, S.-J., Jung, W.-Y. (2002) Preparation of activated carbons derived from KOH-impregnated resin. *Carbon*, **40**, 2021–2022.

23. (a) Bandosz, T.J., Jagiello, J., Schwarz, J.A. (1992) Comparison of methods to assess surface acidic groups on activated carbons. *Anal. Chem.*, **64**, 891–895; (b) Salame, I.I., Bandosz, T.J.

(2001) Surface chemistry of activated carbons: combining the results of temperature-programmed desorption, Boehm, and potentiometric titrations. *J. Coll. Inter. Sci.*, **240**, 252–258.

24. Zhao, X., Zhang, Q., Chen, C.-M., Zhang, B., Reiche, S., Wang, A., Zhang, T., Schlögl, R., Sheng Su, D. (2012) Aromatic sulfide, sulfoxide, and sulfone mediated mesoporous carbon monolith for use in supercapacitor. *Nano Energ.*, **1**, 624–630.

25. (a) Inagaki, N., Narushima, K., Hashimoto, H., Tamura, K. (2007) Implantation of amino functionality into amorphous carbon sheet surfaces by NH_3 plasma. *Carbon*, **45**, 797–804; (b) Lahaye, J., Nansé, G., Bagreev, A., Strelko, V. (1999) Porous structure and surface chemistry of nitrogen-containing carbons from polymers. *Carbon*, **37**, 585–590; (c) Strelko, V.V., Kuts, V.S., Thrower, P.A. (2000) On the mechanism of possible influence of heteroatoms of nitrogen, boron and phosphorus in a carbon matrix on the catalytic activity of carbons in electron transfer reactions. *Carbon*, **38**, 1499–1503; (d) Biniak, S., Szymański, G., Siedlewski, J., Świtkowski, A. (1997) The characterization of activated carbons with oxygen and nitrogen surface groups. *Carbon*, **35**, 1799–1810.

26. Montes-Morán, M.A., Suárez, D., Menéndez, J.A., Fuente, E. (2004) On the nature of basic sites on carbon surfaces: an overview. *Carbon*, **42**, 1219–1225.

27. (a) Corma, A., Iborra, S., Velty, A. (2007) Chemical routes for the transformation of biomass into chemicals. *Chem. Rev.*, **107**, 2411–2502; (b) Zhou, C.H., Xia, X., Lin, C.X., Tong, D.S., Beltramini, J. (2011) Catalytic conversion of lignocellulosic biomass to fine chemicals and fuels. *Chem. Soc. Rev.*, **40**, 5588–5617; (c) Gallezot, P. (2012) Conversion of biomass to selected chemical products. *Chem. Soc. Rev.*, **41**, 1538–1558.

28. (a) Himmel, M.E., Ding, S.Y., Johnson, D.K., Adney, W.S., Nimlos, M.R., Brady, J.W., Foust, T.D. (2007) Biomass recalcitrance: Engineering plants and enzymes for biofuels production. *Science*, **315**, 804–807; (b) Zhao, X., Zhang, L., Liu, D. (2012) Biomass recalcitrance. Part I: the chemical compositions and physical structures affecting the enzymatic hydrolysis of lignocellulose. *Biofuels Bioprod. Biorefining – Biofpr.*, **6**, 465–482; (c) DeMartini, J.D., Pattathil, S., Miller, J.S., Li, H., Hahn, M.G., Wyman, C.E. (2013) Investigating plant cell wall components that affect biomass recalcitrance in poplar and switchgrass. *Energ. Environ. Sci.*, **6**, 898–909.

29. (a) Climent, M.J., Corma, A., Iborra, S. (2011) Converting carbohydrates to bulk chemicals and fine chemicals over heterogeneous catalysts. *Green Chem.*, **13**, 520–540; (b) Chatterjee, C., Pong, F., Sen, A. (2015) Chemical conversion pathways for carbohydrates. *Green Chem.*, **17**, 40–71.

30. (a) Hu, L., Lin, L., Wu, Z., Zhou, S., Liu, S. (2015) Chemocatalytic hydrolysis of cellulose into glucose over solid acid catalysts. *Appl. Catal. B. Environ.*, **174–175**, 225–243; (b) Wang, J., Xi, J., Wang, Y. (2015) Recent advances in the catalytic production of glucose from lignocellulosic biomass. *Green Chem.*, **17**, 737–751.

31. Butera, G., De Pasquale, C., Maccotta, A., Alonzo, G., Conte, P. (2011) Thermal transformation of micro-crystalline cellulose in phosphoric acid. *Cellulose*, **18**, 1499–1507.

32. Kabyemela, B.M., Adschiri, T., Malaluan, R.M., Arai, K. (1997) Kinetics of glucose epimerization and decomposition in subcritical and supercritical water. *Ind. Eng. Chem. Res.*, **36**, 1552–1558.

33. (a) Lanzafame, P., Temi, D.M., Perathoner, S., Spadaro, A.N., Centi, G. (2012) Direct conversion of cellulose to glucose and valuable intermediates in mild reaction conditions over solid acid catalysts. *Catal. Today*, **179**, 178–184; (b) Rinaldi, R., Schüth, F. (2009) Acid hydrolysis of cellulose as the entry point into biorefinery schemes. *ChemSusChem*, **2**, 1096–1107; (c) Huang, Y.-B., Fu, Y. (2013) Hydrolysis of cellulose to glucose by solid acid catalysts. *Green Chem.*, **15**, 1095–1111.

34. Toda, M., Takagaki, A., Okamura, M., Kondo, J.N., Hayashi, S., Domen, K., Hara, M. (2005) Green Chem.: Biodiesel made with sugar catalyst. *Nature*, **438**, 178–178.

35. Hara, M., Yoshida, T., Takagaki, A., Takata, T., Kondo, J.N., Hayashi, S., Domen, K. (2004) A carbon material as a strong protonic acid. *Angew. Chem. Int. Ed.*, **43**, 2955–2958.

36. Hara, M. (2010) Biomass conversion by a solid acid catalyst. *Energ. Environ. Sci.*, **3**, 601–607.

37. (a) Liu, M., Jia, S., Gong, Y., Song, C., Guo, X. (2013) Effective hydrolysis of cellulose into glucose over sulfonated sugar-derived carbon in an ionic liquid. *Ind. Eng. Chem. Res.*, **52**, 8167–8173;

(b) Guo, H., Qi, X., Li, L., Smith Jr, R.L. (2012) Hydrolysis of cellulose over functionalized glucose-derived carbon catalyst in ionic liquid. *Bioresource Technol.*, **116**, 355–359; (c) Qi, X., Lian, Y., Yan, L., Smith Jr, R.L. (2014) One-step preparation of carbonaceous solid acid catalysts by hydrothermal carbonization of glucose for cellulose hydrolysis. *Catal. Commun.*, **57**, 50–54.

38. Wu, Y., Fu, Z., Yin, D., Xu, Q., Liu, F., Lu, C., Mao, L. (2010) Microwave-assisted hydrolysis of crystalline cellulose catalyzed by biomass char sulfonic acids. *Green Chem.*, **12**, 696–700.

39. Fan, J., De Bruyn, M., Budarin, V.L., Gronnow, M.J., Shuttleworth, P.S., Breeden, S., Macquarrie, D.J., Clark, J.H. (2013) Direct microwave-assisted hydrothermal depolymerization of cellulose. *J. Am. Chem. Soc.*, **135**, 11728–11731.

40. Jiang, Y., Li, X., Wang, X., Meng, L., Wang, H., Peng, G., Wang, X., Mu, X. (2012) Effective sac-charification of lignocellulosic biomass over hydrolysis residue derived solid acid under microwave irradiation. *Green Chem.*, **14**, 2162–2167.

41. Onda, A., Ochi, T., Yanagisawa, K. (2008) Selective hydrolysis of cellulose into glucose over solid acid catalysts. *Green Chem.*, **10**, 1033–1037.

42. Pang, J., Wang, A., Zheng, M., Zhang, T. (2010) Hydrolysis of cellulose into glucose over carbons sulfonated at elevated temperatures. *Chem. Commun.*, **46**, 6935–6937.

43. Liu, Z., Fu, X., Tang, S., Cheng, Y., Zhu, L., Xing, L., Wang, J., Xue, L. (2014) Sulfonated magnetic carbon nanotube arrays as effective solid acid catalysts for the hydrolyses of polysaccharides in crop stalks. *Catal. Commun.*, **56**, 1–4.

44. Joo, S.H., Choi, S.J., Oh, I., Kwak, J., Liu, Z., Terasaki, O., Ryoo, R. (2001) Ordered nanoporous arrays of carbon supporting high dispersions of platinum nanoparticles. *Nature*, **412**, 169–172.

45. Kobayashi, H., Komanoya, T., Hara, K., Fukuoka, A. (2010) Water-tolerant mesoporous-carbon-supported ruthenium catalysts for the hydrolysis of cellulose to glucose. *ChemSusChem*, **3**, 440–443.

46. Van de Vyver, S., Peng, L., Geboers, J., Schepers, H., de Clippel, F., Gommes, C.J., Goderis, B., Jacobs, P.A., Sels, B.F. (2010) Sulfonated silica/carbon nanocomposites as novel catalysts for hydrolysis of cellulose to glucose. *Green Chem.*, **12**, 1560–1563.

47. Pang, Q., Wang, L., Yang, H., Jia, L., Pan, X., Qiu, C. (2014) Cellulose-derived carbon bearing -Cl and -SO$_3$H groups as a highly selective catalyst for the hydrolysis of cellulose to glucose. *RSC Adv.*, **4**, 41212–41218.

48. Hu, S., Smith, T.J., Lou, W., Zong, M. (2014) Efficient hydrolysis of cellulose over a novel sucralose-derived solid acid with cellulose-binding and catalytic sites. *J. Agric. Food Chem.*, **62**, 1905–1911.

49. Zhao, X., Wang, J., Chen, C., Huang, Y., Wang, A., Zhang, T. (2014) Graphene oxide for cellulose hydrolysis: how it works as a highly active catalyst? *Chem. Commun.*, **50**, 3439–3442.

50. Foo, G.S., Sievers, C. (2015) Synergistic effect between defect sites and functional groups on the hydrolysis of cellulose over activated carbon. *ChemSusChem*, **8**, 534–543.

51. Zhang, J., Li, J.-b., Wu, S.-B., Liu, Y. (2013) Advances in the catalytic production and utilization of sorbitol. *Ind. Eng. Chem. Res.*, **52**, 11799–11815.

52. (a) Ruppert, A.M., Weinberg, K., Palkovits, R. (2012) Hydrogenolysis goes bio: From carbohydrates and sugar alcohols to platform chemicals. *Angew. Chem. Int. Ed.*, **51**, 2564–2601; (b) Li, Y., Liao, Y., Cao, X., Wang, T., Ma, L., Long, J., Liu, Q., Xua, Y. (2015) Advances in hexitol and ethylene glycol production by one-pot hydrolytic hydrogenation and hydrogenolysis of cellulose. *Biomass Bioenergy*, **74**, 148–161; (c) Vilcocq, L., Cabiac, A., Especel, C., Lacombe, S., Duprez, D. (2012) Sorbitol transformation in aqueous medium: Influence of metal/acid balance on reaction selectivity. *Catal. Today*, **189**, 117–122.

53. Luo, C., Wang, S.A., Liu, H.C. (2007) Cellulose conversion into polyols catalyzed by reversibly formed acids and supported ruthenium clusters in hot water. *Angew. Chem. Int. Ed.*, **46**, 7636–7639.

54. Fukuoka, A., Dhepe, P.L. (2006) Catalytic conversion of cellulose into sugar alcohols. *Angew. Chem. Int. Ed.*, **45**, 5161–5163.

55. Kobayashi, H., Hosaka, Y., Hara, K., Feng, B., Hirosaki, Y., Fukuoka, A. (2014) Control of selectivity, activity and durability of simple supported nickel catalysts for hydrolytic hydrogenation of cellulose. *Green Chem.*, **16**, 637–644.

56. Deng, W., Tan, X., Fang, W., Zhang, Q., Wang, Y. (2009) Conversion of cellulose into sorbitol over carbon nanotube-supported ruthenium catalyst. *Catal. Lett.*, **133**, 167–174.

57. Ding, L.N., Wang, A.Q., Zheng, M.Y., Zhang, T. (2010) Selective transformation of cellulose into sorbitol by using a bifunctional nickel phosphide catalyst. *ChemSusChem*, **3**, 818–821.

58. Van de Vyver, S., Geboers, J., Dusselier, M., Schepers, H., Vosch, T., Zhang, L., Van Tendeloo, G., Jacobs, P.A., Sels, B.F. (2010) Selective bifunctional catalytic conversion of cellulose over reshaped Ni particles at the tip of carbon nanofibers. *ChemSusChem*, **3**, 698–701.

59. Kobayashi, H., Ito, Y., Komanoya, T., Hosaka, Y., Dhepe, P.L., Kasai, K., Hara, K., Fukuoka, A. (2011) Synthesis of sugar alcohols by hydrolytic hydrogenation of cellulose over supported metal catalysts. *Green Chem.*, **13**, 326–333.

60. Zhou, L., Li, Z., Pang, J., Zheng, M., Wang, A., Zhang, T. (2015) Catalytic conversion of Jerusalem artichoke tuber into hexitols using the bifunctional catalyst Ru/(AC-SO$_3$H). *Chin. J. Catal.*, **36**, 1694–1700.

61. Wang, D., Niu, W., Tan, M., Wu, M., Zheng, X., Li, Y., Tsubaki, N. (2014) Pt nanocatalysts supported on reduced graphene oxide for selective conversion of cellulose or cellobiose to sorbitol. *ChemSusChem*, **7**, 1398–1406.

62. Pang, J., Wang, A., Zheng, M., Zhang, Y., Huang, Y., Chen, X., Zhang, T. (2012) Catalytic conversion of cellulose to hexitols with mesoporous carbon supported Ni-based bimetallic catalysts. *Green Chem.*, **14**, 614–617.

63. Xu, J., Wang, A., Wang, X., Su, D., Zhang, T. (2011) Synthesis, characterization, and catalytic application of highly ordered mesoporous alumina-carbon nanocomposites. *Nano Res.*, **4**, 50–60.

64. Kobayashi, H., Matsuhashi, H., Komanoya, T., Hara, K., Fukuoka, A. (2011) Transfer hydrogenation of cellulose to sugar alcohols over supported ruthenium catalysts. *Chem. Commun.*, **47**, 2366–2368.

65. Shrotri, A., Kobayashi, H., Tanksale, A., Fukuoka, A., Beltramini, J. (2014) Transfer hydrogenation of cellulose-based oligomers over carbon-supported ruthenium catalyst in a fixed-bed reactor. *ChemCatChem*, **6**, 1349–1356.

66. Hilgert, J., Meine, N., Rinaldi, R., Schueth, F. (2013) Mechanocatalytic depolymerization of cellulose combined with hydrogenolysis as a highly efficient pathway to sugar alcohols. *Energ. Environ. Sci.*, **6**, 92–96.

67. Kobayashi, H., Yamakoshi, Y., Hosaka, Y., Yabushita, M., Fukuoka, A. (2014) Production of sugar alcohols from real biomass by supported platinum catalyst. *Catal. Today*, **226**, 204–209.

68. Pang, J., Zheng, M., Sun, R., Wang, A.-Q., Wang, X., Zhang, T. (2015) Synthesis of ethylene glycol and terephthalic acid from biomass for producing PET. *Green Chem.*, **18**, 342–359.

69. (a) Xu, S., Zheng, A., Wei, Y., Chen, J., Li, J., Chu, Y., Zhang, M., Wang, Q., Zhou, Y., Wang, J., Deng, F., Liu, Z. (2013) Direct observation of cyclic carbenium ions and their role in the catalytic cycle of the methanol-to-olefin reaction over chabazite zeolites. *Angew. Chem. Int. Ed.*, **52**, 11564–11568; (b) Ng, S.H., Al-Sabawi, M., Wang, J., Ling, H., Zheng, Y., Wei, Q., Ding, F., Little, E. (2015) FCC coprocessing oil sands heavy gas oil and canola oil. 1. Yield structure. *Fuel*, **156**, 163–176.

70. Levy, R.B., Boudart, M. (1973) Platinum-like behavior of tungsten carbide in surface catalysis. *Science*, **181**, 547–549.

71. Ji, N., Zhang, T., Zheng, M., Wang, A., Wang, H., Wang, X., Chen, J.G. (2008) Direct catalytic conversion of cellulose into ethylene glycol using nickel-promoted tungsten carbide catalysts. *Angew. Chem. Int. Ed.*, **47**, 8510–8513.

72. (a) Ji, N., Zhang, T., Zheng, M., Wang, A., Wang, H., Wang, X., Shu, Y., Stottlemyer, A.L., Chen, J.G. (2009) Catalytic conversion of cellulose into ethylene glycol over supported carbide catalysts. *Catal. Today*, **147**, 77–85; (b) Ji, N., Zheng, M., Wang, A., Zhang, T., Chen, J.G. (2012) Nickel-promoted

tungsten carbide catalysts for cellulose conversion: effect of preparation methods. *ChemSusChem*, **5**, 939–944.

73. Zhao, G., Zheng, M., Wang, A., Zhang, T. (2010) Catalytic conversion of cellulose to ethylene glycol over tungsten phosphide catalysts. *Chin. J. Catal.*, **31**, 928–932.

74. (a) Zheng, M.-Y., Wang, A.-Q., Ji, N., Pang, J.-F., Wang, X.-D., Zhang, T. (2010) Transition metal-tungsten bimetallic catalysts for the conversion of cellulose into ethylene glycol. *ChemSusChem*, **3**, 63–66; (b) Tai, Z., Zhang, J., Wang, A., Zheng, M., Zhang, T. (2012) Temperature-controlled phase-transfer catalysis for ethylene glycol production from cellulose. *Chem. Commun.*, **48**, 7052–7054; (c) Liu, Y., Luo, C., Liu, H. (2012) Tungsten trioxide promoted selective conversion of cellulose into propylene glycol and ethylene glycol on a ruthenium catalyst. *Angew. Chem. Int. Ed.*, **51**, 3249–3253; (d) Tai, Z., Zhang, J., Wang, A., Pang, J., Zheng, M., Zhang, T. (2013) Catalytic conversion of cellulose to ethylene glycol over a low-cost binary catalyst of Raney Ni and tungstic acid. *ChemSusChem*, **6**, 652–658.

75. Zhang, Y., Wang, A., Zhang, T. (2010) A new 3D mesoporous carbon replicated from commercial silica as a catalyst support for direct conversion of cellulose into ethylene glycol. *Chem. Commun.*, **46**, 862–864.

76. (a) Pang, J., Zheng, M., Wang, A., Zhang, T. (2011) Catalytic hydrogenation of corn stalk to ethylene glycol and 1,2-propylene glycol. *Ind. Eng. Chem. Res.*, **50**, 6601–6608; (b) Li, C., Zheng, M., Wang, A., Zhang, T. (2012) One-pot catalytic hydrocracking of raw woody biomass into chemicals over supported carbide catalysts: simultaneous conversion of cellulose, hemicellulose and lignin. *Energ. Environ. Sci.*, **5**, 6383–6390; (c) Zhou, L., Pang, J., Wang, A., Zhang, T. (2013) Catalytic conversion of Jerusalem artichoke stalk to ethylene glycol over a combined catalyst of WO_3 and Raney Ni. *Chin. J. Catal.*, **34**, 2041; (d) Pang, J., Zheng, M., Wang, A., Sun, R., Wang, H., Jiang, Y., Zhang, T. (2014) Catalytic conversion of concentrated miscanthus in water for ethylene glycol production. *AIChE J.*, **60**, 2254–2262; (e) Zhou, L., Wang, A., Li, C., Zheng, M., Zhang, T. (2012) Selective production of 1,2-propylene glycol from Jerusalem artichoke tuber using Ni-W2C/AC catalysts. *ChemSusChem*, **5**, 932–938; (f) Fabicovicova, K., Lucas, M., Claus, P. (2015) From microcrystalline cellulose to hard- and softwood-based feedstocks: their hydrogenolysis to polyols over a highly efficient ruthenium-tungsten catalyst. *Green Chem.*, **17**, 3075–3083.

77. (a) Wang, A., Zhang, T. (2013) One-pot conversion of cellulose to ethylene glycol with multifunctional tungsten-based catalysts. *Acc. Chem. Res.*, **46**, 1377–1386; (b) Zheng, M., Pang, J., Wang, A., Zhang, T. (2014) One-pot catalytic conversion of cellulose to ethylene glycol and other chemicals: From fundamental discovery to potential commercialization. *Chin. J. Catal.*, **35**, 602–613.

78. Zhao, G., Zheng, M., Zhang, J., Wang, A., Zhang, T. (2013) Catalytic conversion of concentrated glucose to ethylene glycol with semicontinuous reaction system. *Ind. Eng. Chem. Res.*, **52**, 9566–9572.

79. (a) Zhang, J., Yang, X., Hou, B., Wang, A., Li, Z., Wang, H., Zhang, T. (2014) Comparison of cellobiose and glucose transformation to ethylene glycol. *Chin. J. Catal.*, **35**, 1181–1187; (b) Zhang, J., Hou, B., Wang, A., Li, Z., Wang, H., Zhang, T. (2014) Kinetic study of retro-aldol condensation of glucose to glycolaldehyde with ammonium metatungstate as the catalyst. *AIChE J.*, 3804–3813; (c) Zhang, J., Hou, B., Wang, A., Li, Z., Wang, H., Zhang, T. (2015) Kinetic study of the competitive hydrogenation of glycolaldehyde and glucose on Ru/C with or without AMT. *AIChE J.*, **61**, 224–238.

80. (a) Zheng, M., Wang, A., Ji, N., Pang, J., Wang, X., Zhang, T. (2010) Transition metal–tungsten bimetallic catalysts for the conversion of cellulose into ethylene glycol. *ChemSusChem*, **3**, 63–66; (b) Xiao, Z., Jin, S., Pang, M., Liang, C. (2013) Conversion of highly concentrated cellulose to 1,2-propanediol and ethylene glycol over highly efficient CuCr catalysts. *Green Chem.*, **15**, 891–895.

81. Yabushita, M., Kobayashi, H., Fukuoka, A. (2014) Catalytic transformation of cellulose into platform chemicals. *Appl. Catal. B Environ.*, **145**, 1–9.

82. Deng, W., Liu, M., Zhang, Q., Tan, X., Wang, Y. (2010) Acid-catalysed direct transformation of cellulose into methyl glucosides in methanol at moderate temperatures. *Chem. Commun.*, **46**, 2668–2670.

83. Dora, S., Bhaskar, T., Singh, R., Naik, D.V., Adhikari, D.K. (2012) Effective catalytic conversion of cellulose into high yields of methyl glucosides over sulfonated carbon-based catalyst. *Bioresource Technol.*, **120**, 318–321.

84. (a) Saha, B., Abu-Omar, M.M. (2014) Advances in 5-hydroxymethylfurfural production from biomass in biphasic solvents. *Green Chem.*, **16**, 24–38; (b) Li, H., Zhang, Q., Bhadury, P.S., Yang, S. (2014) Furan-type compounds from carbohydrates via heterogeneous catalysis. *Current Org. Chem.*, **18**, 547–597.

85. Villa, A., Schiavoni, M., Fulvio, P.F., Mahurin, S.M., Dai, S., Mayes, R.T., Veith, G.M., Prati, L. (2013) Phosphorylated mesoporous carbon as effective catalyst for the selective fructose dehydration to HMF. *J. Energ. Chem.*, **22**, 305–311.

86. Daengprasert, W., Boonnoun, P., Laosiripojana, N., Goto, M., Shotipruk, A. (2011) Application of sulfonated carbon-based catalyst for solvothermal conversion of cassava waste to hydroxymethylfurfural and furfural. *Ind. Eng. Chem. Res.*, **50**, 7903–7910.

87. Hu, L., Zhao, G., Tang, X., Wu, Z., Xu, J., Lin, L., Liu, S. (2013) Catalytic conversion of carbohydrates into 5-hydroxymethylfurfural over cellulose-derived carbonaceous catalyst in ionic liquid. *Bioresource Technol.*, **148**, 501–507.

88. Ramachandran, S., Fontanille, P., Pandey, A., Larroche, C. (2006) Gluconic acid: Properties, applications and microbial production. *Food Tech. Bio.*, **44**, 185–195.

89. Biella, S., Prati, L., Rossi, M. (2002) Selective oxidation of D-glucose on gold catalyst. *J. Catal.*, **206**, 242–247.

90. Tan, X., Deng, W., Liu, M., Zhang, Q., Wang, Y. (2009) Carbon nanotube-supported gold nanoparticles as efficient catalysts for selective oxidation of cellobiose into gluconic acid in aqueous medium. *Chem. Commun.*, **7179–7181**.

91. Onda, A., Ochi, T., Yanagisawa, K. (2011) New direct production of gluconic acid from polysaccharides using a bifunctional catalyst in hot water. *Catal. Commun.*, **12**, 421–425.

92. (a) Holladay, J.E., White, J.F., Bozell, J.J., Johnson, D. (2007) Pacific Northwest National Laboratory. *Operated by Battelle for the US Department of Energy*, **PNNL-16983** 1–79; (b) Zhang, X., Tu, M., Paice, M.G. (2011) Routes to potential bioproducts from lignocellulosic biomass lignin and hemicelluloses. *BioEnergy Res.*, **4**, 246–257; (c) Ragauskas, A.J. (2012) Lignin: the new paradigm in biofuels. *Pharm. Biol.*, **50**, 671–671; (d) Hicks, J.C. (2011) Advances in C–O bond transformations in lignin-derived compounds for biofuels production. *J. Phys. Chem. Lett.*, 2280–2287; (e) Calvo-Flores, F.G., Dobado, J.A. (2010) Lignin as renewable raw material. *ChemSusChem*, **3**, 1227–1235.

93. (a) Li, C., Zhao, X., Wang, A., Huber, G.W., Zhang, T. (2015) Catalytic transformation of lignin for the production of chemicals and fuels. *Chem. Rev.*, **115**, 11559–11624; (b) Huber, G.W., Iborra, S., Corma, A. (2006) Synthesis of transportation fuels from biomass: chemistry, catalysts, and engineering. *Chem. Rev.*, **106**, 4044–4098; (c) Zakzeski, J., Bruijnincx, P.C.A., Jongerius, A.L., Weckhuysen, B.M. (2010) The catalytic valorization of lignin for the production of renewable chemicals. *Chem. Rev.*, **110**, 3552–3599.

94. Song, Q., Wang, F., Cai, J., Wang, Y., Zhang, J., Yu, W., Xu, J. (2013) Lignin depolymerization (LDP) in alcohol over nickel-based catalysts via a fragmentation-hydrogenolysis process. *Energ. Environ. Sci.*, **6**, 994–1007.

95. Ji, N., Wang, X., Weidenthaler, C., Spliethoff, B., Rinaldi, R. (2015) Iron(II) disulfides as precursors of highly selective catalysts for hydrodeoxygenation of dibenzyl ether into toluene. *ChemCatChem*, **7**, 960–966.

96. (a) Busetto, L., Fabbri, D., Mazzoni, R., Salmi, M., Torri, C., Zanotti, V. (2011) Application of the Shvo catalyst in homogeneous hydrogenation of bio-oil obtained from pyrolysis of white poplar: New mild upgrading conditions. *Fuel*, **90**, 1197–1207; (b) Ben, H., Mu, W., Deng, Y., Ragauskas, A.J. (2013) Production of renewable gasoline from aqueous phase hydrogenation of lignin pyrolysis oil. *Fuel*, **103**, 1148–1153.

97. Vispute, T.P., Zhang, H., Sanna, A., Xiao, R., Huber, G.W. (2010) Renewable chemical commodity feedstocks from integrated catalytic processing of pyrolysis oils. *Science*, **330**, 1222–1227.

98. Song, Q., Wang, F., Xu, J. (2012) Hydrogenolysis of lignosulfonate into phenols over heterogeneous nickel catalysts. *Chem. Commun.*, **48**, 7019–7021.

99. Ye, Y., Zhang, Y., Fan, J., Chang, J. (2012) Selective production of 4-ethylphenolics from lignin via mild hydrogenolysis. *Bioresource Technol.*, **118**, 648–651.

100. Ruiz, P.E., Frederick, B.G., De Sisto, W.J., Austin, R.N., Radovic, L.R., Leiva, K., García, R., Escalona, N., Wheeler, M.C. (2012) Guaiacol hydrodeoxygenation on MoS_2 catalysts: Influence of activated carbon supports. *Catal. Commun.*, **27**, 44–48.

101. Xu, X., Li, Y., Gong, Y., Zhang, P., Li, H., Wang, Y. (2012) Synthesis of palladium nanoparticles supported on mesoporous N-doped carbon and their catalytic ability for biofuel upgrade. *J. Am. Chem. Soc.*, **134**, 16987–16990.

102. Zhao, C., Kou, Y., Lemonidou, A.A., Li, X., Lercher, J.A. (2009) Highly selective catalytic conversion of phenolic bio-oil to alkanes. *Angew. Chem. Int. Ed.*, **48**, 3987–3990.

103. Zhao, C., He, J., Lemonidou, A.A., Li, X., Lercher, J.A. (2011) Aqueous-phase hydrodeoxygenation of bio-derived phenols to cycloalkanes. *J. Catal.*, **280**, 8–16.

104. Zhao, C., Lercher, J.A. (2012) Selective hydrodeoxygenation of lignin-derived phenolic monomers and dimers to cycloalkanes on Pd/C and HZSM-5 catalysts. *ChemCatChem*, **4**, 64–68.

105. Li, Z., Garedew, M., Lam, C.H., Jackson, J.E., Miller, D.J., Saffron, C.M. (2012) Mild electrocatalytic hydrogenation and hydrodeoxygenation of bio-oil derived phenolic compounds using ruthenium supported on activated carbon cloth. *Green Chem.*, **14**, 2540–2549.

106. Ma, R., Hao, W., Ma, X., Tian, Y., Li, Y. (2014) Catalytic ethanolysis of Kraft lignin into high-value small-molecular chemicals over a nanostructured α-molybdenum carbide catalyst. *Angew. Chem. Int. Ed.*, **53**, 7310–7315.

107. Huber, G.W., Iborra, S., Corma, A. (2006) Synthesis of transportation fuels from biomass: chemistry, catalysts, and engineering. *Chem. Rev.*, **106**, 4044–4098.

108. (a) De, S., Saha, B., Luque, R. (2015) Hydrodeoxygenation processes: advances on catalytic transformations of biomass-derived platform chemicals into hydrocarbon fuels. *Bioresource Technol.*, **178**, 108–118; (b) Bohre, A., Dutta, S., Saha, B., Abu-Omar, M.M. (2015) Upgrading furfurals to drop-in biofuels: an overview. *ACS Sus. Chem. Eng.*, **3**, 1263–1277; (c) Deneyer, A., Renders, T., Van Aelst, J., Van den Bosch, S., Gabriels, D., Sels, B.F. (2015) Alkane production from biomass: chemo-, bio- and integrated catalytic approaches. *Curr. Opinion Chem. Biol.*, **29**, 40–48; (d) Eleni, F.I. (2010) Review of C-C coupling reactions in biomass exploitation processes. *Current Org. Synth.*, **7**, 587–598; (e) Serrano-Ruiz, J.C., Dumesic, J.A. (2011) Catalytic routes for the conversion of biomass into liquid hydrocarbon transportation fuels. *Energ. Environ. Sci.*, **4**, 83–99.

109. Faba, L., Díaz, E., Ordóñez, S. (2013) Improvement on the catalytic performance of Mg–Zr mixed oxides for furfural–acetone aldol condensation by supporting on mesoporous carbons. *ChemSusChem*, **6**, 463–473.

110. Zapata, P.A., Faria, J., Pilar Ruiz, M., Resasco, D.E. (2012) Condensation/hydrogenation of biomass-derived oxygenates in water/oil emulsions stabilized by nanohybrid catalysts. *Top. Catal.*, **55**, 38–52.

111. Anbarasan, P., Baer, Z.C., Sreekumar, S., Gross, E., Binder, J.B., Blanch, H.W., Clark, D.S., Toste, F.D. (2012) Integration of chemical catalysis with extractive fermentation to produce fuels. *Nature*, **491**, 235–239.

112. (a) Li, G., Li, N., Wang, Z., Li, C., Wang, A., Wang, X., Cong, Y., Zhang, T. (2012) Synthesis of high-quality diesel with furfural and 2-methylfuran from hemicellulose. *ChemSusChem*, **5**, 1958–1966; (b) Li, G., Li, N., Yang, J., Wang, A., Wang, X., Cong, Y., Zhang, T. (2013) Synthesis of renewable diesel with the 2-methylfuran, butanal and acetone derived from lignocellulose. *Bioresource Technol.*, **134**, 66–72.

113. Huang, Y.-B., Yang, Z., Dai, J.-J., Guo, Q.-X., Fu, Y. (2012) Production of high-quality fuels from lignocellulose-derived chemicals: a convenient C-C bond formation of furfural, 5-methylfurfural and aromatic aldehyde. *RSC Adv.*, **2**, 11211–11214.

114. Liu, D., Chen, E.Y.X. (2013) Diesel and alkane fuels from biomass by organocatalysis and metal–acid tandem catalysis. *ChemSusChem*, **6**, 2236–2239.

115. Sutton, A.D., Waldie, F.D., Wu, R., Schlaf, M., 'Pete' Silks, L.A., Gordon, J.C. (2013) The hydrodeoxygenation of bioderived furans into alkanes. *Nat. Chem.*, **5**, 428–432.

116. Liu, D., Chen, E.Y.X. (2014) Integrated catalytic process for biomass conversion and upgrading to C12 furoin and alkane fuel. *ACS Catal.*, **4**, 1302–1310.

117. Wegenhart, B.L., Yang, L., Kwan, S.C., Harris, R., Kenttämaa, H.I., Abu-Omar, M.M. (2014) From furfural to fuel: synthesis of furoins by organocatalysis and their hydrodeoxygenation by cascade catalysis. *ChemSusChem*, **7**, 2742–2747.

118. Xing, R., Subrahmanyam, A.V., Olcay, H., Qi, W., van Walsum, G.P., Pendse, H., Huber, G.W. (2010) Production of jet and diesel fuel range alkanes from waste hemicellulose-derived aqueous solutions. *Green Chem.*, **12**, 1933–1946.

119. Chatterjee, M., Matsushima, K., Ikushima, Y., Sato, M., Yokoyama, T., Kawanami, H., Suzuki, T. (2010) Production of linear alkane via hydrogenative ring opening of a furfural-derived compound in supercritical carbon dioxide. *Green Chem.*, **12**, 779–782.

120. Pholjaroen, B., Li, N., Yang, J.F., Li, G., Wang, W.T., Wang, A.Q., Cong, Y., Wang, X.D., Zhang, T. (2014) Production of renewable jet fuel range branched alkanes with xylose and methyl isobutyl ketone. *Ind. Eng. Chem. Res.*, **53**, 13618–13625.

121. Alonso, D.M., Bond, J.Q., Serrano-Ruiz, J.C., Dumesic, J.A. (2010) Production of liquid hydrocarbon transportation fuels by oligomerization of biomass-derived C9 alkenes. *Green Chem.*, **12**, 992.

122. (a) Corma, A., de la Torre, O., Renz, M. (2011) High-quality diesel from hexose- and pentose-derived biomass platform molecules. *ChemSusChem*, **4**, 1574–1577; (b) Corma, A., de la Torre, O., Renz, M., Villandier, N. (2011) Production of high-quality diesel from biomass waste products. *Angew. Chem. Int. Ed.*, **50**, 2375–2378.

123. Li, G., Li, N., Li, S., Wang, A., Cong, Y., Wang, X., Zhang, T. (2013) Synthesis of renewable diesel with hydroxyacetone and 2-methyl-furan. *Chem. Commun. (Camb)*, **49**, 5727–5729.

124. Li, G., Li, N., Yang, J., Li, L., Wang, A., Wang, X., Cong, Y., Zhang, T. (2014) Synthesis of renewable diesel range alkanes by hydrodeoxygenation of furans over Ni/Hβ under mild conditions. *Green Chem.*, **16**, 594–599.

125. Yang, J., Li, N., Li, G., Wang, W., Wang, A., Wang, X., Cong, Y., Zhang, T. (2013) Solvent-free synthesis of C10 and C11 branched alkanes from furfural and methyl isobutyl ketone. *ChemSusChem*, **6**, 1149–1152.

126. Mascal, M., Dutta, S., Gandarias, I. (2014) Hydrodeoxygenation of the Angelica lactone dimer, a cellulose-based feedstock: simple, high-yield synthesis of branched C7–C10 gasoline-like hydrocarbons. *Angew. Chem. Int. Ed.*, **53**, 1854–1857.

3

Nanoporous Carbon/Nitrogen Materials and their Hybrids for Biomass Conversion

Hui Su, Hong-Hui Wang, Tian-Jian Zhao, and Xin-Hao Li

School of Chemistry and Chemical Engineering, Shanghai Jiao Tong University, China

3.1 Introduction

With the rapid development of materials science, nanoporous carbon/nitrogen materials, including carbon nitrides, conjugated polymers and nitrogen-doped carbons, have attracted extensive attention in wide applications in energy-storage devices, catalysis, separation, and molecular electronic devices [1–4]. To be specific, large-scale catalysis transformation containing biomass conversion is a key process for achieving chemicals, pharmaceuticals and industrial compounds [5–7]. All of these applications require high stability, ease of separation, high surface area and, most importantly, highly efficient active sites in one catalyst. With regards to a heterogeneous catalyst, the high thermal/chemical stability of carbon/nitrogen materials sufficiently ensures the repeatable utilization of catalysts under critical reaction conditions. In order to increase the diffusion and pre-adsorption of reactants to the surface of catalysts, nanoporous structures were introduced into these carbon/nitrogen materials, which resulted in a high surface area and also large numbers of low-coordinated sites [8].

Besides the high stability, which is also the typical feature of traditional carbon materials, nitrogen atoms in nitrogen/carbon materials can change the electron distribution by donating electrons, thereby adjusting the electronic structure of the carbon frameworks. Hence, the band gap of carbon material was opened accordingly, and this led to a boost in the catalytic performance for hydrogen photosynthesis, and selective oxidation reactions

[9–12]. As a result, nanoporous carbon/nitrogen materials are preferential candidates for varieties of important industrial catalysis reactions. Likewise, other heteroatoms (e.g., boron, phosphorus, sulfur) were introduced into the carbon lattice via many synthetic strategies.

The catalytic activity of pristine carbon/nitrogen materials is essentially limited to the intrinsic features of the carbon and nitrogen elements, which is the main hurdle for catalytic reactions. The introduction of functional nanoparticles, with particular focus on metal nanoparticles in carbon/nitrogen materials systems, becomes a direct and excellent means of further boosting catalytic performance. Moreover, nanoporous carbon/nitrogen materials, along with tunable pore size and processable surface groups, are also preferred to serve as metal-nanoparticle supports for low-cost and flexible structures and properties. Previously, metal nanoparticles at carbon nitride-based semiconductive materials were accessed to activate metal sites for H_2-generation reactions, hydrogenation reactions, and carbon–carbon coupling reactions [13–15]. As a result, the concept of the Mott–Schottky effect is introduced to describe the mutual electron interaction between metal nanoparticles and supports [16]. Carbon/nitrogen materials with a moderate band gap and a relatively lower work function ensure the construction of a rectifying contact between the metal particles and supports, which is indispensable for Mott–Schottky catalysts. With regards to metal–ligands in homogeneous catalysis, nanoporous carbon/nitrogen materials can modulate the delocalized electron state by using different precursors and synthesis methods, which plays a role of 'solid ligand' to activate metal particles through the Schottky barrier at the metal–support interface. Such an efficient approach is capable of prompting the activity of metal particles for catalytic reactions in remarkable fashion. Carbon/nitrogen material hybrids with functional particles permit the fabrication of efficient and sustainable catalysts, which will pave the way to the development of 'green' chemical processes.

Within the huge 'family' of catalytic transformation, biomass materials – as promising feedstocks or precursors – have been converted into fuels, fine chemicals, and functional materials due to their abundant resources and specific structures [17–19]. Through the chemical or biochemical conversion of biomasses such as lignocellulose, cellulose, starch and triglyceride, high hydrogen-content platform chemicals (e.g., alcohols and formic acid) have emerged as renewable alternatives to mineral precursors [20]. These sustainable biomass-derived resources may not only be used to generate hydrogen as clean energy, but also serve as the building blocks to prompt advances in green chemistry [7,21–23]. For example, furfural can be obtained by the hydrolysis of cellulose or hemicelluloses, and starting from furfural as a platform molecule other value-added fuels and fuel additives may be created via hydrodeoxygenation (HDO) or hydrogenation processes [5,7]. As a result, to architectural development of an efficient and green catalytic protocol has been the mainstream tendency for chemical conversions from biomass-derived feedstocks. During recent years, noble metal-based homogeneous catalysts have been discovered that are capable of transforming biomass-derived alcohol to esters [24]. Yet, the challenge remains to develop heterogeneous catalysts for similar processes. Recently, Li's group carried out a series of intriguing and fundamental investigations whereby carbon nitride-based Mott–Schottky catalysts were successfully applied to the dehydrogenation of formic acid and the transfer hydrogenation of unsaturated compounds [15,25]. Accordingly, carbon/nitrogen materials and their hybrids, all of which have copious physicochemical properties and high-performance, provide a great opportunity to create sustainable biomass conversion.

The aim of this chapter is to provide an overview of the related concepts and principles regarding the application of carbon/nitrogen materials and their hybrids materials as effective heterogeneous catalysts in the field of biomass conversions. Attention will also be focused on the dehydrogenation of formic acid, the transfer hydrogenation reaction with biomass molecules, and the commercial production of industrial chemicals derived from biomass raw materials.

3.2 Dehydrogenation of Formic Acid

In order to meet the demands of sustainable development, hydrogen (H_2) has been widely regarded as one of the most green and environmental energy carriers all the time, especially in the application of polymer electrolyte membrane (PEM) fuel cells in electric automobiles [26,27]. Recently, with non-toxicity, high-energy density, and excellent chemical and thermal stability, formic acid, as one type of major biomass, has been a hot research topic in H_2 storage [28–30]. Two possible reaction pathways may occur during the decomposition of formic acid, namely the dehydrogenation and dehydration reactions, respectively, as follows:

$$HCOOH\ (l) \rightarrow H_2\ (g) + CO_2\ (g)\ \Delta G_{298\ k} = -35.0\ KJ\ mol^{-1}$$

$$HCOOH\ (l) \rightarrow H_2O\ (g) + CO\ (g)\ \Delta G_{298\ k} = -14.9\ KJ\ mol^{-1}$$

It is well known that carbon monoxide (CO) is a fatal poison to the catalysts of most chemical reactions [31,32], but this can be avoided by using different catalysts or adjusting the reaction conditions, such as pH value and temperature of the reaction system.

The dehydrogenation of formic acid through heterogeneous catalysis has been developed very rapidly during recent years because of the advantages of catalyst recycling and reusing. Heterogeneous catalysts usually consist of noble metal nanoparticles as catalytic active sites, with semiconductive support materials as cocatalyst [16], according to which the catalysts can be divided into several categories such as mono/bi/tri-metallic [33–40]-based composites, along with activated carbon materials [35], grapheme [33,37], and mesoporous carbon nitride [13] acting as supporting materials. Heterocatalysts can also be further designated into core–shell nanostructures to improve the catalytic activity of the reaction of formic acid decomposition [38].

3.2.1 Mono-Metallic Nanoparticle/Carbon–Nitrogen Nanocomposites: Metal-Support Effect

Li and coworkers designed a new series of heterogeneous catalysts based on the Mott–Schottky effect which showed a greatly enhanced catalytic performance for the catalytic decomposition of formic acid [13]. The Mott–Schottky effect is a basic concept in solid physics. A rectifying metal–semiconductor heterojunction should be constructed for Mott–Schottky catalysts, where the rectifying contact depends on the work function of the metal and electronic structure of the semiconductor. For a Mott–Schottky catalyst (Figure 3.1a and b), a current flows through the interface of the metal and semiconductor until the work functions of the metal and semiconductor become the same [16]. Band

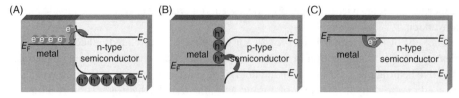

Figure 3.1 *Schematic view of typical (a) rectifying metal–n-type semiconductor contact. (b) Rectifying metal–p-type semiconductor. (c) Metal–semiconductor ohmic contact. Reproduced by permission from Ref. [16]. Copyright 2013, Royal Society of Chemistry. (See color plate section for the color representation of this figure.)*

bending of the semiconductor part could more or less modify the redox ability of the catalysts. More importantly, an electron enrichment of the metal side – and thus its catalytic activity for specific reactions – is programmable by changing the semiconductor support, by following the metal–ligand coupling effect in homogeneous catalysis.

While the work function of noble metal Pd is located exactly between the band structure of carbon nitride (g-C$_3$N$_4$ or CN) with a bandgap around 2.7 eV [41] (Figure 3.2), electrons will transfer Pd nanoparticles (NPs) and thus enrich its electron density. Hence, it is possible to activate Pd NPs by constructing Pd/g-C$_3$N$_4$ heterojunctions. The turnover frequency (TOF) of the dehydrogenation of formic acid over Pd@CN can reach as high as 50 mol H$_2$ mol^{-1} Pd h^{-1} at a temperature as low as 15 °C, which can be raised to 71 mol H$_2$ mol^{-1} Pd h^{-1} under visible light because of the photocatalytic activity of carbon nitride.

Considering that the position of the conduction band has a great influence on the process of electron transference as to the Mott–Schottky effect, Li and coworkers further tuned the electronic band structure of carbon nitride by changing the temperature of the synthesis [42]. It was found that, by lowering the condensation temperature from 500 °C to 300 °C, the absorption band edge in the UV-visible absorption spectra of the support material CN/SiO$_2$ samples showed clear blue shifts. The band gap of the as-obtained CN/SiO$_2$ samples was widened, while the conduction band was elevated accordingly. Hence, when embedded with Pd NPs, the nanocomposite catalyst Pd@CN/SiO$_2$-300 exhibited an exceedingly high catalytic activity towards the reaction of formic acid decomposition, the TOF of which can

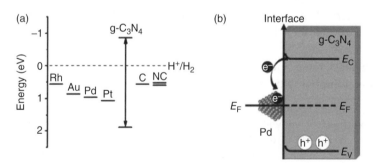

Figure 3.2 *(a) Work functions of typical metals and carbon, and band structures of carbon nitride and N-doped carbon (NC). (b) Schematic view of a Mott–Schottky-type Pd@CN contact (E$_F$: work function; E$_C$: conduction band; E$_V$: valence band). Reproduced with permission from Ref. [16]. Copyright 2013, Royal Society of Chemistry.*

be increased up to 1119 mol H_2 mol^{-1} Pd h^{-1} at 323K, which was comparable to that of the best heterogeneous and homogeneous catalysts reported [35,43].

Although carbon nitride is a good support with abundant in-built amino functional groups in favor of stabilizing Pd NPs embedded on the surface, one intrinsic defect that is apparent is the relatively low specific surface area, of approximately 10 m^2 g^{-1}. (*Note*: bulk carbon nitrides derived from different precursors such as cyanamide, dicyandiamide, urea, and melamine have variable surface areas.) [41]. It is well known that the specific surface area of the support has a significant influence on the promotion of catalytic performance of the hybrids. Xu and colleagues selected an elaborate nanoporous carbon, Maxsorb MSC-30, which has a very high surface area in excess of 3000 m^2 g^{-1}, to support the Pd NPs [35]. The experimental results obtained showed that Pd NPs@MSC-30 may be the best heterogeneous catalyst for the dehydrogenation of formic acid identified to date, with TOF values of 750 mol H_2 mol^{-1} Pd h^{-1} at 298K, and 2623 mol H_2 mol^{-1} Pd h^{-1} at 323K.

3.2.2 Bimetallic Nanoparticle/Carbon–Nitrogen Nanocomposites

With some other elements incorporated, such as Au, Ag, and Co, the catalytic surface of Pd NPs can be modified and higher activities and selectivities achieved. Zhou and coworkers synthesized a series of alloy catalysts such as Pd–Au/C and Pd–Ag/C, which overcame the CO-poisoning caused by the dehydration reaction and produced high-quality hydrogen efficiently in the presence of CeO_2 $(H_2O)_x$, and among which the catalyst Pd–Au/C showed the highest catalytic activity with a TOF of 113 mol H_2 mol^{-1} Pd h^{-1} [36]. It is well known that one of the most important applications of hydrogen is the fuel cell. Zhang and coworkers developed a bimetallic PtAu NP catalyst embedded onto various supports as fuel cell electrocatalysts for formic acid in-situ oxidation ($HCOOH \rightarrow CO_2 + 2H^+ + 2e^-$) [37]. As shown in Figure 3.3, compared with the commercial electrocatalyst Etek-Pt/C, both PtAu/graphene and PtAu/carbon black (CB) exhibited an onset potential of formic acid oxidation of 170 mV, which was much lower than the value of 300 mV for Etek-Pt/C. The peak current densities on PtAu/graphene and PtAu/CB were 2.310 A mg^{-1} Pt and 1.682 A mg^{-1} Pt, respectively, both of which were much more higher than that on Etek-Pt/C (0.182 A mg^{-1} Pt). Moreover, in contrast to PtAu/CB, graphene in PtAu/graphene greatly suppressed the evolution of CO during formic acid oxidation. All of the above results indicate that PtAu/graphene would serve as an excellent electrocatalyst towards formic acid oxidation.

Zhang and coworkers synthesized monodispersed binary AgPd alloy NPs with a tunable size and composition that could be easily controlled by changing the molar ratio of the precursors via a facile approach [40]. The $Ag_{42}Pd_{58}$ alloy NPs with the size of 2.2 nm were found to possess the highest activity for the dehydrogenation of formic acid, and the initial TOF reached 382 mol H_2 mol^{-1} Pd h^{-1} at 323 K in 1 M formic acid aqueous solution, without any additive.

3.2.3 Trimetallic Nanoparticle/Carbon–Nitrogen Nanocomposites

Cobalt (Co), with its relatively low cost and excellent catalytic activity towards a large number of chemical reactions, is potentially a favorite catalytic element, although the fact that it is easily etched by acids limits its application. Wang and coworkers improved the

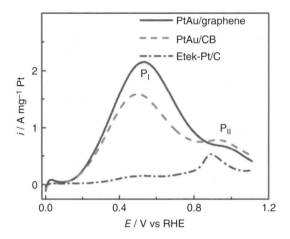

Figure 3.3   *Formic acid oxidation at a scan rate of 50 mV s⁻¹ on PtAu/graphene, PtAu/CB, and commercial Etek-Pt/C in N₂-saturated 0.5 M H₂SO₄ + 0.5 M HCOOH. Reproduced with permission from Ref. [37]. Copyright 2011, American Chemical Society.*

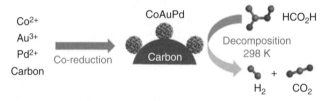

Figure 3.4   *Preparation and application of CoAuPd/C nanocatalyst for formic acid decomposition at 298K. Reproduced with permission from Ref. [40]. Copyright 2013, Wiley. (See color plate section for the color representation of this figure.)*

acid-resistant properties of Co NPs by introducing noble metals (Au and Pd) to produce CoAuPd alloy NPs [39]. The experimental results showed that the TOF at 298K of the trimetallic catalyst $Co_{0.30}Au_{0.35}Pd_{0.35}/C$ was twice of that of the catalyst $Au_{0.5}Pd_{0.5}/C$, without the addition of Co (Figure 3.4).

3.2.4   Core–Shell Nanostructure/Carbon–Nitrogen Nanocomposites

Core–shell nanostructures have always been research 'hotspots' because of their tunable structures in relation to their properties and wide range of applications in sensors, catalysis, biology, and materials chemistry [44]. It is well known that the inner core of one metal element can play a great influence on the external shell of another metal element, so as to endow special physical and chemical properties in a bimetallic core–shell structure, because of the differences in work function, electronegativity, and so on [45,46]. When two metals with different Fermi levels come into contact, a charge will flow between them until the Fermi levels of the two metals are the same (this is reflected in their work functions).

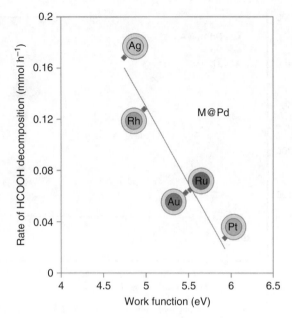

Figure 3.5 *Correlation with the work function of the M core, where M = fcc (111) Ag, Rh, Au, Ru and Pt or hexagonal close-packed (hcp) (0001) Ru. Ag, with the largest difference in work function in relation to Pd, gives the strongest electron promotion to the Pd shell. Reproduced with permission from Ref. [38]. Copyright 2011, Nature.*

As shown in Figure 3.5, theoretical calculations demonstrate that Ag@Pd core–shell NPs would display highest activity. The corresponding TOF was 125 mol H_2 mol^{-1} Pd h^{-1} (based on the moles of Pd active sites) at 293K [38].

Huang and coworkers prepared a core–shell structured PdAu@Au/C (Figure 3.6) by simultaneous reduction of the precursors mixed solution, specifically without the use of any stabilizing agents [47]. For comparison, Pd/C and Au/C were synthesized at the same time. The novel nanostructure and synergistic effect between Pd and Au endowed a high catalytic ability for formic acid decomposition, with the evolution of H_2 and CO_2 over PdAu@Au/C being ninefold that of the other two contrast samples.

3.2.5 Reduction of Carbon Dioxide to Formic Acid Using Carbon/Nitrogen Materials

As the industrial civilization of human society becomes highly developed, environmental issues are becoming increasingly serious, one of these issues being the 'greenhouse effect'. As carbon dioxide (CO_2) is the main greenhouse gas, the conversion of CO_2 into other useful organics such as formic acid is of great significance. Zhao and coworkers achieved the reduction of CO_2 into formic acid by means of electrochemistry [48], by preparing a high-performance catalyst Pd-PANI/CNT in situ through reducing a mixed solution of H_2PdCl_4, aniline and multi-walled carbon nanotubes (MWCNTs) with sodium citrate as a

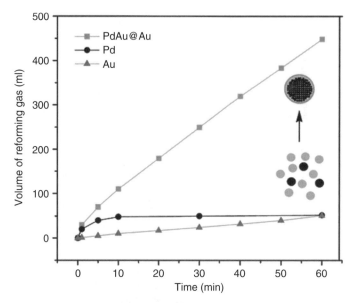

Figure 3.6 *Volume change of reforming gas with time for 60 mg of the synthesized catalyst in 5 ml of solution containing 6.64 M formic acid and 6.64 M sodium formate at a reaction temperature of 92 °C. Reproduced with permission from Ref. [48]. Copyright 2010, American Chemical Society.*

protective stabilizer (Figure 3.7). The results showed that at the low overpotential (–0.8 V), the concentration of both $[CO_2]$ and $[HCO_3^-]$ (from the hydrolysis of carbon dioxide: $CO_2 + H_2O \rightarrow H^+ + HCO_3^- \rightarrow 2H^+ + CO_3^{2-}$) could contribute to the formation of formate, and the corresponding Faradaic efficiency reached the highest value, that is, 83%.

In terms of electrocatalysis, a higher overpotential means more electric energy consumed, which does not meet the demands of sustainable development. So, under the precondition of high catalytic performance, a lowering of the overpotential is required by employing new catalysts and changing the reaction conditions. Min Xiao-Quan and coworkers dispersed Pd NPs on CB, and achieved high mass activities (50–80 mA $HCOO^-$ formed per mg Pd), while the overpotential was less than 0.2 V [49]. A suggested mechanism for this process is illustrated in Figure 3.8.

Whilst it is common sense that metal elements have a better catalytic performance than non-metal elements, it is the low cost and abundant availability of non-metal elements that leads to their being favored by research groups. Thus, there is a trend that metal-free catalysis – mainly carbo-catalysis – will become a research hotspot. Zhang and coworkers designed a nitrogen-doped CNT catalyst to reduce CO_2 to formate electrochemically [50]. Formate, as the dominant electrolysis product at PEI-NCNT/GC electrodes, had an 85% yield, while the Faradaic efficiency reached a maximum at a potential of −1.8 V.

Currently, many chemicals and energy are derived from fossil energy, which is non-renewable, and consequently several new energy sources have been explored [51]. Biomass, as one of the renewable sources, has multiple advantages such as large reserves and low toxicity. As a result, many research groups have explored the possibility of biomass-derived energy to replace petroleum-based energy [52].

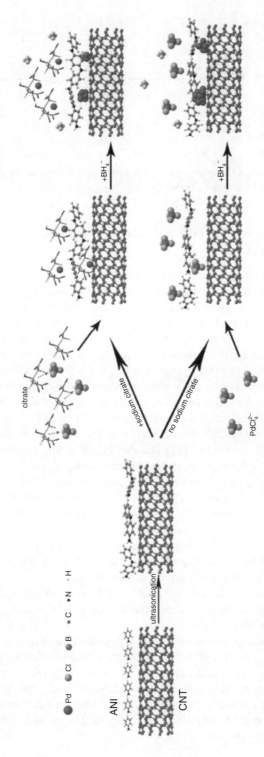

Figure 3.7 *Proposed mechanisms for the synthesis of Pd-PANI/CNT catalysts. Reproduced with permission from Ref. [49]. Copyright 2015, Wiley. (See color plate section for the color representation of this figure.)*

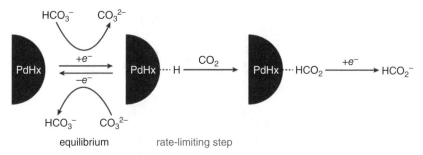

equilibrium   rate-limiting step

Figure 3.8   *Electrohydrogenation mechanism for CO_2 reduction on Pd/C. Reproduced with permission from Ref. [50]. Copyright 2015, American Chemical Society.*

3.3   Transfer Hydrogenation of Unsaturated Compounds from Formic Acid

Hydrogenation is vital in catalytic reactions that may be used to synthesize important chemicals such as pharmaceuticals. Although, in conventional methods, hydrogen is chosen as the hydrogen source it is difficult to handle due to its dangerous properties. In addition, as high-pressure hydrogen gas would generate a high cost of related devices on an industrial scale, it would be more effective to seek an alternative reagent to hydrogen.

Formic acid, as biomass, has the potential to serve as a new hydrogen source, based on its characteristics of low cost, low toxicity, high stability, and high energy density [13,39,53]. Formic acid is mainly produced by the hydrogenation of waste carbon and biomass processes. Most importantly, the active energy of formic acid decomposition is relativity low, and it can be easily dehydrogenated even under mild conditions with transition metals as catalysts. This, in turn, means that formic acid is an excellent environmental source of hydrogen for hydrogenation reactions [13]. Another benefit is that hydrogen gas released by the dehydrogenation of formic acid in-situ can participate in reduction reactions, and can meet the concepts of green chemistry, with features of lower costs and environment friendliness. Moreover, liquid hydrogen sources can alleviate the hazards associated with high pressures and reduce the costs and complexities of devices in experiments [39].

Li and coworkers have succeeded in conducting the hydrogenation of nitrobenzene (denoted as NB) into aniline, with formic acid as hydrogen source [14]. The catalyst was prepared by loading Pd NPs onto carbon nitride, and a Mott–Schottky heterojunction interface was created which could efficiently facilitate the hydrogenation of nitrobenzene [16]. The yield of aniline was up to 99%, and the TOF was 1183 mol (NB) mol^{-1} per active site, which was higher than in other studies (Figure 3.9). A series of compounds was also hydrogenated, with good yields.

Another study was conducted in which the hydrogenation reactions of olefins and unsaturated biomass were processed using the same strategy of the Mott–Schottky effect [14,25]. In these investigations, a variety of unsaturated carbon–carbon bonds (e.g., ethylbenzene) were successfully hydrogenated in high yields, with formic acid as the hydrogen source. The ability to rapidly and automatically separate the product from water, the mild conditions employed, and the inclusion of catalysts and a 'green' hydrogen source, all met with the concept of green chemistry (Figure 3.10).

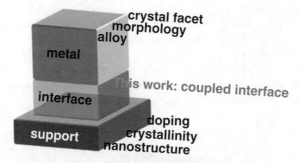

Figure 3.9 *Procedures or parameters for varying the catalytic performance of supported noble metal nanocrystals. A highly coupled interface is the third aspect to be considered for the design of highly efficient catalysts.*

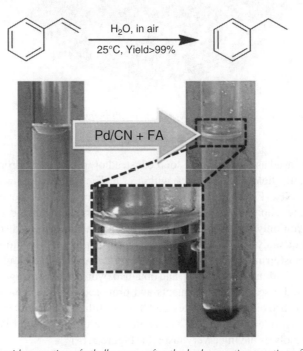

Figure 3.10 *The rapid separation of ethylbenzene after the hydrogenation reaction of styrene. After the reaction, the as-formed ethylbenzene was separated automatically from the water phase, within 10 min. The solid catalyst was precipitated automatically at the bottom and easily separated from the oil phase by filtration or decantation. The reaction conditions were: 10 mmol styrene, 250 ml H_2O, 30 mmol formic acid (FA), and 500 mg Pd/CN, at 298K. (See color plate section for the color representation of this figure.)*

The same strategy was also employed by others. γ-Valerolactone (GVL) is a very important biomass molecule which can be converted into valuable chemicals such as sulcatol (WS75624B; Steganacin) [54]. Following an often-used pathway, levulinic acid (LA) was applied as substrate, and the whole reaction was conducted under relatively harsh conditions requiring a high temperature and high hydrogen pressure. Recently, Sanny Verma and coworkers succeeded in synthesizing GVL under very mild conditions where biomass

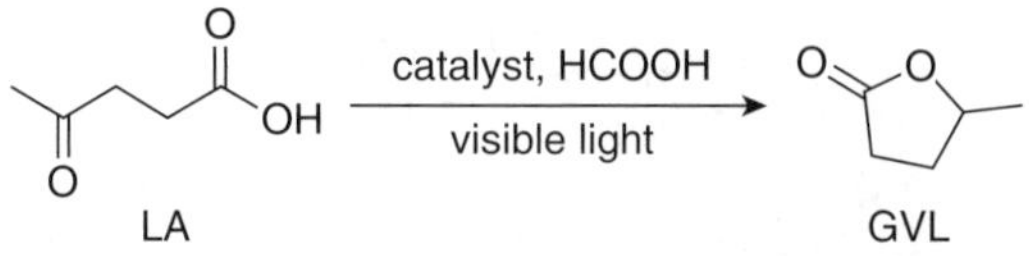

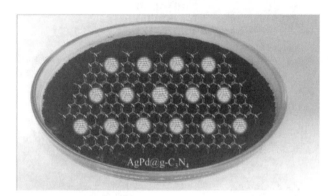

Figure 3.11 *Screening of catalysts for the hydrogenation of levulinic acid (LA) to γ-valerolactone (GVL). Reproduced with permission from Ref. [55]. Copyright 2016, Wiley.*

Figure 3.12 *Structure of AgPd@g-C₃N₄. Reproduced with permission from Ref. [55]. Copyright 2016, Wiley.*

formic acid was used as the hydrogen source instead of traditional hydrogen (Figure 3.11) [55]; moreover, the yield was up to 98% within 12 h. The catalyst used was $g\text{-}C_3N_4$ loaded with Ag and Pd NPs (Figure 3.12), the catalytic ability of which was enhanced by visible light. The yield was up to 98% and the entire process was characterized as 'green'.

Formic acid not only serves as a simple hydrogen source, but can also be used for the immediate hydrogenolysis of benzylic alcohols [56]. Supaporn Sawadjoon and coworkers succeeded in transferring the hydrogenolysis of primary, secondary, and tertiary benzylic alcohols with a palladium catalyst [57], and obtained hydrocarbons in high yields. The same authors conducted a series of experiments and proposed a mechanism whereby formate anions were adsorbed onto open Pd sites, and this was followed by a reversible protonation. A rate-limiting hydride transfer with chemisorbed hydrogen to obtain Pd sites led to the alcohol hydrogenolysis having a very fast rate (Figure 3.13).

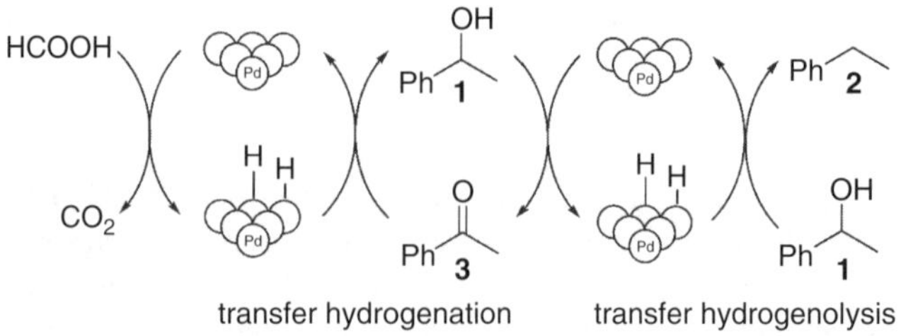

Figure 3.13 *Proposed possible mechanism of disproportion and transfer hydrogenation. Reproduced with permission from Ref. [57]. Copyright 2013, American Chemical Society.*

3.4 Synthesis of High-Value-Added Chemicals from Biomass

Biomass feedstock may be converted into abundant energy sources and applied in areas such as physical chemistry. Hence, the molecular synthesis of biomass into other high-value chemicals is a hot topic that is currently being investigated by many research groups.

Bimetallic catalysts have been proven more highly active for some catalytic syntheses than monometallic metals, due to synergistic effects [58]. When Sanny Verma and coworkers successfully prepared the bimetallic catalyst AgPd@g-C_3N_4, its performance in converting 2-methoxy-4-methylphenol into 4-hydroxy-3-methoxybenzaldehyde (vanillin), a potential future biofuel, was considered to be very efficient (Figure 3.14) [59]. Indeed, the yield of 99% was much higher than with monometallic metal Pd@g-C_3N_4, coupled with the ability to use formic acid as the hydrogen source, proved valuable for the synthesis of biofuels on an industrial scale.

One very important method of catalyst modification involves doping [60]. Chen and coworkers synthesized a protonated vanadium-doped g-C_3N_4 that would transform a biomass platform of molecular fructose into 2,5-diformylfuran (DFF) in a one-pot process (Figure 3.15) [61]. The protonated g-C_3N_4 was able to convert fructose into 5-hydroxymethylfurfural (HMF) under acidic conditions, and the metal sites had an important effect on aerobic oxidation. Somewhat ingeniously, the same authors combined the protonation of g-C_3N_4 and doping by vanadium to create V-g-C_3N_4 (H^+), which

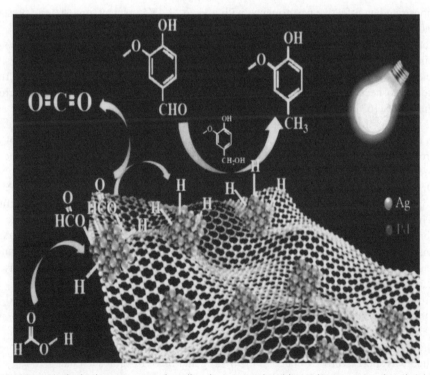

Figure 3.14 *The hydrodeoxygenation of vanillin, depicting a plausible mechanism. Reproduced with permission from Ref. [59]. Copyright 2016, Royal Society of Chemistry.*

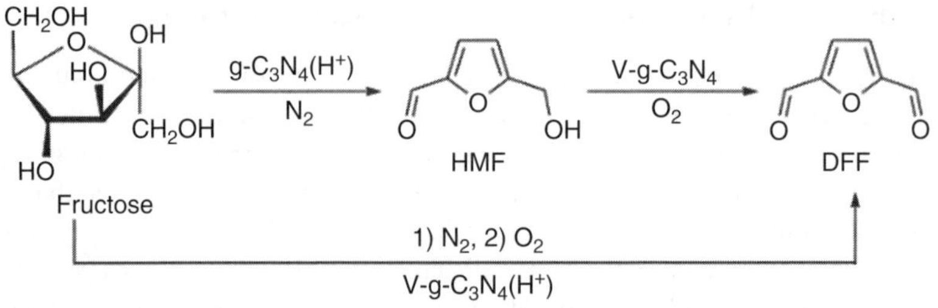

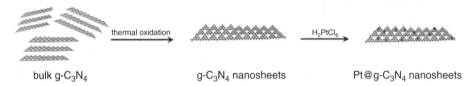

Figure 3.15 *Conversion of fructose into 5-hydroxymethylfurfural (HMF) and 2,5-diformylfuran (DFF), catalyzed by functionalized g-C₃N₄. Reproduced with permission from Ref. [61]. Copyright 2014, Wiley.*

bulk g-C₃N₄   g-C₃N₄ nanosheets   Pt@g-C₃N₄ nanosheets

Figure 3.16 *Schematic illustration of the fabrication of Pt@CN nanosheets. Reproduced with permission from Ref. [63]. Copyright 2016, Nature.*

demonstrated an effective performance with a yield of 45%, even at atmospheric pressure of molecular oxygen.

Furfuryl alcohol is an important biomass-derived chemical intermediate used in the synthesis of other chemicals such as resins, plasticizers and lysine [62]. Noble metal NPs are efficient in the hydrogenation reaction. Chen and coworkers synthesized graphitic carbon nitride nanosheets using a simple, 'green' method that involved thermal oxidation etching (Figure 3.16) [63]. The nanosheet, which was 3 nm thick and had a very high surface area ($142 \text{ m}^2 \text{ g}^{-1}$), provided a good support for loading Pt NPs with an average size only about 4.25 nm. Rather surprisingly, the sheets showed an excellent performance in terms of the hydrogenation of furfural alcohol into furfuryl alcohol, with a selectivity of 99% and no deep hydrogenation products being generated.

One very important industrial reaction is that of esterification. Rather ingeniously, Panagiotopoulou and coworkers performed a catalytic transfer hydrogenation (CTH) of furfural into furfural alcohol, by using 2-propanol as a hydrogen source, followed by in-situ esterification [64]. The catalytic activity exhibited a significant increase with an empirical scale of Lewis acid strength (Figure 3.17) [65]. A high furfural alcohol yield of 97% was obtained at 180 °C, and the combination of Ru/C and Lewis acid resulted in a significant increase in furfural conversion.

Supercritical carbon dioxide (scCO₂) as an effective solvent was always selected for use in catalytic reactions [66,67]. Chatterjee and coworkers recently combined scCO₂ and H₂O as the solvent applied to the hydrogenation of HMF with a Pd/C catalyst (Figure 3.18) [68]. Although a very high yield (100%) of DMF was obtained at 80 °C within 2 h, the method was successfully extended to the hydrogenation of furfural such that a 100% yield of 2-methylfuran was obtained. When the reaction time was shortened to 10 min, furfural was completely converted to 2-methylfuran.

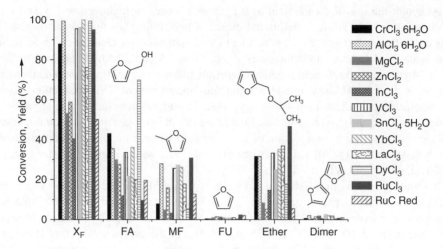

Figure 3.17 *Effect of the addition of Ru/C catalyst on furfural conversion and product yield for the indicated homogeneous Lewis acid catalysts. Reproduced with permission from Ref. [65]. Copyright 2014, Wiley. (See color plate section for the color representation of this figure.)*

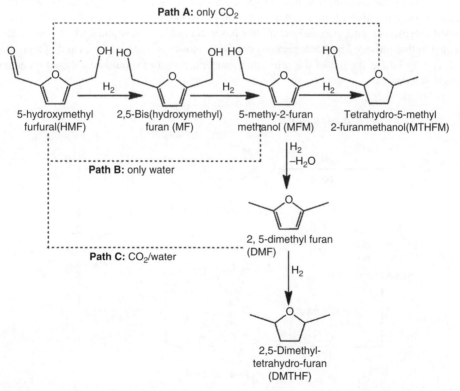

Figure 3.18 *Reaction pathway of 5-hydroxymethylfurfural (HMF) hydrogenation in supercritical CO_2. Reproduced with permission from Ref. [68]. Copyright 2014, Royal Society of Chemistry.*

Although the use of formic acid as a hydrogen source to hydrogenate LA and form GVL was discussed above, a traditional means of performing this reaction also exists. In this case, the catalyst was Ru/C, which had shown potential for application on an industrial scale. Abdelrahman and coworkers showed that LA could be hydrogenated to GVL via a 4-hydroxypentanoic acid-mediated pathway, followed by acid-catalyzed dehydration (Figure 3.19) [69]. At low temperatures, the rate-limiting step of GVL production was the intramolecular esterification of 4-hydroxypentanoic acid, whereas under high-temperature conditions it was the mass transfer that limited the rate of hydrogenation. Abdelrahman and colleagues employed a combination of Ru/C for the hydrogenation, and a strongly acidic catalyst, Amberlyst-15 [70], to increase the yield of GVL. Indeed, even at a low temperature (303K) the yield of GVL was 81%.

It should be noted that Selva and coworkers designed a liquid triphase system that included an aqueous phase, an organic phase and an ionic liquid, and the whole was applied to the hydrogenation and dehydration of LA to GVL [71]. The entire reaction was conducted at 100–150 °C and 35 atm of H_2, in the presence of either Ru/C or a homogeneous Ru precursor. This liquid triphase system showed three advantages: (i) the desired product, GVL, was obtained with quantitative conversion and 100% selectivity; (ii) the product could be created by a simple phase-separation method; and (iii) the catalyst activity could be preserved for in-situ recycles, without any loss of metal (Figure 3.20).

Studies on the dehydrogenation of formic acid and reduction of CO_2 to formic acid over heterocatalysts either by traditional means or electrochemical methods, have been developed extensively, and the catalytic ability based on noble metals and carbon materials is equal to that provided by homogeneous catalysts. However, the high costs and cycling stability may inhibit the use of the procedure and future studies should be focused on actual applications.

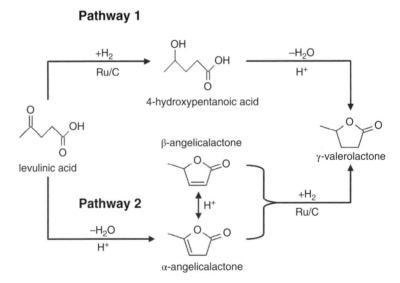

Figure 3.19 *Pathways for the hydrogenation of levulinic acid to γ-valerolactone. Reproduced with permission from Ref. [69]. Copyright 2014, American Chemical Society.*

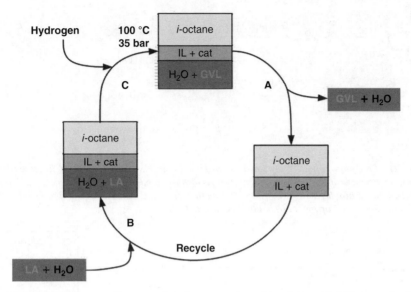

Figure 3.20 *Multiphase recycling procedure for conversion of levulinic acid (LA) to γ-valerolactone (GVL). Reproduced with permission from Ref. [71]. Copyright 2013, American Chemical Society.*

3.5 Metal-Free Catalyst: Graphene Oxide for the Conversion of Fructose

The use of metal-free catalysts remains a 'hotpot' in catalyst research, due to their low cost and unique physical and chemical properties. Specifically, in the past graphene oxide had functioned as a precursor in the production of grapheme, with one of the most widespread strategies for preparing graphene oxide being Hummers' method, which was based on the direct oxidation of graphite under strong acid conditions.

Lv and coworkers reported that graphene oxide exhibited a good performance in the synthesis of 2,5-diformylfuran (DFF) from fructose (Figure 3.21) [72]. In this case, graphene oxide was prepared by Hummers' method and the objective product was obtained in a yield of 53% by a one-step process. In this reaction, graphene oxide with a rich amount of oxygen-containing groups on the surface could stimulate the dehydration reaction of fructose. The built-in carboxylic acid groups with the unpaired electrons at the edge of the graphene oxide layers played a key role in promoting the selective dehydrogenation of fructose to 2,5-diformylfuran [73,74].

Wang and coworkers discovered graphene oxide to be an efficient acid catalyst for the one-step conversion of fructose-based biopolymers into 5-ethoxymethylfurfural (EMF), which could serve as a potential biofuel alternative (Figure 3.22) [75,76]. The solvent was a mixture of ethanol and dimethylsulfoxide (DMSO). 5-Hydroxymethylfurfural (HMF) derived from fructose dehydration could be transformed in-situ into EMF by esterification. When HMF, fructose, sucrose and inulin were chosen as substrates, the yields of EMF were 92%, 71%, 34%, and 66%, respectively.

Again, sulfuric acid-modified graphene oxides could also act as solid acid for the transformation of the biomass 5-(hydroxymethyl)-2-furfural into 5-ethoxymethylfurfural, 5-(ethoxymethyl) furfural diethylacetal, and/or ethyl levulinate by using ethanol as

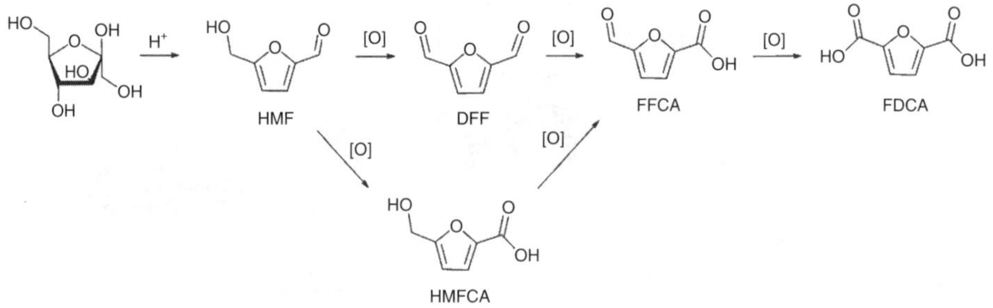

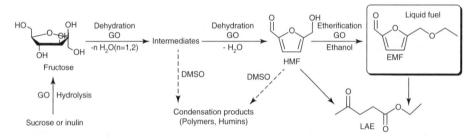

Figure 3.21 *Oxidation products of 5-hydroxymethylfurfural (HMF) derived from fructose dehydration. The byproducts are 2,5-diformylfuran (DFF), 5-hydroxymethyl-2-furancarboxylic acid (HMFCA), 5-formyl-2-furancarboxylic acid (FFCA), and 2,5-furandicarboxylic acid (FDCA). Reproduced with permission from Ref. [72]. Copyright 2016, Royal Society of Chemistry.*

Figure 3.22 *The conversion of carbohydrates to 5-ethoxymethylfurfural (EMF), catalyzed by graphene oxide (GO). Reproduced with permission from Ref. [76]. Copyright 2013, Royal Society of Chemistry.*

solvent (Figure 3.23). The cooperative effects of the sulfonic acid groups and carboxylic acid groups ensured high yields of the target productions [5-(ethoxymethyl) furfural diethylacetal, and ethyl levulinate], which could be as high as 100% under optimized conditions [77].

3.6   Conclusions and Outlook

Nanoporous carbon/nitrogen materials and their hybrids have been proven to activate functional particles to a significant degree, taking metal NPs for example through the Mott–Schotty heterojunction for the dehydrogenation of formic acid, the transfer hydrogenation of unsaturated bond, and the production of high-value chemicals. The support effect on the basis of carbon/nitrogen materials causes an adjustment of the electron density of a metal center to a large extent, which directly triggers an enhancement of catalytic activity in catalysis systems. The programmable band structure of carbon/nitrogen materials offers a favorable routine to rationally engineer superb high-performance catalysts for specific reactions. The introduction and modification of different metal particles will open the door towards a rich chemistry of biomass conversion. Moreover, carbon/nitrogen materials are easily applicable to mass production, and it can be envisioned that a combination of an efficient Mott–Schotty effect and the diversity of metal sites would provide further

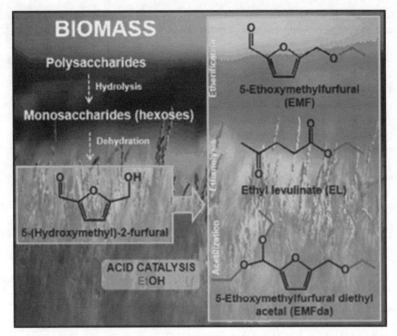

Figure 3.23 *Conversion of HMF into products for biofuels applications. Reproduced with permission from Ref. [77]. Copyright 2014, Wiley. (See color plate section for the color representation of this figure.)*

possibilities of developing biomass conversion systems towards environment-friendly and economic processes.

References

1. Wang, X.C., Maeda, K., Thomas, A., Takanabe, K., Xin, G., Carlsson, J.M., Domen, K., Antonietti, M. (2009) A metal-free polymeric photocatalyst for hydrogen production from water under visible light. *Nat. Mater.*, **8** (1), 76–80.
2. Hao, L., Ning, J., Luo, B., Wang, B., Zhang, Y.B., Tang, Z.H., Yang, J.H., Thomas, A., Zhi, L.J. (2015) Structural evolution of 2D microporous covalent triazine-based framework toward the study of high-performance supercapacitors. *J. Am. Chem. Soc.*, **137** (1), 219–225.
3. Wang, X.R., Li, X.L., Zhang, L., Yoon, Y., Weber, P.K., Wang, H.L., Guo, J., Dai, H.J. (2009) N-Doping of graphene through electrothermal reactions with ammonia. *Science*, **324** (5928), 768–771.
4. Li, X.H., Kurasch, S., Kaiser, U., Antonietti, M. (2012) Synthesis of monolayer-patched graphene from glucose. *Angew. Chem. Int. Ed.*, **51** (38), 9689–9692.
5. Zhou, C.H., Xia, X., Lin, C.X., Tong, D.S., Beltramini, J. (2011) Catalytic conversion of lignocellulosic biomass to fine chemicals and fuels. *Chem. Soc. Rev.*, **40** (11), 5588–5617.
6. Wang, C.T., Wang, L., Zhang, J., Wang, H., Lewis, J.P., Xiao, F.S. (2016) Product selectivity controlled by zeolite crystals in biomass hydrogenation over a palladium catalyst. *J. Am. Chem. Soc.*, **138** (25), 7880–7883.
7. Climent, M.J., Corma, A., Iborra, S. (2014) Conversion of biomass platform molecules into fuel additives and liquid hydrocarbon fuels. *Green Chem.*, **16** (2), 516–547.

8. Li, X.H., Wang, X.C., Antonietti, M. (2012) Solvent-free and metal-free oxidation of toluene using O-2 and g-C3N4 with nanopores: nanostructure boosts the catalytic selectivity. *ACS Catal.*, **2** (10), 2082–2086.

9. Li, X.H., Chen, J.S., Wang, X.C., Sun, J.H., Antonietti, M. (2011) Metal-free activation of dioxygen by graphene/g-C3N4 nanocomposites: functional dyads for selective oxidation of saturated hydrocarbons. *J. Am. Chem. Soc.*, **133** (21), 8074–8077.

10. Li, X.H., Antonietti, M. (2013) Polycondensation of boron- and nitrogen-codoped holey graphene monoliths from molecules: carbocatalysts for selective oxidation. *Angew. Chem. Int. Ed.*, **52** (17), 4572–4576.

11. Yang, S.L., Peng, L., Huang, P.P., Wang, X.S., Sun, Y.B., Cao, C.Y., Song, W.G. (2016) Nitrogen, phosphorus, and sulfur co-doped hollow carbon shell as superior metal-free catalyst for selective oxidation of aromatic alkanes. *Angew. Chem. Int. Ed.*, **55** (12), 4016–4020.

12. Zhang, J.S., Sun, J.H., Maeda, K., Domen, K., Liu, P., Antonietti, M., Fu, X.Z., Wang, X.C. (2011) Sulfur-mediated synthesis of carbon nitride: Band-gap engineering and improved functions for photocatalysis. *Energy Environ. Sci.*, **4** (3), 675–678.

13. Cai, Y.Y., Li, X.H., Zhang, Y.N., Wei, X., Wang, K.X., Chen, J.S. (2013) Highly efficient dehydrogenation of formic acid over a palladium-nanoparticle-based Mott-Schottky photocatalyst. *Angew. Chem. Int. Ed.*, **52** (45), 11822–11825.

14. Li, X.H., Cai, Y.Y., Gong, L.H., Fu, W., Wang, K.X., Bao, H.L., Wei, X., Chen, J.S. (2014) Photochemically engineering the metal-semiconductor interface for room-temperature transfer hydrogenation of nitroarenes with formic acid. *Chem. Eur. J.*, **20** (50), 16732–16737.

15. Li, X.H., Baar, M., Blechert, S., Antonietti, M. (2013) Facilitating room-temperature Suzuki coupling reaction with light: Mott-Schottky photocatalyst for C-C-coupling. *Sci. Rep.*, **3** (4), 1743.

16. Li, X.H., Antonietti, M. (2013) Metal nanoparticles at mesoporous N-doped carbons and carbon nitrides: functional Mott-Schottky heterojunctions for catalysis. *Chem. Soc. Rev.*, **42** (16), 6593–6604.

17. Vardon, D.R., Franden, M.A., Johnson, C.W., Karp, E.M., Guarnieri, M.T., Linger, J.G., Salm, M.J., Strathmann, T.J., Beckham, G.T. (2015) Adipic acid production from lignin. *Energy Environ. Sci.*, **8** (2), 617–628.

18. Chen, Y.Z., Cai, G.R., Wang, Y.M., Xu, Q., Yu, S.H., Jiang, H.L. (2016) Palladium nanoparticles stabilized with N-doped porous carbons derived from metal-organic frameworks for selective catalysis in biofuel upgrade: the role of catalyst wettability. *Green Chem.*, **18** (5), 1212–1217.

19. Mao, L.-B., Gao, H.-L., Yao, H.-B., Liu, L., Cölfen, H., Liu, G., Chen, S.-M., Li, S.-K., Yan, Y.-X., Liu, Y.-Y., Yu, S.-H. (2016) Synthetic nacre by predesigned matrix-directed mineralization. *Science*, **354** (6308), 107–110.

20. Straathof, A.J.J. (2014) Transformation of biomass into commodity chemicals using enzymes or cells. *Chem. Rev.*, **114** (3), 1871–1908.

21. Nielsen, M., Kammer, A., Cozzula, D., Junge, H., Gladiali, S., Beller, M. (2011) Efficient hydrogen production from alcohols under mild reaction conditions. *Angew. Chem. Int. Ed.*, **50** (41), 9593–9597.

22. Boddien, A., Mellmann, D., Gartner, F., Jackstell, R., Junge, H., Dyson, P.J., Laurenczy, G., Ludwig, R., Beller, M. (2011) Efficient dehydrogenation of formic acid using an iron catalyst. *Science*, **333** (6050), 1733–1736.

23. Lu, R., Lu, F., Chen, J.Z., Yu, W.Q., Huang, Q.Q., Zhang, J.J., Xu, J. (2016) Production of diethyl terephthalate from biomass-derived muconic acid. *Angew. Chem. Int. Ed.*, **55** (1), 249–253.

24. Nielsen, M., Junge, H., Kammer, A., Beller, M. (2012) Towards a green process for bulk-scale synthesis of ethyl acetate: efficient acceptorless dehydrogenation of ethanol. *Angew. Chem. Int. Ed.*, **51** (23), 5711–5713.

25. Gong, L.H., Cai, Y.Y., Li, X.H., Zhang, Y.N., Su, J., Chen, J.S. (2014) Room-temperature transfer hydrogenation and fast separation of unsaturated compounds over heterogeneous catalysts in an aqueous solution of formic acid. *Green Chem.*, **16** (8), 3746–3751.

26. Turner, J.A. (2004) Sustainable hydrogen production. *Science*, **305** (5686), 972–974.
27. Schlapbach, L., Zuttel, A. (2001) Hydrogen-storage materials for mobile applications. *Nature*, **414** (6861), 353–358.
28. Johnson, T.C., Morris, D.J., Wills, M. (2010) Hydrogen generation from formic acid and alcohols using homogeneous catalysts. *Chem. Soc. Rev.*, **39** (1), 81–88.
29. Enthaler, S., von Langermann, J., Schmidt, T. (2010) Carbon dioxide and formic acid – the couple for environmental-friendly hydrogen storage? *Energy Environ. Sci.*, **3** (9), 1207–1217.
30. Singh, A.K., Singh, S., Kumar, A. (2016) Hydrogen energy future with formic acid: a renewable chemical hydrogen storage system. *Catal. Sci. Tech.*, **6** (1), 12–40.
31. Ye, T.-N., Lv, L.-B., Li, X.-H., Xu, M., Chen, J.-S. (2014) Strongly veined carbon nanoleaves as a highly efficient metal-free electrocatalyst. *Angew. Chem. Int. Ed.*, **53** (27), 6905–6909.
32. Ji, X., Lee, K.T., Holden, R., Zhang, L., Zhang, J., Botton, G.A., Couillard, M., Nazar, L.F. (2010) Nanocrystalline intermetallics on mesoporous carbon for direct formic acid fuel cell anodes. *Nat. Chem.*, **2** (4), 286–293.
33. Yang, J., Tian, C., Wang, L., Fu, H. (2011) An effective strategy for small-sized and highly-dispersed palladium nanoparticles supported on graphene with excellent performance for formic acid oxidation. *J. Mater. Chem.*, **21** (10), 3384–3390.
34. Guo, L.T., Cai, Y.Y., Ge, J.M., Zhang, Y.N., Gong, L.H., Li, X.H., Wang, K.X., Ren, Q.Z., Su, J., Chen, J.S. (2013) Multifunctional Au-Co@CN nanocatalyst for highly efficient hydrolysis of ammonia borane. *ACS Catal.*, **5** (1), 388–392.
35. Zhu, Q.-L., Tsumori, N., Xu, Q. (2014) Sodium hydroxide-assisted growth of uniform Pd nanoparticles on nanoporous carbon MSC-30 for efficient and complete dehydrogenation of formic acid under ambient conditions. *Chem. Sci.*, **5** (1), 195–199.
36. Zhou, X., Huang, Y., Xing, W., Liu, C., Liao, J., Lu, T. (2008) High-quality hydrogen from the catalyzed decomposition of formic acid by Pd-Au/C and Pd-Ag/C. *Chem. Commun.*, 3540–3542.
37. Zhang, S., Shao, Y., Liao, H.-g., Liu, J., Aksay, I. A., Yin, G., Lin, Y. (2011) Graphene decorated with PtAu alloy nanoparticles: facile synthesis and promising application for formic acid oxidation. *Chem. Mater.*, **23** (5), 1079–1081.
38. Tedsree, K., Li, T., Jones, S., Chan, C.W.A., Yu, K.M.K., Bagot, P.A.J., Marquis, E.A., Smith, G.D.W., Tsang, S.C.E. (2011) Hydrogen production from formic acid decomposition at room temperature using a Ag-Pd core-shell nanocatalyst. *Nat. Nano*, **6** (5), 302–307.
39. Wang, Z.-L., Yan, J.-M., Ping, Y., Wang, H.-L., Zheng, W.-T., Jiang, Q. (2013) An efficient CoAuPd/C catalyst for hydrogen generation from formic acid at room temperature. *Angew. Chem. Int. Ed.*, **52** (16), 4406–4409.
40. Zhang, S., Metin, Ö., Su, D., Sun, S. (2013) Monodisperse AgPd alloy nanoparticles and their superior catalysis for the dehydrogenation of formic acid. *Angew. Chem. Int. Ed.*, **52** (13), 3681–3684.
41. Wang, X., Maeda, K., Thomas, A., Takanabe, K., Xin, G., Carlsson, J.M., Domen, K., Antonietti, M. (2009) A metal-free polymeric photocatalyst for hydrogen production from water under visible light. *Nat. Mater.*, **8** (1), 76–80.
42. Wang, H.-H., Zhang, B., Li, X.-H., Antonietti, M., Chen, J.-S. (2016) Activating Pd nanoparticles on sol-gel prepared porous g-C_3N_4/SiO_2 via enlarging the Schottky barrier for efficient dehydrogenation of formic acid. *Inorg. Chem. Front.*, **3** (9), 1124–1129.
43. Boddien, A., Loges, B., Junge, H., Beller, M. (2008) Hydrogen generation at ambient conditions: application in fuel cells. *ChemSusChem*, **1** (8-9), 751–758.
44. Gawande, M.B., Goswami, A., Asefa, T., Guo, H., Biradar, A.V., Peng, D.-L., Zboril, R., Varma, R.S. (2015) Core-shell nanoparticles: synthesis and applications in catalysis and electrocatalysis. *Chem. Soc. Rev.*, **44** (21), 7540–7590.
45. Tao, F., Grass, M.E., Zhang, Y., Butcher, D.R., Renzas, J.R., Liu, Z., Chung, J.Y., Mun, B.S., Salmeron, M., Somorjai, G.A. (2008) Reaction-driven restructuring of Rh-Pd and Pt-Pd core-shell nanoparticles. *Science*, **322** (5903), 932–934.

46. Kobayashi, H., Yamauchi, M., Kitagawa, H., Kubota, Y., Kato, K., Takata, M. (2008) Hydrogen absorption in the core/shell interface of Pd/Pt nanoparticles. *J. Am. Chem. Soc.*, **130** (6), 1818–1819.

47. Huang, Y., Zhou, X., Yin, M., Liu, C., Xing, W. (2010) Novel PdAu@Au/C core−shell catalyst: superior activity and selectivity in formic acid decomposition for hydrogen generation. *Chem. Mater.*, **22** (18), 5122–5128.

48. Zhao, C., Yin, Z., Wang, J. (2015) Efficient electrochemical conversion of CO$_2$ to HCOOH using Pd-polyaniline/CNT nanohybrids prepared in situ. *ChemElectroChem*, **2** (12), 1974–1982.

49. Min, X., Kanan, M.W. (2015) Pd-catalyzed electrohydrogenation of carbon dioxide to formate: high mass activity at low overpotential and identification of the deactivation pathway. *J. Am. Chem. Soc.*, **137** (14), 4701–4708.

50. Zhang, S., Kang, P., Ubnoske, S., Brennaman, M.K., Song, N., House, R.L., Glass, J.T., Meyer, T.J. (2014) Polyethylenimine-enhanced electrocatalytic reduction of CO$_2$ to formate at nitrogen-doped carbon nanomaterials. *J. Am. Chem. Soc.*, **136** (22), 7845–7848.

51. Daigneault, A., Sohngen, B., Sedjo, R. (2012) Economic approach to assess the forest carbon implications of biomass energy. *Environ. Sci. Tech.*, **46** (11), 5664–5671.

52. Ruppert, A.M., Weinberg, K., Palkovits, R. (2012) Hydrogenolysis goes bio: from carbohydrates and sugar alcohols to platform chemicals. *Angew. Chem. Int. Ed.*, **51** (11), 2564–2601.

53. Boddien, A., Mellmann, D., Gärtner, F., Jackstell, R., Junge, H., Dyson, P.J., Laurenczy, G., Ludwig, R., Beller, M. (2011) Efficient dehydrogenation of formic acid using an iron catalyst. *Science*, **333** (6050), 1733–1736.

54. Michel, C., Zaffran, J., Ruppert, A.M., Matras-Michalska, J., Jedrzejczyk, M., Grams, J., Sautet, P. (2014) Role of water in metal catalyst performance for ketone hydrogenation: a joint experimental and theoretical study on levulinic acid conversion into gamma-valerolactone. *Chem. Commun.*, **50** (83), 12450–12453.

55. Verma, S., Baig, R.B.N., Nadagouda, M.N., Varma, R.S. (2016) Sustainable strategy utilizing biomass: visible-light-mediated synthesis of γ-valerolactone. *ChemCatChem*, **8** (4), 690–693.

56. Schlaf, M. (2006) Selective deoxygenation of sugar polyols to [small alpha],[small omega]-diols and other oxygen content reduced materials – a new challenge to homogeneous ionic hydrogenation and hydrogenolysis catalysis. *Dalton Trans.*, **39**, 4645–4653.

57. Sawadjoon, S., Lundstedt, A., Samec, J.S.M. (2013) Pd-catalyzed transfer hydrogenolysis of primary, secondary, and tertiary benzylic alcohols by formic acid: a mechanistic study. *ACS Catal.*, **3** (4), 635–642.

58. Hu, C., Mu, X., Fan, J., Ma, H., Zhao, X., Chen, G., Zhou, Z., Zheng, N. (2016) Interfacial effects in PdAg bimetallic nanosheets for selective dehydrogenation of formic acid. *ChemNanoMater.*, **2** (1), 28–32.

59. Verma, S., Nasir Baig, R.B., Nadagouda, M.N., Varma, R.S. (2016) Visible light mediated upgrading of biomass to biofuel. *Green Chem.*, **18** (5), 1327–1331.

60. Wang, W., Ding, G., Jiang, T., Zhang, P., Wu, T., Han, B. (2013) Facile one-pot synthesis of VxOy@C catalysts using sucrose for the direct hydroxylation of benzene to phenol. *Green Chem.*, **15** (5), 1150–1154.

61. Chen, J., Guo, Y., Chen, J., Song, L., Chen, L. (2014) One-step approach to 2,5-diformylfuran from fructose by proton- and vanadium-containing graphitic carbon nitride. *ChemCatChem*, **6** (11), 3174–3181.

62. Corma, A., Iborra, S., Velty, A. (2007) Chemical routes for the transformation of biomass into chemicals. *Chem. Rev.*, **107** (6), 2411–2502.

63. Chen, X., Zhang, L., Zhang, B., Guo, X., Mu, X. (2016) Highly selective hydrogenation of furfural to furfuryl alcohol over Pt nanoparticles supported on g-C(3)N(4) nanosheets catalysts in water. *Sci. Rep.*, **6**, 28558.

64. Saha, B., Bohn, C.M., Abu-Omar, M.M. (2014) Zinc-assisted hydrodeoxygenation of biomass-derived 5-hydroxymethylfurfural to 2,5-dimethylfuran. *ChemSusChem*, **7** (11), 3095–3101.

65. Panagiotopoulou, P., Martin, N., Vlachos, D.G. (2015) Liquid-phase catalytic transfer hydrogenation of furfural over homogeneous Lewis acid–Ru/C catalysts. *ChemSusChem*, **8** (12), 2046–2054.

66. Hiyoshi, N., Mine, E., Rode, C.V., Sato, O., Shirai, M. (2006) Low temperature hydrogenation of tetralin over supported rhodium catalysts in supercritical carbon dioxide solvent. *Appl. Catal. A: General*, **310**, 194–198.

67. Hou, A., Chen, B., Dai, J., Zhang, K. (2010) Using supercritical carbon dioxide as solvent to replace water in polyethylene terephthalate (PET) fabric dyeing procedures. *J. Clean. Prod.*, **18** (10–11), 1009–1014.

68. Chatterjee, M., Ishizaka, T., Kawanami, H. (2014) Hydrogenation of 5-hydroxymethylfurfural in supercritical carbon dioxide-water: a tunable approach to dimethylfuran selectivity. *Green Chem.*, **16** (3), 1543–1551.

69. Abdelrahman, O.A., Heyden, A., Bond, J.Q. (2014) Analysis of kinetics and reaction pathways in the aqueous-phase hydrogenation of levulinic acid to form γ-valerolactone over Ru/C. *ACS Catal.*, **4** (4), 1171–1181.

70. Boz, N., Degirmenbasi, N., Kalyon, D.M. (2015) Esterification and transesterification of waste cooking oil over Amberlyst 15 and modified Amberlyst 15 catalysts. *Appl. Catal. B: Environ.*, **165**, 723–730.

71. Selva, M., Gottardo, M., Perosa, A. (2013) Upgrade of biomass-derived levulinic acid via Ru/C-catalyzed hydrogenation to γ-valerolactone in aqueous–organic–ionic liquids multiphase systems. *ACS Sustain. Chem. Eng.*, **1** (1), 180–189.

72. Lv, G., Wang, H., Yang, Y., Deng, T., Chen, C., Zhu, Y., Hou, X. (2016) Direct synthesis of 2,5-diformylfuran from fructose with graphene oxide as a bifunctional and metal-free catalyst. *Green Chem.*, **18** (8), 2302–2307.

73. Wang, H., Kong, Q., Wang, Y., Deng, T., Chen, C., Hou, X., Zhu, Y. (2014) Graphene oxide catalyzed dehydration of fructose into 5-hydroxymethylfurfural with isopropanol as cosolvent. *ChemCatChem*, **6** (3), 728–732.

74. Su, C., Acik, M., Takai, K., Lu, J., Hao, S.-j., Zheng, Y., Wu, P., Bao, Q., Enoki, T., Chabal, Y.J., Ping Loh, K. (2012) Probing the catalytic activity of porous graphene oxide and the origin of this behaviour. *Nat. Commun.*, **3**, 1298.

75. Lew, C.M., Rajabbeigi, N., Tsapatsis, M. (2012) One-pot synthesis of 5-(ethoxymethyl)furfural from glucose using Sn-BEA and Amberlyst catalysts. *Ind. Eng. Chem. Res.*, **51** (14), 5364–5366.

76. Wang, H., Deng, T., Wang, Y., Cui, X., Qi, Y., Mu, X., Hou, X., Zhu, Y. (2013) Graphene oxide as a facile acid catalyst for the one-pot conversion of carbohydrates into 5-ethoxymethylfurfural. *Green Chem.*, **15** (9), 2379–2383.

77. Antunes, M.M., Russo, P.A., Wiper, P.V., Veiga, J.M., Pillinger, M., Mafra, L., Evtuguin, D.V., Pinna, N., Valente, A.A. (2014) Sulfonated graphene oxide as effective catalyst for conversion of 5-(hydroxymethyl)-2-furfural into biofuels. *ChemSusChem*, **7**, 804–812.

4

Recent Developments in the Use of Porous Carbon Materials for Cellulose Conversion

Abhijit Shrotri, Hirokazu Kobayashi, and Atsushi Fukuoka

Institute for Catalysis, Hokkaido University, Japan

4.1 Introduction

The conversion of structural carbohydrates present in lignocellulose produces monomeric sugars, which are useful for the synthesis of important industrial chemicals [1]. Cellulose is the main component of lignocellulose (30–50%), followed by hemicellulose (10–40%), and lignin (5–30%) [2,3]. Cellulose is a homopolymer of glucose, whereas hemicellulose is a heteropolymer containing a mixture of C5 and C6 sugars such as glucose, xylose, mannose, galactose, and arabinose [4]. The hydrolysis of glycosidic bonds interlinking the monomer units in cellulose and hemicellulose is the primary method for conversion of these carbohydrate polymers. In this reaction, cellulose selectively produces glucose (Scheme 4.1) [5], and xylan-type hemicellulose yields xylose as a main product [6]. Together, glucose and xylose represent the largest fraction of organic molecules produced from biomass.

The versatility of glucose as a platform chemical is well-recognized, and it can be converted to many industrially useful chemicals and fuels (Figure 4.1). The enzymatic conversion of glucose to fuel-based ethanol is industrially viable, and many commercial facilities are operating worldwide [7,8]. The conversion of glucose to 5-hydroxy-methylfurfural (5-HMF) through isomerization to fructose and subsequent dehydration is one of the most widely researched topics in platform chemical synthesis [9]. Other important applications of glucose include hydrogenation to produce sorbitol [10], hydrogenolysis to ethylene glycol [11], and fermentation to organic acids such as succinic acid [12] and

Nanoporous Catalysts for Biomass Conversion, First Edition. Edited by Feng-Shou Xiao and Liang Wang.
© 2018 John Wiley & Sons Ltd. Published 2018 by John Wiley & Sons Ltd.

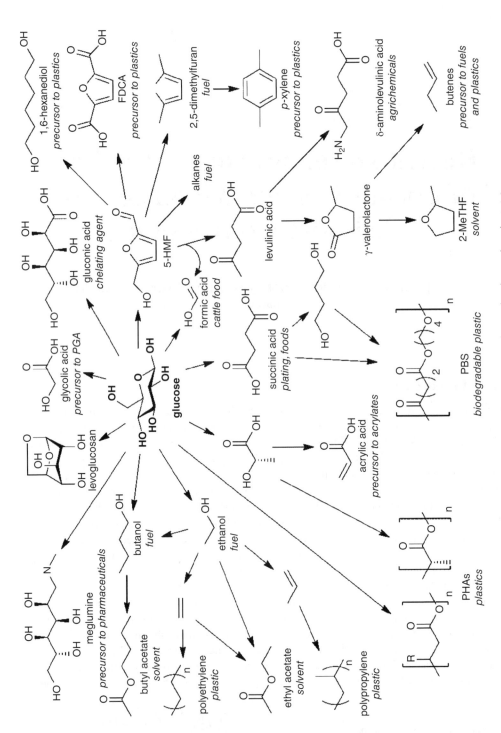

Figure 4.1 Applications of cellulose-derived glucose for chemical synthesis [5].

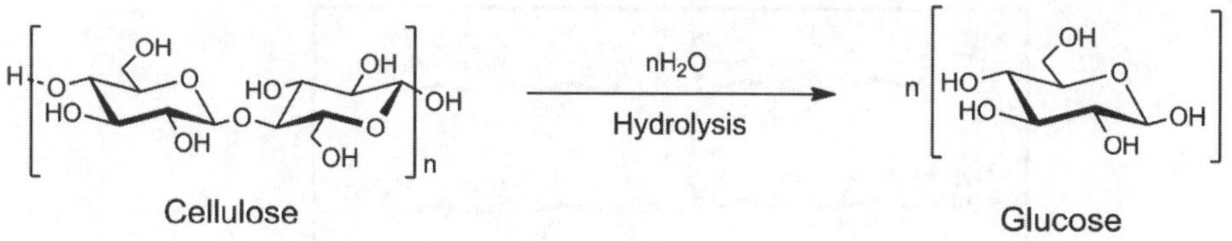

Scheme 4.1   *Hydrolysis of cellulose to glucose.*

gluconic acid [13]. The usefulness of glucose for the synthesis of chemicals calls for a strong focus on research for the selective synthesis of glucose from cellulose present in lignocellulosic biomass.

The conversion of cellulose to glucose is a major challenge that has many unresolved issues. Although cellulose itself is homogeneous, lignocellulose contains many other components, which degrade rapidly during reactions. These chemicals are contaminants for the downstream process of glucose conversion [14]. Furthermore, glucose is prone to degradation in the reaction conditions used for cellulose hydrolysis [15], and it is essential therefore that the cellulose hydrolysis reaction is selective and utilizes mild reaction conditions in order to suppress the degradation of products. Finally, the hydrolysis process needs to be cost-effective and environmentally friendly in order to be adopted for commercialization [16]. For this reason, the catalyst should be cheap and easily recyclable to avoid processing costs. The use of mild reaction conditions also reduces the capital cost needed for high-pressure and corrosion-resistant equipment.

In this chapter, attention is focused on the catalytic conversion of cellulose over a functionalized carbon catalyst. Studies in which carbon is used as a support for other active species, where hydrolysis is not achieved on the carbon surface, is beyond the scope of this chapter [17–19]. Following a brief overview of catalytic cellulose hydrolysis, the synthetic methods and properties of functionalized carbon catalysts are discussed. Carbon catalysts containing strong (sulfonic acid) and weak acid (carboxyl and phenolic) groups used for cellulose hydrolysis are discussed separately for clarity, and this is followed by a discussion on the mechanistic aspects of cellulose hydrolysis on carbon catalysts.

4.2   Overview of Catalytic Cellulose Hydrolysis

A discussion of the structure and chemical properties of cellulose is pivotal to understanding the mechanism of cellulose hydrolysis. Cellulose is a polymer of β-D-glucose monomer units present in a $^4C_1$ chair conformation [20]. The monomers are linked by β-1,4-glycosidic bonds to form a linear polymer chain [2]. The three hydroxyl groups present in β-D-glucose units are located in the equatorial position [21], and form multiple intra- and inter-molecular hydrogen bonds with adjoining monomer units. The inter-molecular bonds are present across adjacent chains (Figure 4.2). The hydrogen-bonding interaction assembles these linear chains into ribbon-like structures [22] that are stacked together to form microfibrils held together by van der Waals forces [23]. The microfibrils are densely packed, and the ordered arrangement of cellulose chains induces crystallinity in the microfibrils. The crystalline structure restricts the access of reactants,

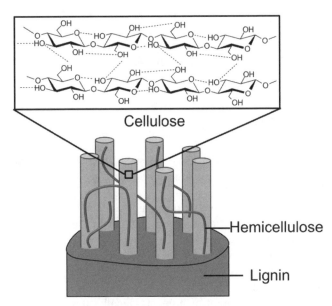

Figure 4.2   *Schematic of lignocellulose showing the three components and molecular structure of cellulose with hydrogen bonding.*

and does not allow the formation of larger products within the structure by steric hindrance [24]. These features make cellulose highly recalcitrant and protect it against chemical attacks. Native cellulose contains a small amount of amorphous domains that are disordered and are more susceptible to undergoing hydrolysis reactions. Often, pretreatment is used to convert the crystalline cellulose to its amorphous form to increase its reactivity [25,26].

Mineral acid processes were the first to be used to successfully hydrolyze cellulose to glucose for small- and medium-scale operations. Seminal studies on the hydrolysis of cellulose for commercial applications used H_2SO_4 and HCl as acid catalysts. The Scholler process [27] and the Madison process [28] each used dilute H_2SO_4 (0.5 wt%) to hydrolyze cellulose, while the Bergius process [29] and the Noguchi process [30] each used high concentrations of HCl in liquid or gaseous form as the catalyst. The Hokkaido process [31] used concentrated H_2SO_4 to achieve a glucose yield of up to 90%. These processes are difficult to commercialize in today's age, as problems related to the separation of spent acid, corrosion and the disposal of neutralization waste have prevented any of these technologies from achieving commercial success [32]. Moreover, mineral acids cause the dehydration of glucose to produce unwanted impurities. Typically, stronger mineral acids show a higher rate of glucose degradation in comparison to weak organic acids [15].

The mechanism of acid-catalyzed hydrolysis of cellulose has been the subject of much investigation over many decades. Seamen reported that cellulose hydrolysis is a pseudo-homogeneous first-order reaction with respect to the concentration of H_3O^+ [33]. The reaction is initiated by the protonation of either the glycosidic bond oxygen (Figure 4.3, path 1) or the pyran ring oxygen (Figure 4.3, path 1), though the former mechanism is more widely accepted. After protonation, a carbocation is formed as the glycosidic bond is cleaved, which then reacts with water to complete the reaction [34]. The glycosidic oxygen atom is only weakly basic, and therefore the likelihood of protonation increases with the

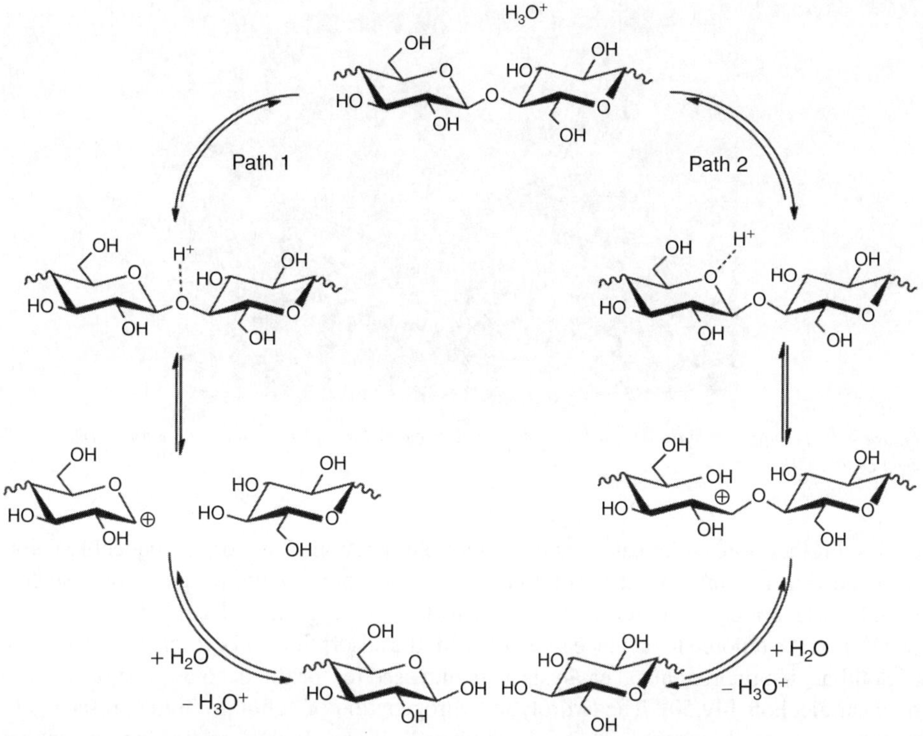

Figure 4.3 *Mechanism of acid hydrolysis of cellulose [32].*

use of a strong acid [35]. Recent investigations have shown that hydrogen bonding can assist the protonation of the glycosidic oxygen by weak acids [36,37].

The enzymatic hydrolysis of cellulose is currently a developing field, with some commercial plants operating in the USA [38]. The hydrolysis relies on three enzymes from the cellulase family [39]: (i) endoglucanase, that randomly cleaves the β-1,4-glycosidic bonds in the amorphous domain of cellulose; (ii) cellobiohydrolase, which depolymerizes the crystalline cellulose to produce cellobiose; and (iii) β-glucosidase, which hydrolyzes the soluble oligomers to glucose (Figure 4.4) [5,40]. The enzymes can produce high yields of glucose under pH-controlled conditions at low reaction temperatures. The enzymatic hydrolysis mechanism is highly sophisticated, and different from the mineral acid approach [41]. The enzymes first bind with the cellulose molecules through CH-π hydrogen bonds and hydrophobic interactions, using the aromatic amino acid residues [42,43]. The adsorption of cellulose is followed by cleavage of the β-1,4 glycosidic bond through the inversion or retention mechanism, using acid–base pairs as the active site [44,45]. The high cost of enzyme synthesis, the long reaction time and the poor recovery of dissolved enzymes are the major drawbacks of this process. Recently, some research groups have reported the use of immobilized enzymes to overcome the poor recovery [46].

The emergence of heterogeneous catalysts for cellulose hydrolysis is a product of the poor recovery and recyclability of mineral acid catalysts and enzymes [47,48]. Solid catalysts can be easily recovered from the product solution and used for multiple reaction

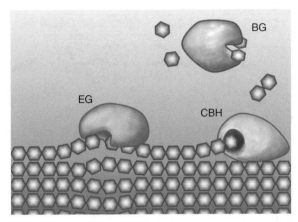

Figure 4.4 *Enzymatic hydrolysis of cellulose. EG: endoglucanase; CBH: cellobiohydrolase; BG: β-glucosidase [5].*

cycles. Furthermore, solid catalysts offer multiple functionalities for subsequent reactions of glucose in a single system. Solid catalysts have unique physical properties, such as a high surface area, an ordered structure and functional groups. Therefore, a solid-acid catalyst can be tailored to enhance its activity for the adsorption and hydrolysis of cellulose.

Solid-acid-supported metal catalysts were first used for the direct conversion of cellulose to sugar alcohols [49,50]. The hydrolytic hydrogenation of cellulose in the presence of a carbon-supported Pt catalyst selectively produced sorbitol. In this reaction, a rapid conversion of the hemiacetal group in glucose and the reducing terminal of intermediate oligomers to alcohol prevents the formation of byproducts. Sorbitol is a feedstock for isosorbide [51], which is used to treat glaucoma and also as a precursor to engineering plastics, such as Durabio$^{®}$ [5,52].

The development of solid-acid catalysts for the hydrolysis of cellulose to glucose is more challenging due to the low activity of the catalyst and the rapid degradation of glucose at higher temperatures. Traditional solid acids such as sulfonic acid resins and zeolites are not useful for the hydrolysis of cellulose [53]. Functionalized carbons have been the most successful solid catalyst for cellulose hydrolysis, due to their simple synthesis and high catalytic activity. Other catalysts reported for the hydrolysis of cellulose to glucose include Cs-substituted heteropoly acids [54,55], sulfonated synthetic polymers [56,57], layered metal oxides [58], and sulfonated silicas [59–61].

4.3   Functionalized Carbon Catalyst for Cellulose Hydrolysis

4.3.1   Synthesis and Properties of Carbon Catalysts

Carbon is a fascinating material that has been used for catalytic applications for many decades [62]. Traditional carbon catalysts such as coke, carbon black and activated carbon are prepared from the thermal treatment of organic compounds [63]. The organic source for carbon, and the treatment method, dictate the structure and physicochemical properties of the materials. Mesoporous and microporous carbon catalysts with ordered structures are prepared with inorganic materials as the cast, using the hard template method. Popular examples of such materials are CMK-3 [64], which is prepared from the carbonization of carbohydrates on the surface of SBA-15, and zeolite-templated carbon (ZTC) [65], which is prepared by the chemical vapor deposition (CVD) of light hydrocarbons on three-dimensional, large-pore zeolites. Graphene [66] and graphene oxide [67] are also used in catalytic applications, and are typically obtained by the exfoliation and oxidation of graphite.

The catalytic activity of carbon materials depends on their surface and their bulk properties. The activity of a carbon catalyst is mostly derived from the functional groups present in the basal planes, or on the edges and defects on the carbon surface. The active sites on a carbon catalyst can be either acidic (COOH, SO_3H) or basic (NH_2, pyridine, quinone). These functional groups are introduced during the synthesis of carbon materials and post-synthetic modification. The characterization of surface functional groups on the carbon surface has been covered extensively [68–70].

4.3.2 Sulfonated Carbon Catalyst for Cellulose Hydrolysis

Sulfonated carbon catalysts possess a high acidic strength due to the protons on the sulfonic groups, which dissociate easily. Seminal studies on use of sulfonated carbon catalysts for the hydrolysis of cellulose were independently reported by Suganuma *et al.* [53] and Onda *et al.* [71], who reported that carbon catalysts prepared by post-synthetic sulfonation were remarkably active for the hydrolysis of β-1,4-glycosidic bonds in cellulose, under mild reaction conditions (373-453K) [72,73]. By comparison, conventional solid-acid catalysts such as proton-form zeolites, Nafion, γ-alumina and niobic acid were not active. Other strong acid catalysts such as Amberlyst-15, sulfated zirconia and H_2SO_4 itself were able to hydrolyze cellulose; however, the selectivity for the sugar products was very low due to the formation of byproducts.

The sulfonic acid catalyst used by Suganuma *et al.* was prepared via a partial carbonization of carbohydrates, including glucose and cellulose as substrate [53]. After sulfonation, the catalyst exhibited SO_3H (1.9 mmol g^{-1}), COOH (0.4 mmol g^{-1}), and OH (2.0 mmol g^{-1}) groups (Figure 4.5). This catalyst was effective for the conversion of crystalline cellulose and eucalyptus flakes at a low substrate-to-catalyst (S/C) ratio of 0.08, achieving more than 99% conversion within 6 h of reaction time. The majority of products formed were soluble glucan with degrees of polymerization of 2–10 units. The presence of highly acidic sulfonic groups on the surface of the catalyst was reported as the active sites for hydrolysis of cellulose. Interestingly, the apparent activation energy for the hydrolysis of crystalline cellulose in the presence of the carbon catalyst was only 110 kJ mol^{-1} between 343 and 373K. This value was smaller than that reported for the hydrolysis of cellulose using sulfuric acid (170 kJ mol^{-1}) under optimal condition. The activity of the carbon catalyst was explained by the ability of the carbon materials to adsorb cellulosic molecules, namely β-1,4-glucans. However, the catalyst was not porous and the measured Brunauer–Emmett–Teller (BET) surface area of the catalyst was only 2 m^2 g^{-1}, and it is well known that adsorption is promoted in porous, high-surface-area materials. Suganuma *et al.* argued that the effective surface area under hydrothermal conditions was as high as

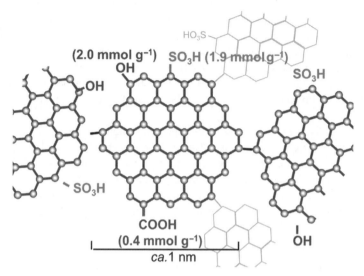

Figure 4.5   *Structure of the sulfonated carbon catalyst. Reprinted with permission from Ref. [74]. Copyright 2008, American Chemical Society.*

560 m^2 g^{-1}, as determined by the adsorption–desorption isotherm of H$_2$O vapor on the catalyst. The ability of the catalyst to adsorb large amounts of water would enable an easy penetration of the hydrophilic reactants to the internal SO$_3$H groups.

Onda *et al.* [71] prepared a solid-acid catalyst (AC-SO$_3$H) by the sulfonation of activated carbon, which contained 1.63 mmol g^{-1} of total acidic functional groups. This catalyst was more active for the hydrolysis of amorphous cellulose in comparison to typical proton-form zeolites, sulfated zirconia, and Amberlyst-15. More importantly, a complete hydrolysis of cellulose to glucose was achieved and a maximum glucose yield of 40.5% was obtained with 95% selectivity among water-soluble products. In contrast, selectivity for the formation of glucose was low in the case of other solid-acid catalysts due to the leaching of SO$_4{}^{2-}$ that caused the degradation of glucose to yield byproducts.

Many other groups have also reported the use of sulfonated carbon catalysts for hydrolysis, with improved yields of products. Wu *et al.* [75] used microwave irradiation that improves the selectivity of glucose formation in comparison to conventional heating. The use of an ionic liquid as the reaction solvent also improves the yield of glucose in the presence of a sulfonated carbon catalyst [76,77]. Ionic liquids such as 1-butyl-3-methyl imidazolium chloride can dissolve cellulose, which promotes contact between the catalyst and the reactant. Other reports have included the use of sulfonated carbon/silica composites [59], sulfonated CMK-3 [78] and sulfonated carbon spheres [79] as active catalysts for cellulose hydrolysis.

The stability of sulfonated carbon catalysts under hydrothermal conditions is very important for their application in industrial processes. The sulfonated carbon can lose its acid sites when the carbon–sulfur bond is hydrolyzed, releasing an SO$_4{}^{2-}$ ion [80]. Several research groups have concluded that the sulfonic groups on the carbon surface are stable at the low temperatures (333–423K) [53,71,76,81,82]. However, hydrothermal treatment above 433K can remove up to 90% of the sulfonic groups, rendering the catalyst inactive after only a few reaction cycles [83,84].

4.3.3   Oxygenated Carbon Catalyst for Cellulose Hydrolysis

Oxygenated functional groups present on the carbon surface are more resistant to deactivation under hydrothermal conditions, and therefore are suitable for cellulose hydrolysis. The present authors' groups first reported that carbon catalysts lacking any strong acid sulfonic groups can also catalyze the hydrolysis of cellulose [85,86]. CMK-3, as the carbon catalyst containing $410\,\mu mol\,g^{-1}$ of oxygenated functional groups, hydrolyzed amorphous cellulose to produce soluble oligosaccharides. However, further hydrolysis of the oligosaccharides to glucose did not occur on unmodified CMK-3 under mild reaction conditions.

The activity of oxygenated carbon catalysts can be increased by the introduction of a large number of oxygenated functional groups. An alkali-activated carbon called K26, containing $0.88\,mmol\,g^{-1}$ oxygenated functional groups, showed the highest conversion (60%) and glucose yield (36%) after the hydrolysis of ball-milled cellulose at 503K. Other alkali-activated carbons also showed good activity for cellulose conversion. By comparison, CMK-3 and carbon blacks BP2000 and XC72 could not produce glucose yields of more than 12%. The number of oxygenated functional groups calculated by the Boehm titration [70,87] was in direct correlation with the yield of glucose obtained after the hydrolysis reaction (Figure 4.6). Removing the oxygenated function groups partially by the heat treatment of K26 over the temperature range 673 to 1273K led to a decrease in catalyst activity and confirmed the role of oxygenated functional groups as active sites.

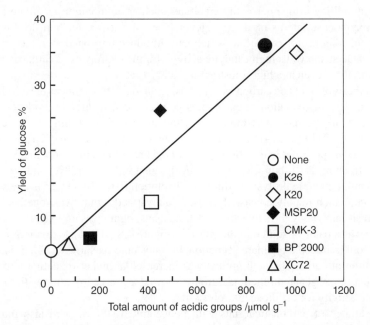

Figure 4.6 *Correlation between total amount of weak acid functional groups on the catalyst surface and the glucose yield. Reproduced with permission from Ref. [38]. Copyright 2016, Springer.*

The hydrolysis of cellulose over a carbon catalyst is a solid–solid reaction that is limited by the contact between cellulose and catalyst particles. Contact can be increased by the physical adsorption of cellulose onto the catalyst surface [88], although cellulose molecules are insoluble in conventional solvents and therefore its molecular adsorption is not possible. Alternatively, solid cellulose can be brought into close contact with a carbon catalyst by ball-milling them together in a pre-treatment termed 'mix-milling'. The mix-milling of cellulose and K26 can provide a 90% yield of soluble oligomers and glucose after a subsequent hydrolysis reaction of 20 min 453K – a dramatic increase from a 13% yield when the cellulose and catalyst are milled separately [89,90]. This high reactivity is purely a result of enhanced solid–solid contact, as the use of a soluble substrate or soluble catalyst did not provide any change in catalytic activity. Mix-milling enhanced the rate constant for the hydrolysis of cellulose to oligomers, from 0.013 to 0.17 h^{-1}, a 13-fold rise [90]. Once the dissolved oligomers were formed, physical contact was lost between the catalyst and substrate, which reduced the rate of oligomer hydrolysis to form glucose. The addition of a small amount of acid (0.012% HCl, pH 2.5) promoted oligomer hydrolysis and produced an 88% yield of glucose under the same reaction conditions (20 min, 453K). These results showed that the hydrolysis of cellulose occurs over weakly acidic function groups, and that the number of oxygenated function groups and accessibility of β-1,4-glycosidic bonds are both important for high hydrolysis activity.

Carboxylic acid groups are the most active for cellulose hydrolysis; however, the selective introduction of a functional group on a carbon catalyst is difficult as conventional oxidation methods introduce all three types of functional group (carboxylic acids, lactones, and phenols). In this respect, mechanocatalytic oxidation was used to introduce large numbers of carboxylic acid groups onto the carbon surface [91]. With this method, activated carbon was oxidized by milling it together with persulfates such as ammonium persulfate and potassium peroxymonosulfate as oxidizing agents. The solvent-free method does not require sulfuric acid solution for oxidation, which can introduce unwanted sulfonic groups. The prepared catalyst showed a high catalytic activity for the hydrolysis of cellulose to glucose after mix-milling, even at a high substrate-to-catalyst ratio.

The synthesis of oxidized carbon catalysts using air-oxidation reduces the cost of catalyst preparation. The air-oxidation of eucalyptus powder at 573K directly produces a catalyst named E-carbon, which has an aromatic structure and contains 2.1 mmol g^{-1} of carboxylic groups (Figure 4.7) [92]. Air-oxidation was essential for the formation of an aromatic structure, as a catalyst prepared in the absence of oxygen under the same temperature lacked any aromatic structure. E-Carbon hydrolyzed eucalyptus to glucose (78%) and xylose (94%) after mix-milling and hydrolysis in dilute HCl solution (0.012%, pH 2.5). The solid reaction residue, which mainly consisted of used catalyst and lignin, was together converted back to fresh catalyst by the same air-oxidation. Thus, lignin was no longer a contaminant but rather was a catalyst source in this system (Figure 4.8). The air-oxidation of activated carbon also introduced oxygenated functional groups onto its surface [93]. Selectivity for the formation of carboxylic acid groups was increased at higher oxidation temperatures. The number of oxygenated functional groups showed a good correlation with the increase in catalytic activity for cellulose hydrolysis.

The treatment of carbon materials with concentrated sulfuric acid also introduces a large number of oxygenated functional groups (–COOH and –OH), along with sulfonic groups. Katz *et al.* prepared a weak acid carbon catalyst by the surface functionalization

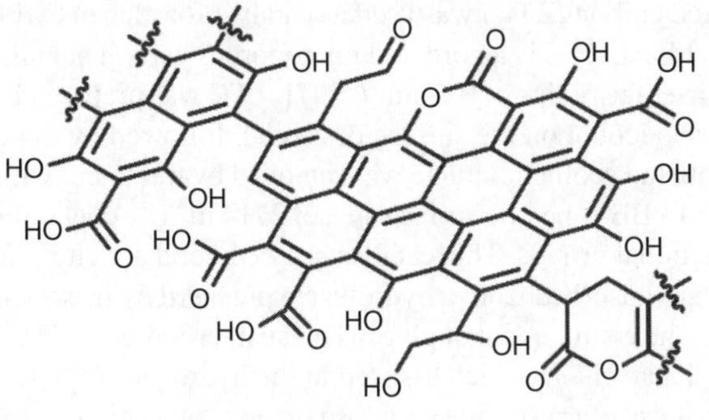

Figure 4.7 *Proposed structure of E-Carbon containing an aromatic framework and weakly acidic functional groups [92].*

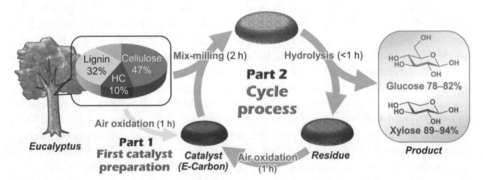

Figure 4.8 *Catalytic cycle for the use of E-Carbon produced by air-oxidation for the hydrolysis of Eucalyptus [92]. HC indicates xylan as a hemicellulose.*

of mesoporous carbon nanoparticles (MCN) with fuming H_2SO_4, followed by the removal of –SO_3H groups by hydrothermal treatment at 473K, repeated five times [94]. The hydrothermally treated catalyst contained 1.53 mmol g^{-1} of acid sites, with less than 10% as sulfonic acid. The remaining sulfonic acid groups were neutralized during hydrolysis with sodium acetate/acetic acid buffer (pH 3.7–4.1) to yield sodium sulfonate salts. The hydrolysis of extracted xylan using this catalyst in buffer solution resulted in a 74.1% yield of xylose after 4 h reaction at 423 K [94]. MCN activated with nitric acid containing only weak acid functional groups (–COOH and –OH) showed a similar activity for xylan hydrolysis under the same reaction conditions.

The post-synthetically modified MCN catalyst was used for the hydrolysis of cello-oligosaccharides adsorbed onto the carbon surface [95]. In order to adsorb β-1,4-glucans onto the surface of MCN, crystalline cellulose was first dissolved in cold concentrated HCl to obtain a hydrolyzate solution. The dissolved glucans were then adsorbed onto MCN, and the residual HCl was washed with water. Hydrolysis of the adsorbed glucans in phosphate buffer at pH 2.0 resulted in a glucose yield of 90%. The activity of the catalyst was enhanced as the pH of the reaction solution was changed from 4.6 to 2.6. The lower activity at pH 4.6 was attributed to the deprotonation of carboxylic groups.

Zeolite-templated carbon (ZTC) was used as catalyst for glucan hydrolysis, without any oxidative treatment [96]. ZTC is an ordered microporous carbon having a narrow pore size distribution, with average radius of 0.6 nm [65,97]. ZTC was prepared by the carbonization of adsorbed furfuryl alcohol on the surface of zeolite, followed by the CVD of propylene. After carbonization, the zeolite template was removed by treatment with 47% hydrofluoric acid. ZTC exhibited a BET specific surface area of $2714\,m^2\,g^{-1}$ containing $870\,mmol\,g^{-1}$ of weakly acidic functional groups. This catalyst showed good activity (73% glucose) for the hydrolysis of glucans adsorbed from a hydrolyzate prepared by dissolving cellulose in cold hydrochloric acid. The confinement of glucans in the micropores of ZTC induced a mechanical strain on the glucan chain, which assisted in the hydrolysis of β-1,4-glycosidic bonds.

The direct hydrolysis of crystalline cellulose on a weakly acidic carbon catalyst is difficult due to a lack of contact between the cellulose and catalyst particles in a dilute aqueous medium. Recently, Katz *et al.* reported that a carbon catalyst prepared by the oxidation of nanoporous carbon MSC-30 with NaOCl can hydrolyze crystalline cellulose [98]. The oxidized MSC-30 catalyst contained $2.2\,mmol\,g^{-1}$ of weakly acidic functional groups. This catalyst, when reacted with crystalline cellulose at low substrate-to-catalyst ratio (1.5 mg cellulose and 20 mg catalyst) afforded 37% glucose after 24 h of reaction at 423 K. The glucose yield was increased to 70% when the reaction was continued for another 3 h at 453 K. Neither the unoxidized MSC-30 catalyst nor a 5 mM H_2SO_4 solution showed any activity for the hydrolysis of crystalline cellulose. The authors concluded that a higher local density of weakly acidic functional groups on the NaOCl-treated MSC-30 catalyst was responsible for its cellulose-hydrolyzing ability.

The use of a hyperbranched polymer enables the preparation of carbon-analog catalysts which selectively bear particular functional groups in high density [99]. When Kakimoto *et al.* synthesized a hyperbranched aromatic poly(ether ketone) with $2.8\,mmol\,g^{-1}$ of accessible carboxylic acid [100], the polymer was capable of hydrolyzing ball-milled cellulose to glucose. The polymer structure may be further improved for cellulose hydrolysis, based on mechanistic studies.

4.3.4   Mechanistic Aspects of Carbon-Catalyzed Cellulose Hydrolysis

Initially, it was believed that adsorption is driven predominantly by interactions between the oxygenated groups of carbon and the hydroxyl groups of cellulose [73,101]. However, the latest results have indicated that the interaction owing to hydrophobic functional group is the major driving force for the adsorption. Katz *et al.* found that the adsorption of cello-oligosaccharides occurs smoothly on MCN in concentrated aqueous HCl solution (pH = 0), where OH–O hydrogen bonding is hindered [102]. To clarify the adsorption site clearly, surface oxygenated groups were removed from an activated carbon (K26) by a thermal treatment at 873–1273K under He flow [103]. The resulting material adsorbed β-1,4-glucans in the same manner, regardless of the content of oxygenated groups. In addition, the formation of a huge number of oxygenated groups on a carbon material decreased the adsorption capacity [93]. These experimental results showed that the major adsorption site is the polycyclic aromatic surface of carbon.

Detailed adsorption parameters were determined for the adsorption of cello-oligosaccharides on K26 [103]. Langmuir-type adsorption isotherms were obtained with

Table 4.1 Langmuir parameters and thermodynamic values for adsorption of cello-oligosaccharides on K26 [103].

Adsorbate	$W_{max}{}^a$ [mg $g_{adsorbent}{}^{-1}$]	$K_{ads}{}^b$ [M^{-1}]	$\Delta H°_{ads}$ [kJ mol^{-1}]	$\Delta S°_{ads}$ [J K^{-1} mol^{-1}]
Glucose	95.2	220	−8.4	+16.5
Cellobiose	412	5660	−14.1	+23.5
Cellotriose	527	181 000	−20.2	+32.2

[a] Adsorption capacity of the substrate.
[b] Adsorption equilibrium constants were measured at 296K.

glucose, cellobiose, and cellotriose. The adsorption equilibrium constant (K_{ads}) was increased exponentially from 220 M^{-1} to 5660 M^{-1} and 181 000 M^{-1} with increasing numbers of glucose unit (Table 4.1). Carbon materials can adsorb longer cello-oligosaccharides very strongly and with a higher adsorption capacity. The temperature-dependence of K_{ads} with the van't Hoff equation indicated a negative enthalpy change ($\Delta H°_{ads}$) and a positive entropy change ($\Delta S°_{ads}$) (Table 4.1). Since $\ln K_{ads}$ is proportional to $-\Delta H°_{ads}$ and $\Delta S°_{ads}$, the adsorption is favored by both enthalpy and entropy. A positive entropy change allows adsorption, even at high temperatures.

The negative value of $\Delta H°_{ads}$ is ascribed to CH-π hydrogen bonds between the CH groups of cellobiose and π electrons of the polycyclic aromatics of carbon. The $\Delta H°_{ads}$ values, divided by number of CH groups on one side of cello-oligosaccharides, were −2.5 to −4.2 kJ mol^{-1} [103], which were in the range of CH-π hydrogen bond energy [104]. Density functional theory (DFT) calculations have also resulted in the formation of CH-π hydrogen bonds. CH groups were directed towards the center of aromatic rings with distances shorter than those expected from London dispersion forces.

The positive $\Delta S°_{ads}$ value is a characteristic of this system as, in general, adsorption decreases the freedom of an adsorbate, mainly translational and rotational entropy. It was proposed that this positive entropy change is ascribed to the hydrophobic interaction. The adsorption of cellobiose onto K26 with hydrophobic groups unbinds the water molecules restricted around the lipophilic surface, which results in an increase of entropy of the water molecules.

The strong adsorption enables a rapid adsorption of long-chain cellulose molecules in the micropores of carbon [96]. Cellulose solubilized in cold concentrated HCl and ZTC were employed in this study. A 1 g sample of ZTC adsorbed 0.80 g of cellulose with an average molecular weight of 3600. The cellulose has a gyration radius of 2.9 nm, which is almost fivefold larger than the pore radius of ZTC, but adsorption in the micropore is completed within a short time of 2 min. A high mobility after adsorption by non-covalent bonds may allow a rapid diffusion of the adsorbate in the pore. Furthermore, ZTC adsorbed cellulose with an average molecular weight of 22 000. In the case of cellulose-hydrolyzing enzymes, the enzyme incorporates long-chain cellulose within its pore and hydrolyzes the molecule. However, this phenomenon has been specific to enzymes, as the large steric hindrance prohibits mimicking of this phenomenon. Hence, this is the first demonstration of the accommodation of long-chain cellulose in an artificial microporous environment. The incorporation of a cellulose chain into ZTC pores follows Rebek's 55% rule, the best host–guest volume relationship [105], which further facilitates the adsorption in addition to a good interaction between polycyclic aromatics and cellulose. The extreme degree of confinement of the micropores provides mechanical strain on the cellulose molecules adsorbed, which increases reactivity of their glycosidic bonds, as described in the previous section.

Substrate specificity has also been observed in the adsorption of sugars in the CH-π system [106]. A metal organic framework (MOF) material bearing pyrene units adsorbed cellobiose and lactose from aqueous solution, while the MOF completely excluded glucose and maltose. Clearly, the material can discriminate between monomer and dimer, as well as between α and β linkages in dimer. The well-defined structure may recognize the number of CH-π hydrogen bonds that can be formed between the pyrene unit and the sugar molecule.

The adsorption of β-1,4-glucan onto the catalyst surface has a synergistic effect on the rate of hydrolysis over catalytic defect sites. Katz *et al.* showed that silica and alumina, both of which are typically inactive catalysts for cellulose hydrolysis, can hydrolyze β-1,4-glucans adsorbed onto its surface [88,107]. This study shows that the hydrolysis of glucans can be achieved when they are constrained to be in close vicinity to functional groups, causing the formation of hydrogen bonds with oxygen atoms on the glucan chains. A linear increase was observed in hydrolysis rate with the number of surface –OH defect sites, which suggested that in the formation of hydrogen bonds was a driving force. Sievers *et al.* reported that the adsorption of glucans near defect sites containing acidic functional groups in polyaromatic carbon structures may cause a conformational change in the structure of the glucose chain [108]. The synergistic effect of adsorption and hydrolysis is plausible, as suggested by reports which examined the correlation between the number of acidic functional groups and catalytic activity [37,78,94].

The synergistic effect of vicinal functional groups enhances the activity of carbon catalysts for cellulose hydrolysis. Oxygenated functional groups, when located adjacent

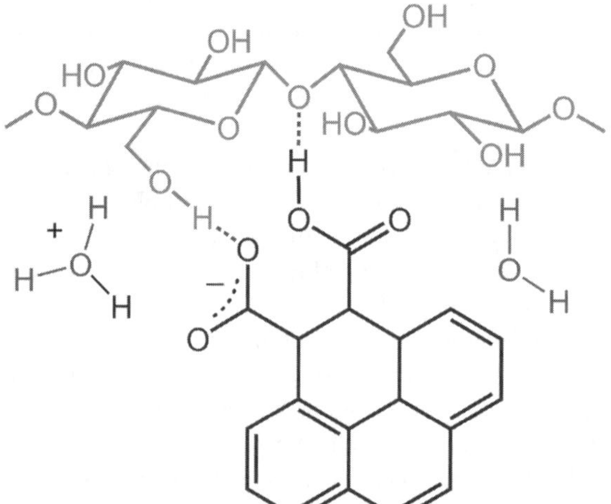

Figure 4.9 *Schematic representation of synergy between adjacent carboxylic acid groups for hydrolysis of β-1,4-glucans.*

to each other (as in phthalic acid and salicylic acid) show a higher frequency factor for the hydrolysis of β-1,4-glycosidic bonds in comparison to *meta-* or *para*-substituted benzoic acids [37]. Moreover, the TOF for these catalysts was higher than that of *o*-chlorobenzoic acid and *o*-trifluoromethylbenzoic acid, despite having a higher pK_a value. In case of the vicinal oxygenated groups, the probability of attack on glycosidic bonds by the carboxylic acid is increased when an adjacent hydroxyl or carboxyl groups forms hydrogen bonds with the glucan chain (Figure 4.9). This insight suggests that the proximity of functional groups may be more important than the strength and abundance of functional groups in the catalyst surface.

4.4 Summary and Outlook

Functionalized carbon catalysts are unique in their activity for cellulose and hemicellulose hydrolysis. This method can be used for purified polysaccharides and real biomass after appropriate pretreatment methods to achieve high yields of monomeric sugars. Both, strong acid sulfonic groups and weak acid carboxylic and phenolic groups are capable of hydrolyzing cellulose to soluble oligosaccharides and glucose. The hydrolysis activity is primarily dependent on the acid strength of the functional group and their relative abundance. However, synergistic effects related to the adsorption of glucans onto the carbon surface through CH-π bonds, as well as hydrophobic interactions, play important roles in hydrolytic activity. The position of functional groups on the catalyst is also important for catalytic activity. One functional group makes a hydrogen bond with the glucan chain and increases the probability of attack on the glycosidic bond with adjacent functional groups. The overall activity for hydrolysis is determined by these three phenomena.

Carbon catalysts are different from enzymes and conventional acid catalysts used in the hydrolysis of cellulose. The ability of carbons to adsorb glucans through CH-π bonds and hydrophobic interactions is unique, and it resembles the mechanism of enzymes. Furthermore, the activity of weak acid functional groups to hydrolyze β-1,4-glucans, especially when located in the vicinal position, contrasts the notion that strong acids are necessary for this reaction. The hydrogen-bonding ability of vicinal functional groups can be compared with enzymes, which utilize a carboxylic acid and carboxylate-conjugated base as active site. However, functionalized carbon catalysts have distinct advantages over conventional acid catalysts and enzymes. Unlike mineral acids, carbon catalysts are easily separable from reaction solutions and can be re-used multiple times. The polyaromatic carbon framework and acidic functional groups are stable under hydrothermal conditions, and the catalyst retains its activity in the presence of impurities produced from lignin, or of unstable degradation products. This contrasts with enzymes, which denature easily under unfavorable reaction conditions. The current major drawback of carbon catalysts for cellulose hydrolysis relates to difficulties in controlling the structure of the catalyst and the active sites during synthesis.

In future, more effort should be focused towards understanding the mechanism of cellulose hydrolysis over carbon catalysts. Additional mechanistic insight would enable the careful design of catalytic active sites for greater hydrolytic activity and product selectivity. Finally, effort should be made to reduce the pretreatment costs and scale-up of the reaction for the commercial synthesis of glucose from cellulose.

References

1. Gruber, P.R., Kamm, M., Kamm, B. (eds) (2006) *Biorefineries: Industrial processes and products: Status quo and future directions*, Wiley-VCH.
2. Klemm, D., Heublein, B., Fink, H.-P., Bohn, A. (2005) Cellulose: Fascinating biopolymer and sustainable raw material. *Angew. Chem. Int. Ed.*, **44** (22), 3358–3393.
3. Ragauskas, A.J., Williams, C.K., Davison, B.H., Britovsek, G., Cairney, J., Eckert, C.A., Frederick, W.J., Hallett, J.P., Leak, D.J., Liotta, C.L., Mielenz, J.R., Murphy, R., Templer, R., Tschaplinski, T. (2006) The path forward for biofuels and biomaterials. *Science*, **311** (5760), 484–489.
4. Aspinall, G.O. (1981) Constitution of plant cell wall polysaccharides, in *Plant Carbohydrates II* (eds W. Tanner and F. Loewus), Springer, Berlin Heidelberg, pp. 3–8.
5. Kobayashi, H., Fukuoka, A. (2013) Synthesis and utilisation of sugar compounds derived from lignocellulosic biomass. *Green Chem.*, **15** (7), 1740–1763.
6. Lavarack, B.P., Griffin, G.J., Rodman, D. (2002) The acid hydrolysis of sugarcane bagasse hemicellulose to produce xylose, arabinose, glucose and other products. *Biomass Bioenergy*, **23** (5), 367–380.
7. Lin, Y., Tanaka, S. (2006) Ethanol fermentation from biomass resources: current state and prospects. *Appl. Microbiol. Biotechnol.*, **69** (6), 627–642.
8. Goldemberg, J. (2007) Ethanol for a sustainable energy future. *Science*, **315** (5813), 808–810.
9. Rosatella, A.A., Simeonov, S.P., Frade, R.F.M., Afonso, C.A.M. (2011) 5-Hydroxymethylfurfural (HMF) as a building block platform: Biological properties, synthesis and synthetic applications. *Green Chem.*, **13** (4), 754–793.
10. Gallezot, P. (2007) Process options for converting renewable feedstocks to bioproducts. *Green Chem.*, **9** (4), 295–302.
11. Zhao, G., Zheng, M., Zhang, J., Wang, A., Zhang, T. (2013) Catalytic conversion of concentrated glucose to ethylene glycol with semicontinuous reaction system. *Ind. Eng. Chem. Res.*, **52** (28), 9566–9572.
12. Zeikus, J.G., Jain, M.K., Elankovan, P. (1999) Biotechnology of succinic acid production and markets for derived industrial products. *Appl. Microbiol. Biotechnol.*, **51** (5), 545–552.
13. Znad, H., Markoš, J., Baleš, V. (2004) Production of gluconic acid from glucose by *Aspergillus niger*: Growth and non-growth conditions. *Process Biochem.*, **39** (11), 1341–1345.
14. Delgenes, J.P., Moletta, R., Navarro, J.M. (1996) Effects of lignocellulose degradation products on ethanol fermentations of glucose and xylose by *Saccharomyces cerevisiae*, *Zymomonas mobilis*, *Pichia stipitis*, and *Candida shehatae*. *Enzyme Microb. Technol.*, **19** (3), 220–225.
15. Mosier, N.S., Ladisch, C.M., Ladisch, M.R. (2002) Characterization of acid catalytic domains for cellulose hydrolysis and glucose degradation. *Biotechnol. Bioeng.*, **79** (6), 610–618.
16. Alonso, D.M., Bond, J.Q., Dumesic, J.A. (2010) Catalytic conversion of biomass to biofuels. *Green Chem.*, **12** (9), 1493–1513.
17. Geboers, J., Van de Vyver, S., Carpentier, K., Blochouse, K., Jacobs, P., Sels, B. (2010) Efficient catalytic conversion of concentrated cellulose feeds to hexitols with heteropoly acids and Ru on carbon. *Chem. Commun.*, **46** (20), 3577–3579.
18. Zhang, Y., Wang, A., Zhang, T. (2010) A new 3D mesoporous carbon replicated from commercial silica as a catalyst support for direct conversion of cellulose into ethylene glycol. *Chem. Commun.*, **46** (6), 862–864.
19. Van de Vyver, S., Geboers, J., Schutyser, W., Dusselier, M., Eloy, P., Dornez, E., Seo, J.W., Courtin, C.M., Gaigneaux, E.M., Jacobs, P.A., Sels, B.F. (2012) Tuning the acid/metal balance of carbon nanofiber-supported nickel catalysts for hydrolytic hydrogenation of cellulose. *ChemSusChem*, **5** (8), 1549–1558.
20. Klemm, D., Schmauder, H.-P., Heinze, T. (2002) Cellulose. *Biopolymers*, **6**, 275–319.
21. Zugenmaier, P. (2001) Conformation and packing of various crystalline cellulose fibers. *Prog. Polym. Sci.*, **26** (9), 1341–1417.

22. Hon, D.N.-S. (1994) Cellulose: a random walk along its historical path. *Cellulose*, **1** (1), 1–25.
23. Jarvis, M. (2003) Chemistry: Cellulose stacks up. *Nature*, **426** (6967), 611–612.
24. Hama, T., Ueta, H., Kouchi, A., Watanabe, N., Tachikawa, H. (2014) Quantum tunneling hydrogenation of solid benzene and its control via surface structure. *J. Phys. Chem. Lett.*, **5** (21), 3843–3848.
25. Mosier, N., Wyman, C., Dale, B., Elander, R., Lee, Y.Y., Holtzapple, M., Ladisch, M. (2005) Features of promising technologies for pretreatment of lignocellulosic biomass. *Bioresource Technol.*, **96** (6), 673–686.
26. Sun, Y., Cheng, J. (2002) Hydrolysis of lignocellulosic materials for ethanol production: a review. *Bioresource Technol.*, **83** (1), 1–11.
27. Faith, W.L. (1945) Development of the Scholler process in the United States. *Ind. Eng. Chem.*, **37** (1), 9–11.
28. Harris, E.E., Beglinger, E. (1946) Madison wood sugar process. *Ind. Eng. Chem.*, **38** (9), 890–895.
29. Bergius, F. (1937) Conversion of wood to carbohydrates. *Ind. Eng. Chem.*, **29** (3), 247–253.
30. Antonoplis, R.A., Blanch, H.W., Freitas, R.P., Sciamanna, A.F., Wilke, C.R. (1983) Production of sugars from wood using high-pressure hydrogen chloride. *Biotechnol. Bioeng.*, **25** (11), 2757–2773.
31. Clausen, E.C., Gaddy, J.L. (1993) Concentrated sulfuric acid process for converting lignocellulosic materials to sugars. US Patent 5,188,673, issued 1993.
32. Rinaldi, R., Schüth, F. (2009) Acid hydrolysis of cellulose as the entry point into biorefinery schemes. *ChemSusChem*, **2** (12), 1096–1107.
33. Saeman, J.F. (1945) Kinetics of wood saccharification – hydrolysis of cellulose and decomposition of sugars in dilute acid at high temperature. *Ind. Eng. Chem.*, **37** (1), 43–52.
34. Bobleter, O. (1994) Hydrothermal degradation of polymers derived from plants. *Prog. Polym. Sci.*, **19** (5), 797–841.
35. Philipp, B., Jacopian, V., Loth, F., Hirte, W., Schulz, G. (1979) Influence of cellulose physical structure on thermohydrolytic, hydrolytic, and enzymatic degradation of cellulose, in *Hydrolysis of Cellulose: Mechanisms of enzymatic and acid catalysts* (eds R.D. Brown, Jr, L. Jurasek) pp. 127–143.
36. Lu, Y., Mosier, N.S. (2007) Biomimetic catalysis for hemicellulose hydrolysis in corn stover. *Biotechnol. Prog.*, **23** (1), 116–123.
37. Kobayashi, H., Yabushita, M., Hasegawa, J., Fukuoka, A. (2015) Synergy of vicinal oxygenated groups of catalysts for hydrolysis of cellulosic molecules. *J. Phys. Chem. C*, **119** (36), 20993–20999.
38. Kobayashi, H., Yabushita, M., Fukuoka, A. (2016) Depolymerization of cellulosic biomass catalyzed by activated carbons, in *Reaction Pathways and Mechanisms in Thermocatalytic Biomass Conversion I* (eds M. Schlaf, Z. Zhang), Springer Singapore, pp. 15–26.
39. Bhat, M.K. (2000) Cellulases and related enzymes in biotechnology. *Biotechnol. Adv.*, **18** (5), 355–383.
40. Tébéka, I.R.M., Silva, A.G.L., Petri, D.F.S. (2009) Hydrolytic activity of free and immobilized cellulase. *Langmuir*, **25** (3), 1582–1587.
41. Zhang, Y.-H.P., Lynd, L.R. (2004) Toward an aggregated understanding of enzymatic hydrolysis of cellulose: Noncomplexed cellulase systems. *Biotechnol. Bioeng.*, **88** (7), 797–824.
42. Chen, W., Enck, S., Price, J.L., Powers, D.L., Powers, E.T., Wong, C.-H., Dyson, H.J., Kelly, J.W. (2013) Structural and energetic basis of carbohydrate–aromatic packing interactions in proteins. *J. Am. Chem. Soc.*, **135** (26), 9877–9884.
43. Zolotnitsky, G., Cogan, U., Adir, N., Solomon, V., Shoham, G., Shoham, Y. (2004) Mapping glycoside hydrolase substrate subsites by isothermal titration calorimetry. *Proc. Natl Acad. Sci. USA*, **101** (31), 11275–11280.
44. Saharay, M., Guo, H., Smith, J.C. (2010) Catalytic mechanism of cellulose degradation by a cellobiohydrolase, CelS. *PLoS One*, **5** (10), e12947.
45. Knott, B.C., Haddad Momeni, M., Crowley, M.F., Mackenzie, L.F., Götz, A.W., Sandgren, M., Withers, S.G. Ståhlberg, J., Beckham, G.T. (2014) The mechanism of cellulose hydrolysis by a two-step,

retaining cellobiohydrolase elucidated by structural and transition path sampling studies. *J. Am. Chem. Soc.*, **136** (1), 321–329.

46. Wang, J., Xi, J., Wang, Y. (2015) Recent advances in the catalytic production of glucose from lignocellulosic biomass. *Green Chem.*, **17** (2), 737–751.

47. Van de Vyver, S., Geboers, J., Jacobs, P.A., Sels, B.F. (2011) Recent advances in the catalytic conversion of cellulose. *ChemCatChem*, **3** (1), 82–94.

48. Yabushita, M., Kobayashi, H., Fukuoka, A. (2014) Catalytic transformation of cellulose into platform chemicals. *Appl. Catal. B Environ.*, **145**, 1–9.

49. Fukuoka, A., Dhepe, P.L. (2006) Catalytic conversion of cellulose into sugar alcohols. *Angew. Chem. Int. Ed.*, **45** (31), 5161–5163.

50. Luo, C., Wang, S., Liu, H. (2007) Cellulose conversion into polyols catalyzed by reversibly formed acids and supported ruthenium clusters in hot water. *Angew. Chem. Int. Ed.*, **46** (40), 7636–7639.

51. Kobayashi, H., Yokoyama, H., Feng, B., Fukuoka, A. (2015) Dehydration of sorbitol to isosorbide over H-beta zeolites with high Si/Al ratios. *Green Chem.*, **17** (5), 2732–2735.

52. Rose, M., Palkovits, R. (2012) Isosorbide as a renewable platform chemical for versatile applications – quo vadis? *ChemSusChem*, **5** (1), 167–176.

53. Suganuma, S., Nakajima, K., Kitano, M., Yamaguchi, D., Kato, H., Hayashi, S., Hara, M. (2008) Hydrolysis of cellulose by amorphous carbon bearing SO_3H, COOH, and OH groups. *J. Am. Chem. Soc.*, **130**, 12787–12793.

54. Okuhara, T. (2002) Water-tolerant solid acid catalysts. *Chem. Rev.*, **102** (10), 3641–3666.

55. Tian, J., Fang, C., Cheng, M., Wang, X. (2011) Hydrolysis of cellulose over $Cs_xH_{3-x}PW_{12}O_{40}$ (x = 1–3) heteropoly acid catalysts. *Chem. Eng. Technol.*, **34** (3), 482–486.

56. Akiyama, G., Matsuda, R., Sato, H., Takata, M., Kitagawa, S. (2011) Cellulose hydrolysis by a new porous coordination polymer decorated with sulfonic acid functional groups. *Adv. Mater.*, **23** (29), 3294–3297.

57. Li, X., Jiang, Y., Shuai, L., Wang, L., Meng, L., Mu, X. (2012) Sulfonated copolymers with SO_3H and COOH groups for the hydrolysis of polysaccharides. *J. Mater. Chem.*, **22** (4), 1283–1289.

58. Takagaki, A., Tagusagawa, C., Domen, K. (2008) Glucose production from saccharides using layered transition metal oxide and exfoliated nanosheets as a water-tolerant solid acid catalyst. *Chem. Commun.*, (42), 5363–5365.

59. Van de Vyver, S., Peng, L., Geboers, J., Schepers, H., De, C.F., Gommes, C.J., Goderis, B., Jacobs, P.A., Sels, B.F. (2010) Sulfonated silica/carbon nanocomposites as novel catalysts for hydrolysis of cellulose to glucose. *Green Chem.*, **12** (9), 1560–1563.

60. Lai, D., Deng, L., Li, J., Liao, B., Guo, Q., Fu, Y. (2011) Hydrolysis of cellulose into glucose by magnetic solid acid. *ChemSusChem*, **4** (1), 55–58.

61. Takagaki, A., Nishimura, M., Nishimura, S., Ebitani, K. (2011) Hydrolysis of sugars using magnetic silica nanoparticles with sulfonic acid groups. *Chem. Lett.*, **40** (10), 1195–1197.

62. Radovic, L.R. (2009) Physicochemical properties of carbon materials: a brief overview, in *Carbon Materials for Catalysis* (eds P. Serp, J.L. Figueiredo), John Wiley & Sons, Hoboken, NJ, pp. 1–44.

63. Lam, E., Luong, J.H.T. (2014) Carbon materials as catalyst supports and catalysts in the transformation of biomass to fuels and chemicals. *ACS Catal.*, **4** (10), 3393–3410.

64. Shin, H.J., Ryoo, R., Kruk, M., Jaroniec, M. (2001) Modification of SBA-15 pore connectivity by high-temperature calcination investigated by carbon inverse replication. *Chem. Commun.*, **279** (4), 349–350.

65. Kyotani, T., Nagai, T., Inoue, S., Tomita, A. (1997) Formation of new type of porous carbon by carbonization in zeolite nanochannels. *Chem. Mater.*, **9** (2), 609–615.

66. Xue, T., Jiang, S., Qu, Y., Su, Q., Cheng, R., Dubin, S., Chiu, C.-Y., Kaner, R., Huang, Y., Duan, X. (2012) Graphene-supported hemin as a highly active biomimetic oxidation catalyst. *Angew. Chem. Int. Ed.*, **51** (16), 3822–3825.

67. Dreyer, D.R., Park, S., Bielawski, C.W., Ruoff, R.S. (2010) The chemistry of graphene oxide. *Chem. Soc. Rev.*, **39** (1), 228–240.

68. Biniak, S., Szymański, G., Siedlewski, J., Świtkowski, A. (1997) The characterization of activated carbons with oxygen and nitrogen surface groups. *Carbon*, **35** (12), 1799–1810.

69. Chingombe, P., Saha, B., Wakeman, R.J. (2005) Surface modification and characterisation of a coal-based activated carbon. *Carbon*, **43** (15), 3132–3143.

70. Lopez-Ramon, M.V., Stoeckli, F., Moreno-Castilla, C., Carrasco-Marin, F. (1999) On the characterization of acidic and basic surface sites on carbons by various techniques. *Carbon*, **37** (8), 1215–1221.

71. Onda, A., Ochi, T., Yanagisawa, K. (2008) Selective hydrolysis of cellulose into glucose over solid acid catalysts. *Green Chem.*, **10** (10), 1033–1037.

72. Onda, A., Ochi, T., Yanagisawa, K. (2009) Hydrolysis of cellulose selectively into glucose over sulfonated activated-carbon catalyst under hydrothermal conditions. *Top. Catal.*, **52** (6), 801–807.

73. Kitano, M., Yamaguchi, D., Suganuma, S., Nakajima, K., Kato, H., Hayashi, S., Hara, M. (2009) Adsorption-enhanced hydrolysis of β-1,4-glucan on graphene-based amorphous carbon bearing SO_3H, COOH, and OH Groups. *Langmuir*, **25** (9), 5068–5075.

74. Hara, M. (2010) Biomass conversion by a solid acid catalyst. *Energy Environ. Sci.*, **3** (5), 601–607.

75. Wu, Y., Fu, Z., Yin, D., Xu, Q., Liu, F., Lu, C., Mao, L. (2010) Microwave-assisted hydrolysis of crystalline cellulose catalyzed by biomass char sulfonic acids. *Green Chem.*, **12** (4), 696–700.

76. Guo, H., Qi, X., Li, L., Smith, R.L. Jr (2012) Hydrolysis of cellulose over functionalized glucose-derived carbon catalyst in ionic liquid. *Bioresource Technol.*, **116**, 355–359.

77. Liu, M., Jia, S., Gong, Y., Song, C., Guo, X. (2013) Effective hydrolysis of cellulose into glucose over sulfonated sugar-derived carbon in an ionic liquid. *Ind. Eng. Chem. Res.*, **52** (24), 8167–8173.

78. Pang, J., Wang, A., Zheng, M., Zhang, T. (2010) Hydrolysis of cellulose into glucose over carbons sulfonated at elevated temperatures. *Chem. Commun.*, **46** (37), 6935–6937.

79. Jiang, Y., Li, X., Cao, Q., Mu, X. (2010) Acid functionalized, highly dispersed carbonaceous spheres: an effective solid acid for hydrolysis of polysaccharides. *J. Nanoparticle Res.*, **13** (2), 463–469.

80. Mo, X., Lotero, E., Lu, C., Liu, Y., Goodwin, J. G. (2008) A novel sulfonated carbon composite solid acid catalyst for biodiesel synthesis. *Catal. Lett.*, **123** (1), 1–6.

81. Ji, J., Zhang, G., Chen, H., Wang, S., Zhang, G., Zhang, F., Fan, X. (2011) Sulfonated graphene as water-tolerant solid acid catalyst. *Chem. Sci.*, **2** (3), 484–487.

82. Daengprasert, W., Boonnoun, P., Laosiripojana, N., Goto, M., Shotipruk, A. (2011) Application of sulfonated carbon-based catalyst for solvothermal conversion of cassava waste to hydroxymethylfurfural and furfural. *Ind. Eng. Chem. Res.*, **50** (13), 7903–7910.

83. Anderson, J.M., Johnson, R.L., Schmidt-Rohr, K., Shanks, B.H. (2014) Solid state NMR study of chemical structure and hydrothermal deactivation of moderate-temperature carbon materials with acidic SO_3H sites. *Carbon*, **74**, 333–345.

84. Van Pelt, A.H., Simakova, O.A., Schimming, S.M., Ewbank, J.L., Foo, G.S., Pidko, E.A., Hensen, E.J.M., Sievers, C. (2014) Stability of functionalized activated carbon in hot liquid water. *Carbon*, **77**, 143–154.

85. Kobayashi, H., Komanoya, T., Hara, K., Fukuoka, A. (2010) Water-tolerant mesoporous-carbon-supported ruthenium catalysts for the hydrolysis of cellulose to glucose. *ChemSusChem*, **3** (4), 440–443.

86. Kobayashi, H., Komanoya, T., Guha, S.K., Hara, K., Fukuoka, A. (2011) Conversion of cellulose into renewable chemicals by supported metal catalysis. *Appl. Catal. A General*, **409–410** (0), 13–20.

87. Boehm, H.P. (2002) Surface oxides on carbon and their analysis: a critical assessment. *Carbon*, **40** (2), 145–149.

88. Gazit, O.M., Katz, A. (2013) Understanding the role of defect sites in glucan hydrolysis on surfaces. *J. Am. Chem. Soc.*, **135** (11), 4398–4402.

89. Kobayashi, H., Yabushita, M., Komanoya, T., Hara, K., Fujita, I., Fukuoka, A. (2013) High-yielding one-pot synthesis of glucose from cellulose using simple activated carbons and trace hydrochloric acid. *ACS Catal.*, **3** (4), 581–587.

90. Yabushita, M., Kobayashi, H., Hara, K., Fukuoka, A. (2014) Quantitative evaluation of ball-milling effects on the hydrolysis of cellulose catalysed by activated carbon. *Catal. Sci. Technol.*, **4** (8), 2312–2317.

91. Shrotri, A., Kobayashi, H., Fukuoka, A. (2016) Mechanochemical synthesis of a carboxylated carbon catalyst and its application in cellulose hydrolysis. *ChemCatChem*, **8** (6), 1059–1064.

92. Kobayashi, H., Kaiki, H., Shrotri, A., Techikawara, K., Fukuoka, A. (2016) Hydrolysis of woody biomass by a biomass-derived reusable heterogeneous catalyst. *Chem. Sci.*, **7** (1), 692–696.

93. Shrotri, A., Kobayashi, H., Fukuoka, A. (2016) Air oxidation of activated carbon to synthesize a biomimetic catalyst for hydrolysis of cellulose. *ChemSusChem*, **9** (11), 1299–1303.

94. Chung, P.-W., Charmot, A., Olatunji-Ojo, O.A., Durkin, K.A., Katz, A. (2014) Hydrolysis catalysis of miscanthus xylan to xylose using weak-acid surface sites. *ACS Catal.*, **4** (1), 302–310.

95. Charmot, A., Chung, P.-W., Katz, A. (2014) Catalytic hydrolysis of cellulose to glucose using weak-acid surface sites on postsynthetically modified carbon. *ACS Sustain. Chem. Eng.*, **2** (12), 2866–2872.

96. Chung, P.-W., Yabushita, M., To, A.T., Bae, Y., Jankolovits, J., Kobayashi, H., Fukuoka, A., Katz, A. (2015) Long-chain glucan adsorption and depolymerization in zeolite-templated carbon catalysts. *ACS Catal.*, **5** (11), 6422–6425.

97. Nishihara, H., Kyotani, T. (2012) Templated nanocarbons for energy storage. *Adv. Mater.*, **24** (33), 4473–4498.

98. To, A.T., Chung, P.-W., Katz, A. (2015) Weak-acid sites catalyze the hydrolysis of crystalline cellulose to glucose in water: importance of post-synthetic functionalization of the carbon surface. *Angew. Chem. Int. Ed.*, **54** (38), 11050–11053.

99. Van de Vyver, S., Thomas, J., Geboers, J., Keyzer, S., Smet, M., Dehaen, W., Jacobs, P.A., Sels, B.F. (2011) Catalytic production of levulinic acid from cellulose and other biomass-derived carbohydrates with sulfonated hyperbranched poly(arylene oxindole)s. *Energy Environ. Sci.*, **4** (9), 3601–3610.

100. Shi, Y., Nabae, Y., Hayakawa, T., Kobayashi, H., Yabushita, M., Fukuoka, A., Kakimoto, M. (2014) Synthesis and characterization of hyperbranched aromatic poly(ether ketone)s functionalized with carboxylic acid terminal groups. *Polym. J.*, **46** (10), 722–727.

101. Bui, S., Verykios, X., Mutharasan, R. (1985) In situ removal of ethanol from fermentation broths. 1. Selective adsorption characteristics. *Ind. Eng. Chem. Process Des. Dev.*, **24** (4), 1209–1213.

102. Chung, P.-W., Charmot, A., Gazit, O.M., Katz, A. (2012) Glucan adsorption on mesoporous carbon nanoparticles: effect of chain length and internal surface. *Langmuir*, **28** (43), 15222–15232.

103. Yabushita, M., Kobayashi, H., Hasegawa, J., Hara, K., Fukuoka, A. (2014) Entropically favored adsorption of cellulosic molecules onto carbon materials through hydrophobic functionalities. *ChemSusChem*, **7** (5), 1443–1450.

104. Tsuzuki, S., Honda, K., Uchimaru, T., Mikami, M., Tanabe, K. (2000) The magnitude of the CH/π interaction between benzene and some model hydrocarbons. *J. Am. Chem. Soc.*, **122** (15), 3746–3753.

105. Mecozzi, S., Rebek, J. Jr (1998) The 55% solution: a formula for molecular recognition in the liquid state. *Chem. Eur. J.*, **4** (6), 1016–1022.

106. Yabushita, M., Li, P., Bernales, V., Kobayashi, H., Fukuoka, A., Gagliardi, L., Farha, O.K., Katz, A. (2016) Unprecedented selectivity in molecular recognition of carbohydrates by a metal-organic framework. *Chem. Commun*, **52** (44), 7094–7097.

107. Gazit, O.M., Charmot, A., Katz, A. (2011) Grafted cellulose strands on the surface of silica: effect of environment on reactivity. *Chem. Commun.*, **47** (1), 376–378.

108. Foo, G.S., Sievers, C. (2015) Synergistic effect between defect sites and functional groups on the hydrolysis of cellulose over activated carbon. *ChemSusChem*, **8** (3), 534–543.

5

Ordered Mesoporous Silica-Based Catalysts for Biomass Conversion

Liang Wang, Shaodan Xu, Xiangju Meng, and Feng-Shou Xiao

Key Laboratory of Applied Chemistry of Zhejiang Province, Department of Chemistry, Zhejiang University, China

5.1 Introduction

Since the pioneer studies conducted by the Mobile Corporation scientists on the synthesis of the M41S family of ordered mesoporous materials during the 1990s [1,2], an ocean of ordered mesoporous silica materials have been created using cationic surfactants, with a wide range of compositions [3–9]. Typically, the inorganic component of silica is assembled around the molecular assemblies of surfactants, and open mesopores are obtained after removing the surfactants by calcination or extraction. Compared with conventional nanomaterials, ordered mesoporous materials have significant advantages for catalysis [10–16], as follows:

- An *ordered mesoporous network*, which has a very uniform mesopore size distribution and an ordered orientation, which allows fine control of the molecular diffusion kinetics.
- A *high surface area*, which plays important roles in the extensive dispersion of active sites on the ordered mesoporous materials.
- A *high pore volume,* which allows an enrichment of the molecules for catalytic transformation.
- A silanol-rich surface, which is favorable for functionalization to introduce catalytically active sites or to modify the surface wettability.

Based on these features, the ordered mesoporous materials have been widely used as supports for a huge number of heterogeneous catalysts [16–24]. As expected, in many cases

Nanoporous Catalysts for Biomass Conversion, First Edition. Edited by Feng-Shou Xiao and Liang Wang.
© 2018 John Wiley & Sons Ltd. Published 2018 by John Wiley & Sons Ltd.

the ordered mesoporous silica-based catalysts exhibited clearly higher activity as well as selectivity than the nonporous catalysts. Recently, ordered mesoporous silica-based catalysts have also attracted much attention in the conversion of biomass feedstocks [25–28], where the mesoporous silica materials act as supports for various metal nanoparticles, acidic, and basic sites, providing unusual catalytic performances in a series of oxidation [29–33], hydrogenation [34–40], esterification [41–45], and hydrodeoxygenation [46–51] reactions. The aim of this chapter is to discuss and evaluate ordered mesoporous silica-based catalysts in the conversion of biomass molecules, where the production of fine chemicals and biofuels are also included.

5.2   Sulfated Ordered Mesoporous Silicas

The acid catalysts have been widely used in biomass conversion, playing important roles in catalyzing the reactions of hydration [52–57], dehydration [58–61], esterification [62–64], and isomerization [65–70]. Conventionally, homogeneous acids (e.g., H_2SO_4 and H_3PO_4) are employed in these reactions, exhibiting high catalytic activity [71,72], but such homogeneous features lead to difficulties in their separation and regeneration from the reaction system. The use of heterogeneous acids could easily solve this problem. Until now, sulfated resins, metal oxides, silicas and aluminosilicate zeolites have been generally used as efficient solid acids [73–77]. Compared to these catalysts, sulfated ordered mesoporous silica exhibits unusual catalytic performances in many reactions, owing to the ordered mesoporous structure [78,79], simple modification of the sulfonic acid group, and controllable acid density.

5.2.1   Conversion of Levulinic Acid to Valerate Esters

With the increased consumption of fossil resources leading to problems of environmental pollution and carbon dioxide emission, biomass conversion has attracted considerable attention as it is regarded as a renewable alternative to fossil fuels [80–85]. The selective conversion of biomass-derived feedstocks is very important for the sustainable production of fuels and fine chemicals [86–90]. However, it should be noted that direct conversion of the natural biomass feedstocks to desirable chemicals is challengeable, mainly because of the multiple compositions and complex reaction pathways involved in the conversion of natural biomass. One efficient route is a rapid pyrolysis and/or acidolysis of the cellulose and lignin feedstocks to various platform molecules, after which the platform molecules can be transformed to desired fuels and chemicals [87,91–93].

Levulinic acid (LA) is produced from the acidolysis of cellulose, which has been recognized as one of the top-10 most promising platform molecules derived from biomass by the U.S. Department of Energy [94–96]. Although the selective conversion of LA has become a 'hot topic' in the field of biomass conversion [97–101], the complex reaction pathways involved create difficulties in the selective formation of desired products. A key to success here is the rational design and synthesis of new catalysts.

LA can easily be transformed into ethyl levulinate (EL) [102,103], γ-valerolactone (GVL) [101,104–106] and 2-methyl tetrahydrofuran (MTHF) [107], all of which have

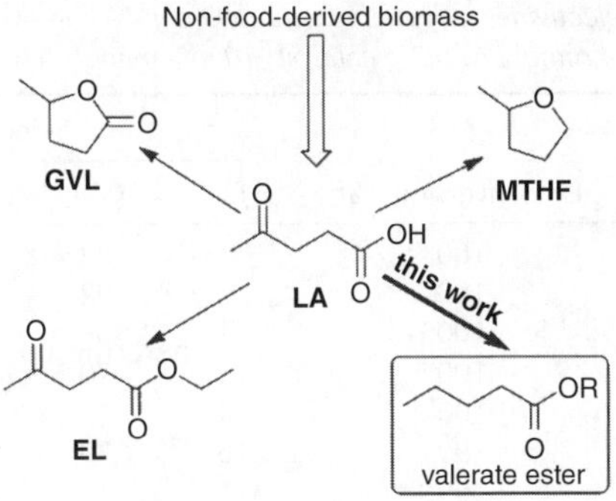

Figure 5.1 *Conversion of levulinic acid (LA) to fuel additives.*

the potential to serve as fuel additives (Figure 5.1) [108]. However, these molecules are still rather unsatisfactory due to their polarity or dissolubility. Recently, ethyl valerate (EV) was proposed as a new oxygenated fuel additive, which could be blended easily with gasoline and diesel [109,110]. Subsequently, a series of studies has demonstrated the significant advantages of EV compared to other cellulose-based molecules when used as fuel additives, including the simple production, good solubility, and high energy density. Until now, many bifunctional heterogeneous catalysts with both metal and acid sites (e.g., Ru/HZSM-5) have been used to produce medium yields of valeric acid (VA) and EV [111–114]. Pan *et al.* systemically studied the catalytic performances of various Ru-based catalysts, including metal oxide-, SiO_2-, and activated carbon-supported Ru nanoparticles (NPs), in the hydrogenation of LA, where maximal selectivities of VA and EV of 15% were obtained over MnO_x-supported Ru catalysts (Table 5.1) [1108]. When ZSM-5 was used as a support, the total yield of VA and VE often reached 29%, indicating the important role of acid sites. Notably, the ordered mesoporous silica of SBA-15 functionalized with different amounts of sulfonic acid group combining Ru sites gave a total yield of VA and VE of 54–90%. Notably, the selectivities towards GVL, a major byproduct in LA conversion, decreased with increasing amounts of acid sites. Notably, the VA and VE yield of sulfated SBA-15-supported Ru catalyst was much higher than for the Ru/ZSM-5 catalyst, though this may have been due to the uniform mesoporous structure of the SBA-15 support, which would favor molecular diffusion on the metal and acid sites. Furthermore, the homogeneous acids of H_2SO_4 and *p*-TsOH, both of which are regarded as very active acid catalysts, produced very low yields of VA and VE, though this may be attributed to the hydrogenation of liquid acid catalyzed by Ru.

5.2.2 One-Pot Conversion of Cellulose into Chemicals

Although a scope of successes has been achieved in the selective conversion of platform chemicals, it is still challengeable for the conversion of original feedstocks (e.g., cellulose) to the desired fuels and fine chemicals [115]. From the point view of 'green' chemistry, manipulation of the tandem catalysis process with multiple reactions would

Table 5.1   Product selectivities in levulinic acid (LA) conversion over different catalysts. Reaction conditions: 4 mmol LA; 0.2 g catalyst; 10 ml ethanol; 6 h; 523K; 4 MPa H_2.[a]

Catalyst	LA conversion (%)	Selectivity (%)			
		EL	GVL	MTHF	VA+VE
Ru/MnO$_x$	100	1	82	1	15
Ru/MgO	100	0	92	0	3
Ru/CeO$_2$	100	1	88	1	6
Ru/TiO$_2$	100	3	80	4	6
Ru/SiO$_2$	100	3	73	9	10
Ru/SBA-15	99	16	70	1	2
Ru/SBA-SO$_3$H-1[b]	99	1	35	1	54
Ru/SBA-SO$_3$H-2[b]	99	1	29	0	63
Ru/SBA-SO$_3$H-3[b]	100	1	14	1	76
Ru/SBA-SO$_3$H-4[b]	100	0	4	1	90
Ru/C + H$_2$SO$_4$	99	45	26	5	8
Ru/C + p-TsOH	100	19	31	4	16

[a]The data are from Ref. [108]. Ru loadings are 5% for all catalysts.
[b]The acid densities for Ru/SBA-SO$_3$H-1, Ru/SBA-SO$_3$H-2, Ru/SBA-SO$_3$H-3, and Ru/SBA-SO$_3$H-4 are 0.17, 0.39, 0.46, and 1.14 mmol g^{-1}, respectively.

be economically and environmentally favorable (Figure 5.2) [114]. Considering that the molecules of the feedstock and intermediates are generally relatively bulky, the ordered mesoporous silica-based acids could serve as efficient catalysts for these reactions.

As a typical reaction route for the conversion of cellulose, the hydrolysis of cellulose to glucose and subsequent conversion to hydroxymethylfurfural (HMF) and LA have been investigated using various acid catalysts [116–120]. In order to achieve a one-pot process, the use of an efficient solid-acid catalyst can play a key role. The hydrolysis of cellulose catalyzed by sulfuric acid has been implemented on a large scale [121–123]. Based on the exception of heterogeneous catalysts, sulfonic acid (–SO$_3$H)-functionalized ordered mesoporous silica (MCM-41) was synthesized and applied in a one-pot conversion of cellulose to chemicals by Wu *et al.* [114]. In this case, the thiol group was modified on MCM-41 by a silane condensation reaction, and subsequent oxidation treatment transformed the thiol into sulfonic acid group (MCM-41-SO$_3$H).

H$_2$SO$_4$ is regarded as very active catalyst for the hydrolysis of cellulose. For example, it took about 22 min to achieve a full conversion of cellulose in proposed reaction conditions over an H$_2$SO$_4$ catalyst, but about 60 min was necessary over the MCM-41-SO$_3$H catalyst. The lower activity of MCM-41-SO$_3$H than that of H$_2$SO$_4$ is reasonably attributed to the limited accessibility between acid sites and cellulose molecules. Interestingly, under the similar conversion of cellulose, MCM-41-SO$_3$H exhibited an even slightly higher LA yield than H$_2$SO$_4$ (Figure 5.3) [114], indicating the good features of both activity and selectivity.

The use of metal catalysts with solid acid can have a significant influence on product selectivity in the conversion of cellulose. For example, the sole Ru/C catalyst catalyzed the hydrogenation of glucose to sorbitol, while the combined Ru/C and WO$_3$ catalysts preferentially produce ethylene glycol. In a one-pot conversion of cellulose over MCM-41-SO$_3$H

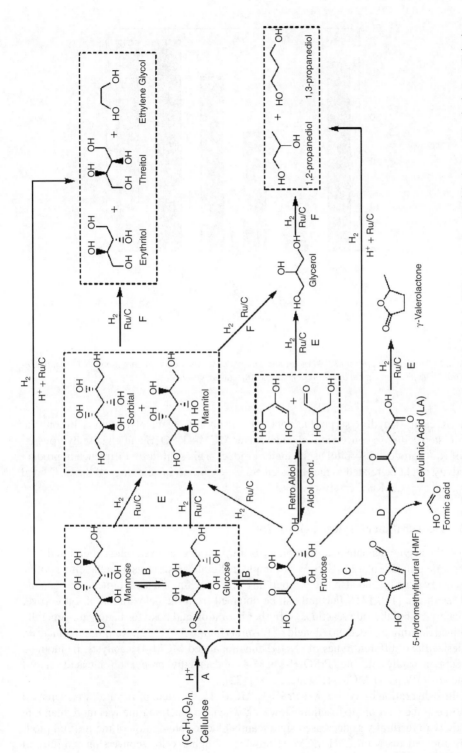

Figure 5.2 Reaction pathways for the conversion of cellulose. The reactions include (A) hydrolysis, (B) isomerization, (C) dehydration, (D) rehydration, (E) hydrogenation, and (F) hydrogenolysis.

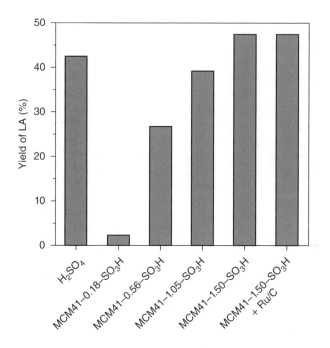

Figure 5.3   *Yield of levulinic acid (LA) over different acid catalysts. Reaction conditions: 0.1 g catalyst or 0.03 mol l⁻¹ H_2SO_4; 0.1 g cellulose; 5 ml H_2O; 473K; 1000 rpm; 4 h.*

and Ru/C catalyst, hydrolysis products of C_2–C_4 alcohols were mainly obtained. In contrast, the two-step cellulose conversion to glucose over MCM-41-SO_3H and glucose hydrogenation of Ru/C produced sorbitol as the major product. This significant difference in product selectivity might be related to the Brønsted acid sites promoting cleavage of the C–C bond in glucose over the Ru/C catalyst.

5.2.3   Dehydration of Xylose to Furfural

Among the chemicals obtained from biomass, furfural has received much attention due to its versatile utilization [124–129]. For example, furfural can serve as a fuel additive with a high energy density, and also act as a building block for furanic acid, dimethyl furan, and ethyl levulinate [130,131]. Furfural can be obtained from the dehydration of xylose over acid catalysts, but the side reactions of further hydration and coke formation are generally combined, leading to a decreased yield of furfural (Figure 5.4) [132]. Kaiprommarata *et al.* synthesized two different types of –SO_3H-functionalized MCM-41 catalysts, including a conventional catalyst denoted $PrSO_3$H-MCM-41 and a methyl propyl sulfonic acid catalyst denoted as $MPrSO_3$H-MCM-41 (Figure 5.5) [132].

In the dehydration of xylose over $PrSO_3$H-MCM-41, the yield of furfural was increased with time at the start of the reaction. However, when the reaction time was more than 8 h, selectivity to furfural was decreased significantly, due to rehydration of the furfural product. In comparison, $MPrSO_3$H-MCM-41 provided a higher xylose conversion and furfural

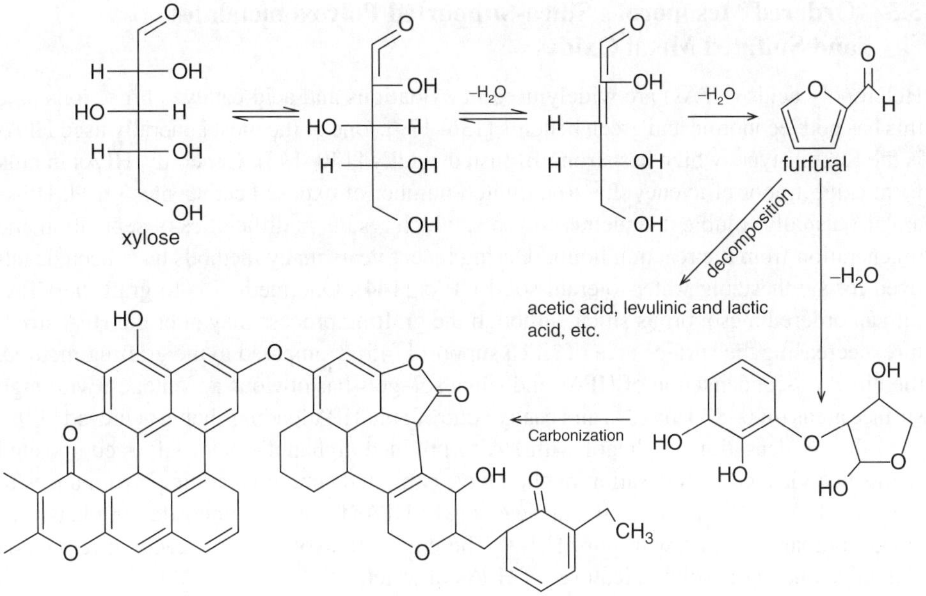

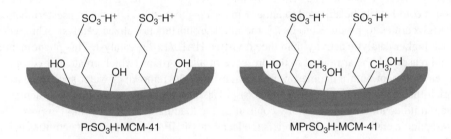

Figure 5.4   *The conversion of xylose to furfural.*

Figure 5.5   *Structures of PrSO₃H-MCM-41 and MPrSO₃H-MCM-41.*

selectivity than PrSO$_3$H-MCM-41. This phenomenon was assigned to the effect of the methyl group on MPrSO$_3$H-MCM-41 and its anti-adsorption effect on furfural at active sites. The acid density was also sensitive to the product yield. It should be emphasized that the balance of furfural formation and hydration played a key role in the yield of furfural, efficiently preventing furfural hydration, which is important for the selective formation of this compound.

Sulfated ordered mesoporous silica has been confirmed as an efficient catalyst in many biomass conversions in the liquid phase, due to their significant advantages of easy synthesis (by grafting) and high acid strength and controllable acid density [78,79,133–135]. The low stability of the organic groups makes these catalysts unsatisfactory for the reactions at high temperature, however. Removal of the coke, which is easily formed in acid-catalyzed biomass conversions, is challenging because these catalysts do not undergo high-temperature calcination.

5.3   Ordered Mesoporous Silica-Supported Polyoxometalates and Sulfated Metal Oxides

Heteropoly acids (HPAs) are widely used in oxidations and acid-catalyzed reactions, and this has both economic and green benefits [136–139]. One of the most generally used HPAs is the Keggin-type, which has strong Brønsted acidity [140–143]. Generally, HPAs in bulk form exhibit poor efficiency due to a limited number of exposed active sites [138]. HPAs are also usually soluble in aqueous solvents, which results in difficulties of separation and regeneration from the reaction liquor. During recent years many methods have been developed for synthesizing water-tolerant solid HPAs [144]. One method is to graft the HPAs into an ordered mesoporous silica, although the grafting process may alter the HPA structure, decreasing the surface area of silica support [145]. Compared to the grafting method, the direct co-condensation of HPAs and silica sol–gels has obvious advantages, with high surface areas (400–800 m^2 g^{-1}) and easily controllable HPA loadings being achieved [146].

In the condensation of phenol with LA to produce diphenolic acid, HPAs composited with silica via a co-condensation method have exhibited similar catalytic performances to the pure HPAs [146]. Importantly, the leaching of HPAs from the composite sample is very slight compared to that with pure HPAs. The complete avoidance of leaching remains a challenge due to the soluble features of HPAs in water.

Adjustment of the wettability of a catalyst surface has emerged as an efficient method for hindering the leaching of HPAs. When Xu *et al.* synthesized Ta$_2$O$_5$ and SiO$_2$ materials that were functionalized with both alkyl groups and a Keggin-type HPA, the alkyl groups could easily be adjusted to either Me- or Ph-groups [147]. In the transesterification of triglycerides to produce biodiesel, the alkyl group-functionalized catalyst exhibited a much higher catalytic activity than the alkyl-free HAPs/Ta$_2$O$_5$ catalyst. This phenomenon could reasonably be attributed to the presence of alkyl groups (which create an hydrophobic environment within the mesopores), by ensuring the removal of water species so as to enrich the organic molecules (Figure 5.6) [147]. The highest activity could be achieved under mild conditions by rationally controlling the loading of HPAs and alkyl groups. The hydrophobic environment also hinders the leaching of HPAs, with constant activities being achieved in recycle tests.

Compared with HPAs, the sulfated metal oxides have good stability in solvents, as also possess features of combined Brønsted and Lewis acidities. The Lewis acidity derives from the metal atoms, while the Brønsted acidity derives from the protons on the surface hydroxyl groups of metal oxides (Figure 5.7) [148]. Generally, the hydroxyl groups on the surface of metal oxides have very weak Brønsted acidity. By applying a sulfation treatment, the S–O bonds on the sulfuric groups could bind strongly to the metal atoms, forming a coordination of the S=O bond with the surface hydroxyl groups on metal oxides, leading to an enhancement of Brønsted acidity. It is well known that a series of metal oxides (e.g., ZrO$_2$, SnO$_2$, TiO$_2$) could be converted to solid-acid catalysts by sulfation treatment [149–156]. These catalysts also suffer from low numbers of exposed acid sites due to their limited surface area, and this greatly hinders the access of acid sites to the reactants. It should also be noted that the leaching of sulfur occurred even at mild temperatures, and consequently the sulfated catalyst was deactivated.

Jiménez-Morales *et al.* impregnated zirconium sulfate into the mesopores of MCM-41, followed by calcination at 750 °C [157]. Thermal decomposition of this salt led to the

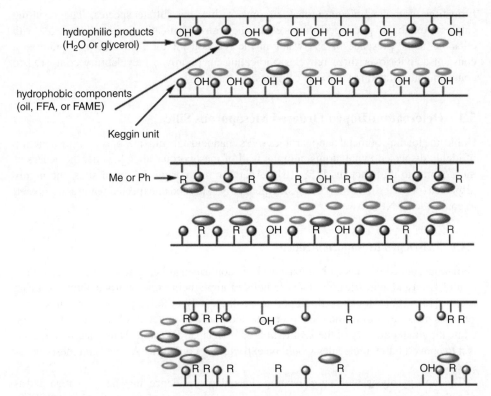

Figure 5.6 *Schematic representation of the influence of surface hydrophobicity on catalytic activity. Top: HPAs/SiO₂-Ta₂O₅. Alkyl-functionalized HPAs/SiO₂-Ta₂O₅ sample synthesized by a co-condensation technique (middle) and a post-synthesis grafting method (bottom). (See color plate section for the color representation of this figure.)*

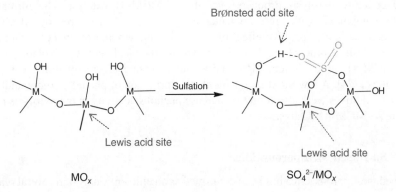

Figure 5.7 *Locations of the Brønsted and Lewis acid sites on metal oxides (MO$_x$) and sulfated metal oxides (SO$_4^{2-}$/MO$_x$).*

formation of monoclinic/tetragonal zirconia with some sulfate species. The resulting catalysts exhibited good performances in the transesterification of sunflower oil with ethanol. In the presence of 5% water in the reaction mixture, the catalyst exhibited a constant activity over three recycles, indicating an improved recyclability compared to conventional sulfated metal oxide catalysts.

5.4 Heteroatom-Doped Ordered Mesoporous Silica

Unlike the loading of metal nanoparticles (NPs)/nanoclusters into mesopores, the impaction of highly dispersed metal atoms onto the wall of mesoporous silica benefits the access of more atoms to the reactants [158–161]. Apart from exposing the metal sites, the highly dispersed (or even single-site) metal sites generally exhibited unexpected features compared to conventional NPs.

5.4.1 Al-Doped Mesoporous Silica

Tetra-coordinated Al sites with silica have been confirmed as being efficient for the production of Brønsted acid sites, the most famous example being microporous aluminosilicate zeolite [162–167]. However, microporous zeolite catalysts failed for some biomass conversions because most of the biomass crude and platform molecules had larger diameters, while the small pore size of the zeolite limited molecular diffusion. Thus, an efficient catalyst to convert bulky molecules would be expected by combining Al sites and mesoporous silicas.

The rapid pyrolysis, with a high heating rate and low residence time, has been regarded as an efficient strategy for obtaining pyrolysis oil and various platform molecules [168,169], though the conversion routes are complex and non-selective. It has been found that, by adding Al-MCM-41 catalyst into the pyrolysis system, it would be possible to adjust product selectivities [170–173]. In the pyrolysis vapors of spruce wood, levoglucosan is completely eliminated, while acetic acid, furfural, and furans become quite important among cellulose pyrolysis products over the Al-MCM-41 catalyst. The formation of phenolic compounds with high molecular masses is greatly hindered by Al-MCM-41. These catalysts have a positive effect on increasing the amount of phenolic compounds, which are desired products and are important in chemical industry. Notably, the type of Al-MCM-41 with different Si/Al ratios and components greatly influences product selectivities [174]. A low Si/Al ratio benefits the formation of phenolic compounds, while Fe- or Cu-additives were also found to increase phenol selectivity in the pyrolysis reaction over the Al-MCM-41 catalyst.

5.4.2 Sn-Doped Mesoporous Silica

Al-doped mesoporous silica has been confirmed as an efficient catalyst in several reactions, but the weak acidity severely hinders its wide application due to the amorphous nature of the mesoporous walls [175,176]. This phenomenon is due to the fact that the Brønsted acid sites of aluminosilicate originate from the protons that are present to neutralize the negative charges of the aluminosilicate walls. Apart from Al atoms, the introduction of Sn atoms into mesoporous silicas may also generate acid sites, but these are of the Lewis type and cannot catalyze reactions requiring Brønsted acid sites. Therefore, the development of alternative Sn catalysts with Brønsted acid sites has attracted much attention, as these may be efficient for the conversion of biomass and perhaps also other classes of reaction [177].

It is well known that small transitional metal nanoclusters (e.g., SnO_2) can form Brønsted acid sites from the surface hydroxyl groups [178]. Therefore, it is possible to design an acidic Sn catalyst with relatively strong Brønsted acidity by loading Sn sites onto the surface of mesoporous silica with ultra-high dispersion, even to form single-site Sn, which would be expected to produce a maximized number of hydroxyl groups on the Sn sites. Thus, a new method was proposed to load single-site Sn onto mesoporous silicas by grafting dimethyldichlorostannane onto SBA-15 (S-Sn-Me), followed by calcination at 600 °C for 3 h to transform the methyl groups into hydroxyl groups (S-Sn-OH) [177]. Using this method, rich hydroxyl groups are formed on the Sn sites. Characterization studies using X-ray diffraction (XRD) and N_2 sorption have shown that the mesoporous structure of SBA-15 is well maintained during the loading of Sn sites with hydroxyl groups.

The high dispersion of Sn on SBA-15 was demonstrated using transmission electron microscopy (TEM) characterizations, whereby the Sn or SnOx clusters were seen to be completely absent in the high-resolution image (Figure 5.8b) [177]. The elemental energy-dispersive X-ray spectroscopy (EDS) map of Sn confirmed the high degree of dispersion. The acidity of S-Sn-OH was investigated using pyridine-adsorption infrared (IR) spectra, with S-Sn-OH exhibiting both bands assigned coordinatively to pyridine and bound to both Brønsted and Lewis acid sites. In contrast, for S-Sn-Me the bands were assigned only to Lewis acid sites. Considering that both the S-Sn-OH and S-Sn-Me samples had the same SBA-15 support and highly dispersed Sn, the difference in acidity between S-Sn-OH and S-Sn-Me could be reasonably attributed to the different groups on the Sn sites, rather than for other reasons. S-Sn-Me with a methyl group exhibited only Lewis acidity, while S-Sn-OH with an hydroxyl group attached to the Sn atoms caused additional Brønsted acidity. In the condensation of glycerol with acetone to produce solketal, which is a model reaction to transform glycerol into valuable products, S-Sn-OH exhibited a much higher solketal yield than did general Sn-doped silica catalysts, a situation that can reasonably be attributed to the Brønsted acidity on S-Sn-OH.

5.5   Ordered Mesoporous Silica-Supported Metal Nanoparticles

Metal nanoparticles (NPs) with tunable composition, size and distinguishable electronic properties compared to the bulky catalyst, have been used widely in a variety of reactions for biomass conversion [179–183]. In order to construct durable and highly dispersed metal NPs, the metal precursors are generally loaded into the mesopores of silicas and transformed into NPs. In this case, the ordered mesoporous silicas are highly favorable as they can offer major advantages owing to their uniform mesopores, controllable pore sizes, and large surface areas, all of which benefit the dispersion of metal NPs and the diffusion of molecules. The fabrication of ordered mesoporous silica-supported metal NP catalysts has become an important topic in the field of biomass conversion [184–186].

Figure 5.8 *(a) The synthesis of S-Sn-OH; (b) Scanning transmission electron microscopy image; (c–e) O, Si, and Sn energy-dispersive X-ray spectroscopy (EDS) elemental maps of S-Sn-OH, respectively. The scale bars in images (b) to (e) are 10, 8, 8, and 8 nm, respectively. (See color plate section for the color representation of this figure.)*

5.5.1 Mesoporous Silica-Supported Pd Nanoparticles

The components of pyrolysis liquids, known as crude bio-oil, include water, acids, aldehydes, unsaturated compounds, and oligomers, and this results in the various undesirable properties of bio-oils [187–189]. Thus, it is necessary to upgrade the pyrolysis oil and to eliminate any undesirable compounds. When upgrading the pyrolysis oils, a major problem is to upgrade the large-molecule oligomers, which are derived from both holocellulose and lignin, and which account for 13.5–27.7% of the pyrolysis oils. One method to eliminate oligomers is the *in-situ* catalytic cracking of pyrolysis vapors following the pyrolysis steps. Taking into consideration that the oligomers have large molecular sizes, SBA-15-supported

Pd NPs rather than zeolite-supported catalysts nanoparticles were applied to the cracking of oligomers [186]. In this process, the pyrolysis oligomers were cracked to minomeric phenols. Of note, any anhydrosugars were almost completely eliminated, and the furan was decarbonylated to form light compounds. The concentration of linear aldehydes was significantly decreased, but the decrease in acid was very slight.

The introduction of acid sites to SBA-15-supported Pd NP catalysts also greatly influences the catalytic performance. For example, in a model reaction of the one-step hydrogenation–esterification (OHE) of furfural and acetic acid bio-oil upgrading, it has been shown that Al-SBA-15-supported Pd NPs favor the OHE reaction, which is reasonably assigned to the synergistic effect between the metal sites and the acid sites [190].

5.5.2   Mesoporous Silica-Supported Pt Nanoparticles

Recently, the utilization of waste from small molecules to create useful alkanes from C7 to C15 has attracted great interest [191], where the normal route is tandem aldol condensation and hydrodeoxygenation. Surplus energy costs and process limitations remain problems in this process, however. For example, cleavage during the reaction causes energy losses, and the separation of products from organic solvents is an energy-costing process. To overcome these limitations, Corma and coworkers developed a new process for the synthesis of C9 to C16 hydrocarbons, avoiding the use of organic solvents and minimizing the side reactions of cleavage (Figure 5.9) [192]. On the basis of this process, the maximal yield of C9 to C16 hydrocarbons could be up to 92.6%, while the yields of cracked products (C3–C8) were less than 2%, indicating the advantage in selectively obtaining alkane products. However, it should be noted that this process involves two individual reactions of alkylation and hydrodeoxygenation, and homogeneous H_2SO_4 was employed in the alkylation reaction.

Wen *et al.* tried to make this process greener by developing a bifunctional catalyst with both acid and metal sites, so as to combine the two individual reactions into a one-pot conversion [191]. In this case, Pt NPs supported on MCM-41 with acid sites were employed. In the model reaction of transforming 2-methylfuran and butanal into alkanes, the 2-methylfuran and butanal go through a hydroalkylation first at the acid sites, followed by dehydroxygenation to produce the alkane fuels (Figure 5.10) [191], where the mesopores of MCM-41 favor the diffusion of molecules between the acid and Pt sites. In this one-pot conversion process, the yield of C_{8+} alkanes could reach 96%. Very importantly, it was shown that the bifunctional catalyst had good recyclability, with constant performances being achieved over four reaction recycles, each of 20 h. These studies demonstrated a typical method for transforming conventional reactions with various reactions into a one-pot conversion, using ordered mesoporous silica-supported catalysts, which was in good agreement with the topic of 'green chemistry'.

5.5.3   Mesoporous Silica-Supported Ni Nanoparticles

Many reaction pathways have been developed for the conversion of biomass via hydrogenation, their success being mainly due to the use of noble metal catalysts such as Pd, Ru, and Pt [193,194]. However, the high cost and limited availability of these precious metals

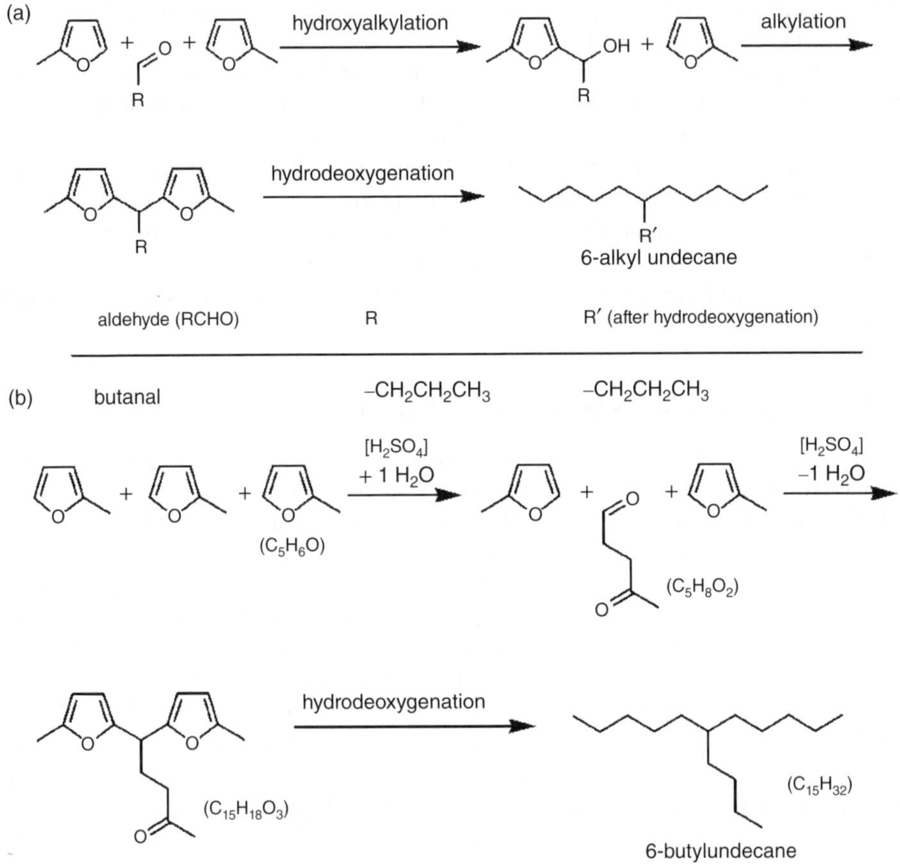

Figure 5.9 *(a) Formation of 1,1-bisylvylalkanes by hydroxyalkylation/alkylation from Sylvan and an aldehyde with subsequent hydrodeoxygenation to 6-alkyl undecane. (b) Formation and hydrodeoxygenation of 5,5-bisylvyl-2-pentanone. Initially, one molecule of Sylvan is hydrolyzed to 4-oxopentanal by sulfuric acid catalysis; two further sylvan molecules are subsequently hydroxyalkylated and alkylated with the intermediary aldehyde.*

have spurred research with non-precious metals having a low cost, such as Co, Ni, and Fe [195,196]. Important applications of these catalysts include the Fischer–Tropsch synthesis using Co and Fe catalysts, and fine-chemical hydrogenations using Ni catalysts [197–199]. Compared to precious metals, the Co-, Fe-, and Ni-based catalysts exhibited lower activities, where higher temperatures and/or hydrogen pressures were necessary. However, due to the significant advantage in the low cost of these metals they are regarded as promising catalysts for biomass conversion. In particular, ordered mesoporous silica-supported Ni catalysts have been widely investigated.

Lercher and coworkers reported mesoporous silica-supported Ni NPs as efficient catalysts for the selective cleavage of ether bonds of (lignin-derived) aromatic ethers and hydrogenation of the oxygen-containing intermediates in the aqueous phase under mild reaction conditions (120 °C, 6 bar H_2) [200]. The C–O bonds of α-O-4 and β-O-4 linkages

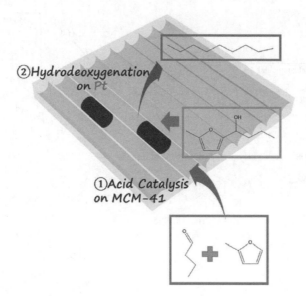

Figure 5.10 *Transformation of 2-methylfuran and butanal into alkanes over a bifunctional Pt/MCM-41 catalyst.*

are cleaved by hydrogenolysis on Ni, while the C–O bond of the 4-O-5 linkage is cleaved via parallel hydrogenolysis and hydrolysis (Figure 5.11). In the conversion of β-O-4 and 4-O-5 ether bonds, C–O bond cleavage is regarded as the rate-determining step. In this case, the reactants competed with hydrogen for active sites, leading to a maximum reaction rate as a function of the H_2 pressure.

Furthermore, Wang and Rinaldi systemically studied the performances of various supported Ni catalysts in the hydrogenation of diphenyl ether, including Ni/ZrO_2, Ni/Al_2O_3, Ni/Al_2O_3-KF, Ni/SBA-15, and Ni/Al-SBA-15 [201]. For example, the Ni/Al-SBA-15 exhibited a full conversion of diphenyl ether with selectivity to cyclohexane at 98%, outperforming other Ni catalysts. This phenomenon was assigned to the acidic support of SBA-15 in the Ni/Al-SBA-15, which benefits cleavage of the C–O bond in the reaction.

In most of these studies it has been shown that the mesopores and the acidity of silica supports are important for catalytic activity and product selectivity, particularly in the conversion of bulky molecules. It is recommended that the selective cleavages of C–O/C–C bonds are usually important steps in the hydrogenation of biomass feedstocks. This exception requires a synergistic effect of the metal sites and acid sites, and the mesoporous channels play an important role in the transfer of molecules between the different active sites.

5.6 Overall Summary and Outlook

Compared to conventional zeolite catalysts, ordered mesoporous silicas with larger surface areas and pore diameters have been widely investigated for the conversion of various biomass molecules, and have exhibited acceptable to excellent performances. Although the mesoporous silica-based catalysts have been shown to be recyclable in biomass conversion reactions, it should be noted that the hydrothermal stability of mesoporous silicas is still

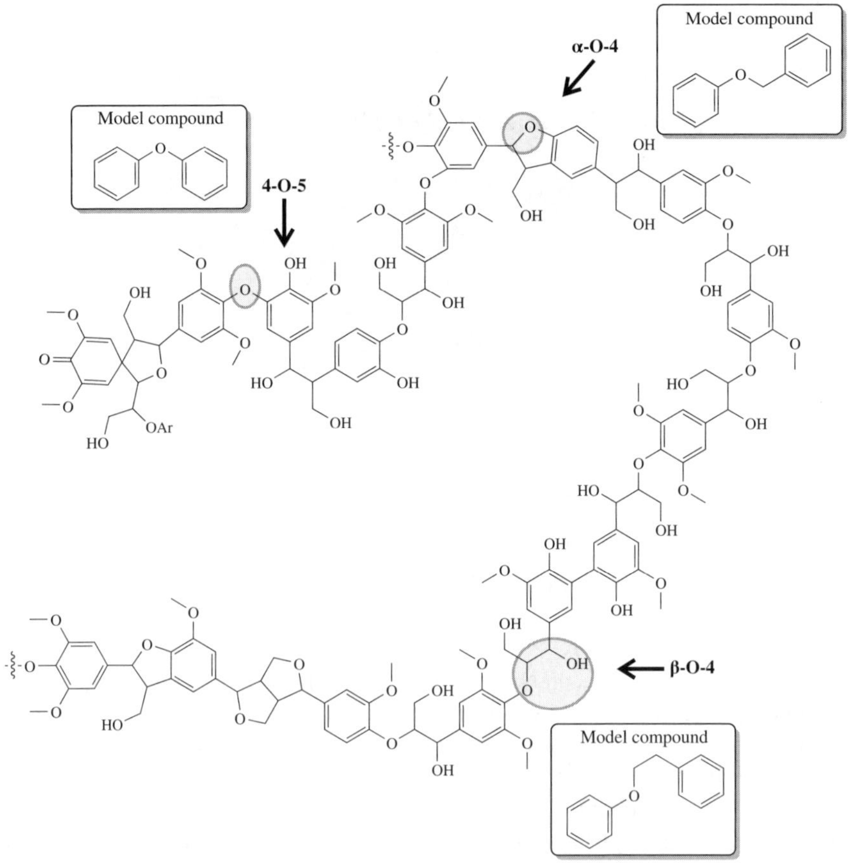

Figure 5.11 *Fragment structure of hardwood lignin.*

unsatisfactory due to the amorphous features of the silicas. To enhance the hydrothermal stability of mesoporous silicas is an old story in the field of ordered mesoporous material synthesis, and much success has been achieved by employing high-temperature syntheses and/or new surfactants, though the stability is still much lower than the practically used catalyst in industry (e.g., zeolite), which renders ordered mesoporous silica catalysts unsuitable for reactions conducted under hydrothermal conditions.

Furthermore, the synthesis of ordered mesoporous silicas generally requires a large amount of surfactants to generate a rich population of mesopores, and the high costs of surfactants represents a challenge to industrial applications. Therefore, it might be favorable to decrease the amount of surfactants in the synthesis and to identify a balance between cost and catalytic performance for potential applications.

Ordered mesoporous silica-based catalysts have significant advantages in mass transfer compared to zeolite catalysts, but show a lack of micropores to control product selectivity (this is known as shape-selective catalysis). It is recommended that catalysts which possess a combination of the advantages of mesoporous silicas and zeolites (e.g., hierarchical zeolites or the physical mixture of mesoporous silicas and zeolites) might be favorable for use in reactions of biomass conversion.

References

1. Beck, J.S., Vartuli, J.C., Roth, W.J., Leonowicz, M.E., Kresge, C.T., Schmitt, K.D., Chu, C.T.W., Olson, D.H., Sheppard, E.W., McCullen, S.B., Higgins, J.B., Schlenker, J.L. (1992) A new family of mesoporous molecular-sieves prepared with liquid-crystal templates. *J. Am. Chem. Soc.*, **114** (27), 10834–10843.
2. Kresge, C.T., Leonowicz, M.E., Roth, W.J., Vartuli, J.C., Beck, J.S. (1992) Ordered mesoporous molecular-sieves synthesized by a liquid-crystal template mechanism. *Nature*, **359** (6397), 710–712.
3. Hoffmann, F., Cornelius, M., Morell, J., Froba, M. (2006) Silica-based mesoporous organic-inorganic hybrid materials. *Angew. Chem. Int. Ed.*, **45** (20), 3216–3251.
4. Davis, M.E. (2002) Ordered porous materials for emerging applications. *Nature*, **417** (6891), 813–821.
5. Yang, P.D., Zhao, D.Y., Margolese, D.I., Chmelka, B.F., Stucky, G.D. (1998) Generalized syntheses of large-pore mesoporous metal oxides with semicrystalline frameworks. *Nature*, **396** (6707), 152–155.
6. Zhao, D.Y., Feng, J.L., Huo, Q.S., Melosh, N., Fredrickson, G.H., Chmelka, B.F., Stucky, G.D. (1998) Triblock copolymer syntheses of mesoporous silica with periodic 50 to 300 angstrom pores. *Science*, **279** (5350), 548–552.
7. Huo, Q.S., Margolese, D.I., Stucky, G.D. (1996) Surfactant control of phases in the synthesis of mesoporous silica-based materials. *Chem. Mater.*, **8** (5), 1147–1160.
8. Asefa, T., MacLachlan, M.J., Coombs, N., Ozin, G.A. (1999) Periodic mesoporous organosilicas with organic groups inside the channel walls. *Nature*, **402** (6764), 867–871.
9. Feng, X., Fryxell, G.E., Wang, L.Q., Kim, A.Y., Liu, J., Kemner, K.M. (1997) Functionalized monolayers on ordered mesoporous supports. *Science*, **276** (5314), 923–926.
10. Yan, J.N., Shi, J.L., Chen, H.R., Zhang, L.X., Li, L. (2003) Catalytic applied prospect of mesoporous materials. *J. Inorg. Mater.*, **18** (4), 725–730.
11. Corma, A. (1997) From microporous to mesoporous molecular sieve materials and their use in catalysis. *Chem. Rev.*, **97** (6), 2373–2420.
12. Sayari, A. (1996) Catalysis by crystalline mesoporous molecular sieves. *Chem. Mater.*, **8** (8), 1840–1852.
13. Soler-illia, G.J.D., Sanchez, C., Lebeau, B., Patarin, J. (2002) Chemical strategies to design textured materials: From microporous and mesoporous oxides to nanonetworks and hierarchical structures. *Chem. Rev.*, **102** (11), 4093–4138.
14. Taguchi, A., Schuth, F. (2005) Ordered mesoporous materials in catalysis. *Micropor. Mesopor. Mater.*, **77** (1), 1–45.
15. Vallet-Regi, M., Balas, F., Arcos, D. (2007) Mesoporous materials for drug delivery. *Angew. Chem. Int. Ed.*, **46** (40), 7548–7558.
16. Ying, J.Y., Mehnert, C.P., Wong, M.S. (1999) Synthesis and applications of supramolecular-templated mesoporous materials. *Angew. Chem. Int. Ed.*, **38** (1-2), 56–77.
17. White, R.J., Luque, R., Budarin, V.L., Clark, J.H., Macquarrie, D.J. (2009) Supported metal nanoparticles on porous materials. Methods and applications. *Chem. Soc. Rev.*, **38** (2), 481–494.
18. Pal, N., Bhaumik, A. (2015) Mesoporous materials: versatile supports in heterogeneous catalysis for liquid phase catalytic transformations. *RSC Adv.*, **5** (31), 24363–24391.
19. Nhut, J.M., Pesant, L., Tessonnier, J.P., Wine, G., Guille, J., Pham-Huu, C., Ledoux, M.J. (2003) Mesoporous carbon nanotubes for use as support in catalysis and as nanosized reactors for one-dimensional inorganic material synthesis. *Appl. Catal. A*, **254** (2), 345–363.

20. Li, C., Zhang, H.D., Jiang, D.M., Yang, Q.H. (2007) Chiral catalysis in nanopores of mesoporous materials. *Chem. Commun.*, (6), 547–558.

21. Perego, C., Millini, R. (2013) Porous materials in catalysis: challenges for mesoporous materials. *Chem. Soc. Rev.*, **42** (9), 3956–3976.

22. Liu, Y., Peng, J.J., Zhai, S.R., Li, J.Y., Mao, J.J., Li, M.J., Qiu, H.Y., Lai, G.Q. (2006) Synthesis of ionic liquid functionalized SBA-15 mesoporous materials as heterogeneous catalyst toward Knoevenagel condensation under solvent-free conditions. *Eur. J. Inorg. Chem.*, (15), 2947–2949.

23. Zhang, H.D., Xiang, S., Xiao, J.L., Li, C. (2005) Heterogeneous enantioselective epoxidation catalyzed by Mn(salen) complexes grafted onto mesoporous materials by phenoxy group. *J. Mol. Catal. A: Chem.*, **238** (1-2), 175–184.

24. Hultman, H.M., de Lang, M., Arends, I.W.C.E., Hanefeld, U., Sheldon, R.A., Maschmeyer, T. (2003) Chiral catalysts confined in porous hosts: 2. Catalysis. *J. Catal.*, **217** (2), 275–283.

25. Dodson, J.R., Cooper, E.C., Hunt, A.J., Matharu, A., Cole, J., Minihan, A., Clark, J.H., Macquarrie, D.J. (2013) Alkali silicates and structured mesoporous silicas from biomass power station wastes: the emergence of bio-MCMs. *Green Chem.*, **15** (5), 1203–1210.

26. Kim, S.H., Huang, Y.L., Sawatdeenarunat, C., Sung, S.W., Linx, V.S.Y. (2011) Selective sequestration of carboxylic acids from biomass fermentation by surface-functionalized mesoporous silica nanoparticles. *J. Mater. Chem.*, **21** (32), 12103–12109.

27. Miao, S.J., Shanks, B.H. (2009) Esterification of biomass pyrolysis model acids over sulfonic acid-functionalized mesoporous silicas. *Appl. Catal. A*, **359** (1-2), 113–120.

28. Wang, L., Zhang, J., Yi, X., Zheng, A., Deng, F., Chen, C., Ji, Y., Liu, F., Meng, X., Xiao, F.-S. (2015) Mesoporous ZSM-5 zeolite-supported Ru nanoparticles as highly efficient catalysts for upgrading phenolic biomolecules. *ACS Catal.*, **5** (5), 2727–2734.

29. Tanev, P.T., Chibwe, M., Pinnavaia, T.J. (1994) Titanium-containing mesoporous molecular-sieves for catalytic-oxidation of aromatic-compounds. *Nature*, **368** (6469), 321–323.

30. Jiao, F., Frei, H. (2009) Nanostructured cobalt oxide clusters in mesoporous silica as efficient oxygen-evolving catalysts. *Angew. Chem. Int. Ed.*, **48** (10), 1841–1844.

31. Jiao, F., Frei, H. (2010) Nanostructured cobalt and manganese oxide clusters as efficient water oxidation catalysts. *Energy Environ. Sci.*, **3** (8), 1018–1027.

32. Liu, J., Yang, H.Q., Kleitz, F., Chen, Z.G., Yang, T.Y., Strounina, E., Lu, G.Q., Qiao, S.Z. (2012) Yolk-shell hybrid materials with a periodic mesoporous organosilica shell: ideal nanoreactors for selective alcohol oxidation. *Adv. Funct. Mater.*, **22** (3), 591–599.

33. Jia, M.J., Seifert, A., Thiel, W.R. (2003) Mesoporous MCM-41 materials modified with oxodiperoxo molybdenum complexes: Efficient catalysts for the epoxidation of cyclooctene. *Chem. Mater.*, **15** (11), 2174–2180.

34. Corma, A., Martinez, A., Martinez-Soria, V. (1997) Hydrogenation of aromatics in diesel fuels on Pt/MCM-41 catalysts. *J. Catal.*, **169** (2), 480–489.

35. Tsung, C.K., Kuhn, J.N., Huang, W.Y., Aliaga, C., Hung, L.I., Somorjai, G.A., Yang, P.D. (2009) Sub-10 nm platinum nanocrystals with size and shape control: catalytic study for ethylene and pyrrole hydrogenation. *J. Am. Chem. Soc.*, **131** (16), 5816–5822.

36. Chen, Y.Y., Qiu, J.S., Wang, X.K., Xiu, J.H. (2006) Preparation and application of highly dispersed gold nanoparticles supported on silica for catalytic hydrogenation of aromatic nitro compounds. *J. Catal.*, **242** (1), 227–230.

37. Huang, W., Kuhn, J.N., Tsung, C.K., Zhang, Y., Habas, S.E., Yang, P., Somorjai, G.A. (2008) Dendrimer templated synthesis of one nanometer Rh and Pt particles supported on mesoporous silica: Catalytic activity for ethylene and pyrrole hydrogenation. *Nano Lett.*, **8** (7), 2027–2034.

38. Kuhn, J.N., Huang, W.Y., Tsung, C.K., Zhang, Y.W., Somorjai, G.A. (2008) Structure sensitivity of carbon-nitrogen ring opening: impact of platinum particle size from below 1 to 5 nm upon pyrrole hydrogenation product selectivity over monodisperse platinum nanoparticles loaded onto mesoporous silica. *J. Am. Chem. Soc.*, **130** (43), 14026.

39. Liu, P.N., Gu, P.M., Wang, F., Tu, Y.Q. (2004) Efficient heterogeneous asymmetric transfer hydrogenation of ketones using highly recyclable and accessible silica-immobilized Ru-TsDPEN catalysts. *Org. Lett.*, **6** (2), 169–172.
40. Vradman, L., Landau, M.V., Herskowitz, M., Ezersky, V., Talianker, M., Nikitenko, S., Koltypin, Y., Gedanken, A. (2003) High loading of short WS2 slabs inside SBA-15: promotion with nickel and performance in hydrodesulfurization and hydrogenation. *J. Catal.*, **213** (2), 163–175.
41. Alvaro, M., Corma, A., Das, D., Fornes, V., Garcia, H. (2005) 'Nafion'-functionalized mesoporous MCM-41 silica shows high activity and selectivity for carboxylic acid esterification and Friedel–Crafts acylation reactions. *J. Catal.*, **231** (1), 48–55.
42. Diaz, I., Marquez-Alvarez, C., Mohino, F., Perez-Pariente, J., Sastre, E. (2000) Combined alkyl and sulfonic acid functionalization of MCM-41-type silica - Part 2. Esterification of glycerol with fatty acids. *J. Catal.*, **193** (2), 295–302.
43. Mbaraka, I.K., Radu, D.R., Lin, V.S.Y., Shanks, B.H. (2003) Organosulfonic acid-functionalized mesoporous silicas for the esterification of fatty acid. *J. Catal.*, **219** (2), 329–336.
44. Mbaraka, I.K., Shanks, B.H. (2005) Design of multifunctionalized mesoporous silicas for esterification of fatty acids. *J. Catal.*, **229** (2), 365–373.
45. Van Rhijn, W.M., De Vos, D.E., Sels, B.F., Bossaert, W.D., Jacobs, P.A. (1998) Sulfonic acid functionalised ordered mesoporous materials as catalysts for condensation and esterification reactions. *Chem. Commun.*, (3), 317–318.
46. Ghampson, I.T., Sepulveda, C., Garcia, R., Fierro, J.L., Escalona, N., DeSisto, W.J. (2012) Comparison of alumina- and SBA-15-supported molybdenum nitride catalysts for hydrodeoxygenation of guaiacol. *Appl. Catal. A*, **435**, 51–60.
47. Kandel, K., Anderegg, J.W., Nelson, N.C., Chaudhary, U., Slowing, I.I. (2014) Supported iron nanoparticles for the hydrodeoxygenation of microalgal oil to green diesel. *J. Catal.*, **314**, 142–148.
48. Selvaraj, M., Shanthi, K., Maheswari, R., Ramanathan, A. (2014) Hydrodeoxygenation of guaiacol over MoO$_3$-NiO/mesoporous silicates: effect of incorporated heteroatom. *Energy Fuels*, **28** (4), 2598–2607.
49. Wang, L., Li, C., Jin, S.H., Li, W.Z., Liang, C.H. (2014) Hydrodeoxygenation of dibenzofuran over SBA-15 supported Pt, Pd, and Ru catalysts. *Catal. Lett.*, **144** (5), 809–816.
50. Wang, L., Zhang, M.M., Zhang, M., Sha, G.Y., Liang, C.H. (2013) Hydrodeoxygenation of dibenzofuran over mesoporous silica COK-12 supported palladium catalysts. *Energy Fuels*, **27** (4), 2209–2217.
51. Zanuttini, M.S., Costa, B.O.D., Querini, C.A., Peralta, M.A. (2014) Hydrodeoxygenation of m-cresol with Pt supported over mild acid materials. *Appl. Catal. A*, **482**, 352–361.
52. Modena, G., Rivetti, F., Scorrano, G., Tonellato, U. (1977) Reactions in moderately concentrated acids. 2. Solvation effects in acid-catalyzed hydration of olefins and acetylenes. *J. Am. Chem. Soc.*, **99** (10), 3392–3395.
53. Hall, R.H., Stern, E.S. (1950) Acid-catalysed hydration of acraldehyde: Kinetics of the reaction and isolation of beta-hydroxypropaldehyde. *J. Chem. Soc.*, **February**, 490–498.
54. Chiang, Y., Kresge, A.J., Zhu, Y. (2002) Flash photolytic generation of *o*-quinone alpha-phenylmethide and *o*-quinone alpha-(*p*-anisyl)methide in aqueous solution and investigation of their reactions in that medium. Saturation of acid-catalyzed hydration. *J. Am. Chem. Soc.*, **124** (4), 717–722.
55. Tomas-Mendivil, E., Cadierno, V., Menendez, M.I., Lopez, R. (2015) Unmasking the action of phosphinous acid ligands in nitrile hydration reactions catalyzed by arene-ruthenium(II) complexes. *Chem. Eur. J.*, **21** (47), 16874–16886.
56. Schroeder, G.K., Johnson, W.H., Huddleston, J.P., Serrano, H., Johnson, K.A., Whitman, C.P. (2012) Reaction of cis-3-chloroacrylic acid dehalogenase with an allene substrate, 2,3-butadienoate: hydration via an enamine. *J. Am. Chem. Soc.*, **134** (1), 293–304.
57. Kore, R., Srivastava, R. (2012) Influence of -SO$_3$H functionalization (N-SO$_3$H or N-R-SO$_3$H, where R = alkyl/benzyl) on the activity of Brønsted acidic ionic liquids in the hydration reaction. *Tetrahedron Lett.*, **53** (26), 3245–3249.

58. Wijaya, Y.P., Kristianto, I., Lee, H., Jae, J. (2016) Production of renewable toluene from biomass-derived furans via Diels-Alder and dehydration reactions: A comparative study of Lewis acid catalysts. *Fuel*, **182**, 588–596.

59. Sairanen, E., Karinen, R., Lehtonen, J. (2014) Comparison of solid acid-catalyzed and autocatalyzed C5 and C6 sugar dehydration reactions with water as a solvent. *Catal. Lett.*, **144** (11), 1839–1850.

60. Phung, T.K., Busca, G. (2015) Diethyl ether cracking and ethanol dehydration: Acid catalysis and reaction paths. *Chem. Eng. J.*, **272**, 92–101.

61. Mahmoud, E., Yu, J.Y., Gorte, R.J., Lobo, R.F. (2015) Diels-Alder and dehydration reactions of biomass-derived furan and acrylic acid for the synthesis of benzoic acid. *ACS Catal.*, **5** (11), 6946–6955.

62. Wu, M., Zhao, Q.Q., Li, J., Su, X.L., Wu, H.Y., Guan, X.X., Zheng, X.C. (2016) Tungstophosphoric acid-based mesoporous materials anchored to MCM-41: characterization and catalytic performance in esterification of levulinic acid with ethanol. *J. Porous Mater.*, **23** (5), 1329–1338.

63. Wu, M., Zhang, X.L., Su, X.L., Li, X.Y., Zheng, X.C., Guan, X.X., Liu, P. (2016) 3D graphene aerogel anchored tungstophosphoric acid catalysts: Characterization and catalytic performance for levulinic acid esterification with ethanol. *Catal. Commun.*, **85**, 66–69.

64. Mutlu, V.N., Yilmaz, S. (2016) Esterification of cetyl alcohol with palmitic acid over WO3/Zr-SBA-15 and Zr-SBA-15 catalysts. *Appl. Catal. A*, **522**, 194–200.

65. Roman-Leshkov, Y., Moliner, M., Labinger, J.A., Davis, M.E. (2010) Mechanism of glucose isomerization using a solid Lewis acid catalyst in water. *Angew. Chem. Int. Ed.*, **49** (47), 8954–8957.

66. Oivanen, M., Kuusela, S., Lonnberg, H. (1998) Kinetics and mechanisms for the cleavage and isomerization of the phosphodiester bonds of RNA by Brønsted acids and bases. *Chem. Rev.*, **98** (3), 961–990.

67. Morterra, C., Cerrato, G., Pinna, F., Signoretto, M., Strukul, G. (1994) On the acid-catalyzed isomerization of light paraffins over a ZRO_2/SO_4 system – the effect of hydration. *J. Catal.*, **149** (1), 181–188.

68. Malhotra, S.K., Ringold, H.J. (1965) Chemistry of conjugate anions and enols. V. Stereochemistry kinetics and mechanism of acid- and enzymatic-catalyzed isomerization of 5-3-keto steroids. *J. Am. Chem. Soc.*, **87** (14), 3228–3236.

69. Harris, J.W., Cordon, M.J., Di Iorio, J.R., Vega-Vila, J.C., Ribeiro, F.H., Gounder, R. (2016) Titration and quantification of open and closed Lewis acid sites in Sn-Beta zeolites that catalyze glucose isomerization. *J. Catal.*, **335**, 141–154.

70. Escobar, M.A., Trofymchuk, O.S., Rodriguez, B.E., Lopez-Lira, C., Tapia, R., Daniliuc, C., Berke, H., Nachtigall, F.M., Santos, L.S., Rojas, R.S. (2015) Lewis acid-enhanced ethene dimerization and alkene isomerization-ESI-MS identification of the catalytically active pyridyldimethoxybenzimidazole nickel(II) hydride species. *ACS Catal.*, **5** (12), 7338–7342.

71. Henschel, H., Kurten, T., Vehkamaki, H. (2016) Computational study on the effect of hydration on new particle formation in the sulfuric acid/ammonia and sulfuric acid/dimethylamine systems. *J. Phys. Chem. A*, **120** (11), 1886–1896.

72. Sathe, M., Gupta, A.K., Kaushik, M.P. (2006) An efficient method for the esterification of phosphonic and phosphoric acids using silica chloride. *Tetrahedron Lett.*, **47** (18), 3107–3109.

73. Venkatesh, K.R., Hu, J.L., Dogan, C., Tierney, J.W., Wender, I. (1995) Sulfated metal-oxides and related solid acids – comparison of protonic acid strengths. *Energy Fuels*, **9** (5), 888–893.

74. Nishiumi, M., Miura, H., Wada, K., Hosokawa, S., Inoue, M. (2010) Recyclable solid ruthenium catalysts supported on metal oxides for the addition of carboxylic acids to terminal alkynes. *Adv. Synth. Catal.*, **352** (17), 3045–3052.

75. Duffy, J.A. (1997) Acid-base reactions of transition metal oxides in the solid state. *J. Am. Ceram. Soc.*, **80** (6), 1416–1420.

76. Kruger, J.S., Nikolakis, V., Vlachos, D.G. (2014) Aqueous-phase fructose dehydration using Brønsted acid zeolites: Catalytic activity of dissolved aluminosilicate species. *Appl. Catal. A*, **469**, 116–123.

77. Caillot, M., Chaumonnot, A., Digne, M., van Bokhoven, J.A. (2014) The variety of Brønsted acid sites in amorphous aluminosilicates and zeolites. *J. Catal.*, **316**, 47–56.

78. Wang, Y.H., Gan, Y.T., Whiting, R., Lu, G.Z. (2009) Synthesis of sulfated titania supported on mesoporous silica using direct impregnation and its application in esterification of acetic acid and n-butanol. *J. Solid State Chem.*, **182** (9), 2530–2534.

79. Pan, H., Wang, J.X., Chen, L., Su, G.H., Cui, J.M., Meng, D.W., Wu, X.L. (2013) Preparation of sulfated alumina supported on mesoporous MCM-41 silica and its application in esterification. *Catal. Commun.*, **35**, 27–31.

80. Alonso, D.M., Bond, J.Q., Dumesic, J.A. (2010) Catalytic conversion of biomass to biofuels. *Green Chem.*, **12** (9), 1493–1513.

81. Gallezot, P. (2012) Conversion of biomass to selected chemical products. *Chem. Soc. Rev.*, **41** (4), 1538–1558.

82. McKendry, P. (2002) Energy production from biomass (part 2): conversion technologies. *Bioresorce Technol.*, **83** (1), 47–54.

83. Stocker, M. (2008) Biofuels and biomass-to-liquid fuels in the biorefinery: catalytic conversion of lignocellulosic biomass using porous materials. *Angew. Chem. Int. Ed.*, **47** (48), 9200–9211.

84. Besson, M., Gallezot, P., Pinel, C. (2014) Conversion of biomass into chemicals over metal catalysts. *Chem. Rev.*, **114** (3), 1827–1870.

85. Jae, J., Tompsett, G.A., Foster, A.J., Hammond, K.D., Auerbach, S.M., Lobo, R.F., Huber, G.W. (2011) Investigation into the shape selectivity of zeolite catalysts for biomass conversion. *J. Catal.*, **279** (2), 257–268.

86. McKendry, P. (2002) Energy production from biomass (part 1): overview of biomass. *Bioresorce Technol.*, **83** (1), 37–46.

87. Czernik, S., Bridgwater, A.V. (2004) Overview of applications of biomass fast pyrolysis oil. *Energy Fuels*, **18** (2), 590–598.

88. Mohan, D., Pittman, C.U., Steele, P.H. (2006) Pyrolysis of wood/biomass for bio-oil: A critical review. *Energy Fuels*, **20** (3), 848–889.

89. Huber, G.W., Iborra, S., Corma, A. (2006) Synthesis of transportation fuels from biomass: Chemistry, catalysts, and engineering. *Chem. Rev.*, **106** (9), 4044–4098.

90. Corma, A., Iborra, S., Velty, A. (2007) Chemical routes for the transformation of biomass into chemicals. *Chem. Rev.*, **107** (6), 2411–2502.

91. Carlson, T.R., Vispute, T.R., Huber, G.W. (2008) Green gasoline by catalytic fast pyrolysis of solid biomass derived compounds. *ChemSusChem*, **1** (5), 397–400.

92. Bridgwater, A.V., Peacocke, G.V.C. (2000) Fast pyrolysis processes for biomass. *Renewable Sustainable Energy Rev.*, **4** (1), 1–73.

93. Bridgwater, A.V. (2012) Review of fast pyrolysis of biomass and product upgrading. *Biomass Bioenergy*, **38**, 68–94.

94. Mascal, M., Nikitin, E.B. (2010) High-yield conversion of plant biomass into the key value-added feedstocks 5-(hydroxymethyl)furfural, levulinic acid, and levulinic esters via 5-(chloromethyl)-furfural. *Green Chem.*, **12** (3), 370–373.

95. Bozell, J.J., Moens, L., Elliott, D.C., Wang, Y., Neuenscwander, G.G., Fitzpatrick, S.W., Bilski, R.J., Jarnefeld, J.L. (2000) Production of levulinic acid and use as a platform chemical for derived products. *Resour. Conserv. Recycling*, **28** (3-4), 227–239.

96. Rackemann, D.W., Doherty, W.O.S. (2011) The conversion of lignocellulosics to levulinic acid. *Biofuel. Bioprod. Biores.*, **5** (2), 198–214.

97. Li, J.M., Jiang, Z.C., Hu, L.B., Hu, C.W. (2014) Selective conversion of cellulose in corncob residue to levulinic acid in an aluminum trichloride-sodium chloride system. *ChemSusChem*, **7** (9), 2482–2488.

98. Luo, W.H., Bruijnincx, P.C.A., Weckhuysen, B.M. (2014) Selective, one-pot catalytic conversion of levulinic acid to pentanoic acid over Ru/H-ZSM5. *J. Catal.*, **320**, 33–41.

99. Bourne, R.A., Stevens, J.G., Ke, J., Poliakoff, M. (2007) Maximising opportunities in supercritical chemistry: the continuous conversion of levulinic acid to gamma-valerolactone in CO_2. *Chem. Commun.*, (44), 4632–4634.

100. Braden, D.J., Henao, C.A., Heltzel, J., Maravelias, C.T., Dumesic, J.A. (2011) Production of liquid hydrocarbon fuels by catalytic conversion of biomass-derived levulinic acid. *Green Chem.*, **13** (7), 1755–1765.

101. Wright, W.R.H., Palkovits, R. (2012) Development of Heterogeneous Catalysts for the Conversion of Levulinic Acid to gamma-Valerolactone. *ChemSusChem*, **5** (9), 1657–1667.

102. Fernandes, D.R., Rocha, A.S., Mai, E.F., Mota, C.J.A., da Silva, V.T. (2012) Levulinic acid esterification with ethanol to ethyl levulinate production over solid acid catalysts. *Appl. Catal. A*, **425**, 199–204.

103. Nandiwale, K.Y., Sonar, S.K., Niphadkar, P.S., Joshi, P.N., Deshpande, S.S., Patil, V.S., Bokade, V.V. (2013) Catalytic upgrading of renewable levulinic acid to ethyl levulinate biodiesel using dodecatungstophosphoric acid supported on desilicated H-ZSM-5 as catalyst. *Appl. Catal. A*, **460**, 90–98.

104. Galletti, A.M.R., Antonetti, C., De Luise, V., Martinelli, M. (2012) A sustainable process for the production of gamma-valerolactone by hydrogenation of biomass-derived levulinic acid. *Green Chem.*, **14** (3), 688–694.

105. Hengne, A.M., Rode, C.V. (2012) Cu-ZrO_2 nanocomposite catalyst for selective hydrogenation of levulinic acid and its ester to gamma-valerolactone. *Green Chem.*, **14** (4), 1064–1072.

106. Yan, Z.P., Lin, L., Liu, S.J. (2009) Synthesis of gamma-valerolactone by hydrogenation of biomass-derived levulinic acid over Ru/C catalyst. *Energy Fuels*, **23** (8), 3853–3858.

107. Zheng, J.L., Zhu, J.H., Xu, X., Wang, W.M., Li, J.W., Zhao, Y., Tang, K.J., Song, Q., Qi, X.L., Kong, D.J., Tang, Y. (2016) Continuous hydrogenation of ethyl levulinate to gamma-valerolactone and 2-methyl tetrahydrofuran over alumina doped Cu/SiO_2 catalyst: the potential of commercialization. *Sci. Rep.*, **2016**, 6.

108. Pan, T., Deng, J., Xu, Q., Xu, Y., Guo, Q.-X., Fu, Y. (2013) Catalytic conversion of biomass-derived levulinic acid to valerate esters as oxygenated fuels using supported ruthenium catalysts. *Green Chem.*, **15** (10), 2967–2974.

109. Dong, L.L., He, L., Tao, G.H., Hu, C.W. (2013) High yield of ethyl valerate from the esterification of renewable valeric acid catalyzed by amino acid ionic liquids. *RSC Adv.*, **3** (14), 4806–4813.

110. Raghavendra, T., Sayania, D., Madamwar, D. (2010) Synthesis of the 'green apple ester' ethyl valerate in organic solvents by *Candida rugosa* lipase immobilized in MBGs in organic solvents: Effects of immobilization and reaction parameters. *J. Mol. Catal. B: Enzym.*, **63** (1-2), 31–38.

111. Dayma, G., Halter, F., Foucher, F., Togbé, C., Mounaim-Rousselle, C., Dagaut, P. (2012) Experimental and detailed kinetic modeling study of ethyl pentanoate (ethyl valerate) oxidation in a jet stirred reactor and laminar burning velocities in a spherical combustion chamber. *Energy Fuels*, **26** (8), 4735–4748.

112. Luo, W., Deka, U., Beale, A.M., van Eck, E.R.H., Bruijnincx, P.C.A., Weckhuysen, B.M. (2013) Ruthenium-catalyzed hydrogenation of levulinic acid: Influence of the support and solvent on catalyst selectivity and stability. *J. Catal.*, **301**, 175–186.

113. Chan-Thaw, C.E., Marelli, M., Psaro, R., Ravasio, N., Zaccheria, F. (2013) New generation biofuels: [gamma]-valerolactone into valeric esters in one pot. *RSC Adv.*, **3** (5), 1302–1306.

114. Wu, Z., Ge, S., Ren, C., Zhang, M., Yip, A., Xu, C. (2012) Selective conversion of cellulose into bulk chemicals over Brønsted acid-promoted ruthenium catalyst: one-pot vs. sequential process. *Green Chem.*, **14** (12), 3336–3343.

115. Le Van Mao, R., Muntasar, A., Petraccone, D., Yan, H.T. (2012) AC3B technology for direct liquefaction of lignocellulosic biomass: new concepts of coupling and decoupling of catalytic/chemical reactions for obtaining a very high overall performance. *Catal. Lett.*, **142** (6), 667–675.

116. Huang, Y.B., Fu, Y. (2013) Hydrolysis of cellulose to glucose by solid acid catalysts. *Green Chem.*, **15** (5), 1095–1111.

117. Kobayashi, H., Komanoya, T., Hara, K., Fukuoka, A. (2010) Water-tolerant mesoporous-carbon-supported ruthenium catalysts for the hydrolysis of cellulose to glucose. *ChemSusChem*, **3** (4), 440–443.

118. Pang, Q., Wang, L.Q., Yang, H., Jia, L.S., Pan, X.W., Qiu, C.C. (2014) Cellulose-derived carbon bearing -Cl and -SO$_3$H groups as a highly selective catalyst for the hydrolysis of cellulose to glucose. *RSC Adv.*, **4** (78), 41212–41218.

119. Van de Vyver, S., Peng, L., Geboers, J., Schepers, H., de Clippel, F., Gommes, C.J., Goderis, B., Jacobs, P.A., Sels, B.F. (2010) Sulfonated silica/carbon nanocomposites as novel catalysts for hydrolysis of cellulose to glucose. *Green Chem.*, **12** (9), 1560–1563.

120. Zhang, Q.H., Benoit, M., Vigier, K.D., Barrault, J., Jegou, G., Philippe, M., Jerome, F. (2013) Pretreatment of microcrystalline cellulose by ultrasounds: effect of particle size in the heterogeneously-catalyzed hydrolysis of cellulose to glucose. *Green Chem.*, **15** (4), 963–969.

121. Kassaye, S., Pant, K.K., Jain, S. (2016) Synergistic effect of ionic liquid and dilute sulphuric acid in the hydrolysis of microcrystalline cellulose. *Fuel Process. Technol.*, **148**, 289–294.

122. Neto, W.P.F., Putaux, J.L., Mariano, M., Ogawa, Y., Otaguro, H., Pasquini, D., Dufresne, A. (2016) Comprehensive morphological and structural investigation of cellulose I and II nanocrystals prepared by sulphuric acid hydrolysis. *RSC Adv.*, **6** (79), 76017–76027.

123. Niu, M.G., Hou, Y.C., Ren, S.H., Wang, W.H., Zheng, Q.T., Wu, W.Z. (2015) The relationship between oxidation and hydrolysis in the conversion of cellulose in NaVO$_3$-H$_2$SO$_4$ aqueous solution with O-2. *Green Chem.*, **17** (1), 335–342.

124. Kubička, D., Kikhtyanin, O. (2015) Opportunities for zeolites in biomass upgrading – Lessons from the refining and petrochemical industry. *Catal. Today*, **243**, 10–22.

125. Chheda, J.N., Roman-Leshkov, Y., Dumesic, J.A. (2007) Production of 5-hydroxymethylfurfural and furfural by dehydration of biomass-derived mono- and poly-saccharides. *Green Chem.*, **9** (4), 342–350.

126. Dias, A.S., Pillinger, M., Valente, A.A. (2005) Dehydration of xylose into furfural over micro-mesoporous sulfonic acid catalysts. *J. Catal.*, **229** (2), 414–423.

127. Karinen, R., Vilonen, K., Niemela, M. (2011) Biorefining: Heterogeneously catalyzed reactions of carbohydrates for the production of furfural and hydroxymethylfurfural. *ChemSusChem*, **4** (8), 1002–1016.

128. Mamman, A.S., Lee, J.M., Kim, Y.C., Hwang, I.T., Park, N.J., Hwang, Y.K., Chang, J.S., Hwang, J.S. (2008) Furfural: Hemicellulose/xylose-derived biochemical. *Biofuel. Bioprod. Biores.*, **2** (5), 438–454.

129. Palmqvist, E., Grage, H., Meinander, N.Q., Hahn-Hagerdal, B. (1999) Main and interaction effects of acetic acid, furfural, and *p*-hydroxybenzoic acid on growth and ethanol productivity of yeasts. *Biotechnol. Bioeng.*, **63** (1), 46–55.

130. Bohre, A., Dutta, S., Saha, B., Abu-Omar, M.M. (2015) Upgrading furfurals to drop-in biofuels: an overview. *ACS Sustain. Chem. Eng.*, **3** (7), 1263–1277.

131. Sain, B., Chaudhuri, A., Borgohain, J.N., Baruah, B.P., Ghose, J.L. (1982) Furfural and furfural-based industrial-chemicals. *J. Sci. Ind. Res.*, **41** (7), 431–438.

132. Kaiprommarat, S., Kongparakul, S., Reubroycharoen, P., Guan, G., Samart, C. (2016) Highly efficient sulfonic MCM-41 catalyst for furfural production: Furan-based biofuel agent. *Fuel*, **174**, 189–196.

133. Chen, X.R., Ju, Y.H., Mou, C.Y. (2007) Direct synthesis of mesoporous sulfated silica-zirconia catalysts with high catalytic activity for biodiesel via esterification. *J. Phys. Chem. C*, **111** (50), 18731–18737.

134. Shao, G.N., Sheikh, R., Hilonga, A., Lee, J.E., Park, Y.H., Kim, H.T. (2013) Biodiesel production by sulfated mesoporous titania-silica catalysts synthesized by the sol-gel process from less expensive precursors. *Chem. Eng. J.*, **215**, 600–607.

135. Thitsartarn, W., Kawi, S. (2011) Transesterification of oil by sulfated Zr-supported mesoporous silica. *Ind. Eng. Chem. Res.*, **50** (13), 7857–7865.

136. Kozhevnikov, I.V., Matveev, K.I. (1983) Homogeneous catalysts based on heteropoly acids (review). *Appl. Catal.*, **5** (2), 135–150.
137. Timofeeva, M.N. (2003) Acid catalysis by heteropoly acids. *Appl. Catal. A*, **256** (1-2), 19–35.
138. Misono, M. (2001) Unique acid catalysis of heteropoly compounds (heteropolyoxometalates) in the solid state. *Chem. Commun.*, (13), 1141–1152.
139. Kozhevnikov, I.V. (1998) Catalysis by heteropoly acids and multicomponent polyoxometalates in liquid-phase reactions. *Chem. Rev.*, **98** (1), 171–198.
140. Mestl, G., Ilkenhans, T., Spielbauer, D., Dieterle, M., Timpe, O., Krohnert, J., Jentoft, F., Knozinger, H., Schlogl, R. (2001) Thermally and chemically induced structural transformations of Keggin-type heteropoly acid catalysts. *Appl. Catal. A*, **210** (1-2), 13–34.
141. Su, F., Ma, L., Guo, Y.H., Li, W. (2012) Preparation of ethane-bridged organosilica group and Keggin-type heteropoly acid co-functionalized ZrO_2 hybrid catalyst for biodiesel synthesis from *Eruca sativa* gars oil. *Catal. Sci. Technol.*, **2** (11), 2367–2374.
142. Zhou, L.L., Wang, L., Zhang, S.J., Yan, R.Y., Diao, Y.Y. (2015) Effect of vanadyl species in Keggin-type heteropoly catalysts in selective oxidation of methacrolein to methacrylic acid. *J. Catal.*, **329**, 431–440.
143. Su, F., Wu, Q.Y., Song, D.Y., Zhang, X.H., Wang, M., Guo, Y.H. (2013) Pore morphology-controlled preparation of ZrO_2-based hybrid catalysts functionalized by both organosilica moieties and Keggin-type heteropoly acid for the synthesis of levulinate esters. *J. Mater. Chem. A*, **1** (42), 13209–13221.
144. Okuhara, T. (2002) Water-tolerant solid acid catalysts. *Chem. Rev.*, **102** (10), 3641–3666.
145. Yue, B., Zhou, Y., Xu, J., Wu, Z., Zhang, X., Zou, Y., Jin, S. (2002) Photocatalytic degradation of aqueous 4-chlorophenol by silica-immobilized polyoxometalates. *Environ. Sci. Technol.*, **36** (6), 1325–1329.
146. Guo, Y., Li, K., Yu, X., Clark, J.H. (2008) Mesoporous $H_3PW_{12}O_{40}$-silica composite: Efficient and reusable solid acid catalyst for the synthesis of diphenolic acid from levulinic acid. *Appl. Catal. B*, **81** (3–4), 182–191.
147. Xu, L., Li, W., Hu, J., Yang, X., Guo, Y. (2009) Biodiesel production from soybean oil catalyzed by multifunctionalized Ta_2O_5/SiO_2-[$H_3PW_{12}O_{40}$/R] (R = Me or Ph) hybrid catalyst. *Appl. Catal. B*, **90** (3–4), 587–594.
148. Wang, L., Xiao, F.S. (2015) Nanoporous catalysts for biomass conversion. *Green Chem.*, **17** (1), 24–39.
149. Arata, K. (2009) Organic syntheses catalyzed by superacidic metal oxides: sulfated zirconia and related compounds. *Green Chem.*, **11** (11), 1719–1728.
150. Brown, A.S.C., Hargreaves, J.S.J. (1999) Sulfated metal oxide catalysts – Superactivity through superacidity? *Green Chem.*, **1** (1), 17–20.
151. Farcasiu, D., Ghenciu, A., Li, J.Q. (1996) The mechanism of conversion of saturated hydrocarbons catalyzed by sulfated metal oxides: Reaction of adamantane on sulfated zirconia. *J. Catal.*, **158** (1), 116–127.
152. Ghenciu, A., Farcasiu, D. (1996) The mechanism of conversion of hydrocarbons on sulfated metal oxides .2. Reaction of benzene on sulfated zirconia. *J. Mol. Catal. A: Chem.*, **109** (3), 273–283.
153. Lu, Q., Xiong, W.M., Li, W.Z., Guo, Q.X., Zhu, X.F. (2009) Catalytic pyrolysis of cellulose with sulfated metal oxides: A promising method for obtaining high yield of light furan compounds. *Biores. Technol.*, **100** (20), 4871–4876.
154. Matsuhashi, H., Tanaka, T., Arata, K. (2001) Measurement of heat of argon adsorption for the evaluation of relative acid strength of some sulfated metal oxides and H-type zeolites. *J. Phys. Chem. B*, **105** (40), 9669–9671.
155. Ohtsuka, H. (2001) The selective catalytic reduction of nitrogen oxides by methane on noble metal-loaded sulfated zirconia. *Appl. Catal. B. Environ.*, **33** (4), 325–333.

156. Pradhan, V.R., Tierney, J.W., Wender, I., Huffman, G.P. (1991) Catalysis in direct coal-liquefaction by sulfated metal-oxides. *Energy Fuels*, **5** (3), 497–507.

157. Jiménez-Morales, I., Santamaría-González, J., Maireles-Torres, P., Jiménez-López, A. (2011) Calcined zirconium sulfate supported on MCM-41 silica as acid catalyst for ethanolysis of sunflower oil. *Appl. Catal. B*, **103** (1–2), 91–98.

158. Eliche-Quesada, D., Merida-Robles, J.M., Rodriguez-Castellon, E., Jimenez-Lopez, A. (2005) Ru, Os and Ru-Os supported on mesoporous silica doped with zirconium as mild thio-tolerant catalysts in the hydrogenation and hydrogenolysis/hydrocracking of tetralin. *Appl. Catal. A*, **279** (1-2), 209–221.

159. Lyu, L., Zhang, L.L., Hu, C. (2015) Enhanced Fenton-like degradation of pharmaceuticals over framework copper species in copper-doped mesoporous silica microspheres. *Chem. Eng. J.*, **274**, 298–306.

160. Carrion, M.C., Manzano, B.R., Jalon, F.A., Eliche-Quesada, D., Maireles-Torres, P., Rodriguez-Castellon, E., Jimenez-Lopez, A. (2005) Influence of the metallic precursor in the hydrogenation of tetralin over Pd-Pt supported zirconium doped mesoporous silica. *Green Chem.*, **7** (11), 793–799.

161. Davis, J.J., Huang, W.Y., Davies, G.L. (2012) Location-tuned relaxivity in Gd-doped mesoporous silica nanoparticles. *J. Mater. Chem.*, **22** (43), 22848–22850.

162. Barata-Rodrigues, P.M., Mays, T.J., Moggridge, G.D. (2003) Structured carbon adsorbents from clay, zeolite and mesoporous aluminosilicate templates. *Carbon*, **41** (12), 2231–2246.

163. Camblor, M.A., Corma, A., Valencia, S. (1998) Synthesis in fluoride media and characterisation of aluminosilicate zeolite beta. *J. Mater. Chem.*, **8** (9), 2137–2145.

164. Liu, Y., Pinnavaia, T.J. (2002) Assembly of hydrothermally stable aluminosilicate foams and large-pore hexagonal mesostructures from zeolite seeds under strongly acidic conditions. *Chem. Mater.*, **14** (1), 3.

165. Liu, Y., Zhang, W.Z., Pinnavaia, T.J. (2000) Steam-stable aluminosilicate mesostructures assembled from zeolite type Y seeds. *J. Am. Chem. Soc.*, **122** (36), 8791–8792.

166. Liu, Y., Zhang, W.Z., Pinnavaia, T.J. (2001) Steam-stable MSU-S aluminosilicate mesostructures assembled from zeolite ZSM-5 and zeolite beta seeds. *Angew. Chem. Int. Ed.*, **40** (7), 1255.

167. Sakthivel, A., Huang, S.J., Chen, W.H., Lan, Z.H., Chen, K.H., Kim, T.W., Ryoo, R., Chiang, A.S.T., Liu, S.B. (2004) Replication of mesoporous aluminosilicate molecular sieves (RMMs) with zeolite framework from mesoporous carbons (CMKs). *Chem. Mater.*, **16** (16), 3168–3175.

168. Garcia-Perez, M., Wang, X.S., Shen, J., Rhodes, M.J., Tian, F.J., Lee, W.J., Wu, H.W., Li, C.Z. (2008) Fast pyrolysis of oil mallee woody biomass: Effect of temperature on the yield and quality of pyrolysis products. *Ind. Eng. Chem. Res.*, **47** (6), 1846–1854.

169. Wang, D., Czernik, S., Montane, D., Mann, M., Chornet, E. (1997) Biomass to hydrogen via fast pyrolysis and catalytic steam reforming of the pyrolysis oil or its fractions. *Ind. Eng. Chem. Res.*, **36** (5), 1507–1518.

170. Adam, J., Blazso, M., Meszaros, E., Stocker, M., Nilsen, M.H., Bouzga, A., Hustad, J.E., Gronli, M., Oye, G. (2005) Pyrolysis of biomass in the presence of Al-MCM-41 type catalysts. *Fuel*, **84** (12-13), 1494–1502.

171. Antonakou, E., Lappas, A., Nilsen, M.H., Bouzga, A., Stocker, M. (2006) Evaluation of various types of Al-MCM-41 materials as catalysts in biomass pyrolysis for the production of bio-fuels and chemicals. *Fuel*, **85** (14-15), 2202–2212.

172. Coriolano, A.C.F., Oliveira, A.A.A., Bandeira, R.A.F., Fernandes, V.J., Araujo, A.S. (2015) Kinetic study of thermal and catalytic pyrolysis of Brazilian heavy crude oil over mesoporous Al-MCM-41 materials. *J. Therm. Anal. Calorim.*, **119** (3), 2151–2157.

173. Liu, W.W., Hu, C.W., Yang, Y., Tong, D.M., Zhu, L.F., Zhang, R.N., Zhao, B.H. (2013) Study on the effect of metal types in (Me)-Al-MCM-41 on the mesoporous structure and catalytic behavior during the vapor-catalyzed co-pyrolysis of pubescens and LDPE. *Appl. Catal. B. Environ.*, **129**, 202–213.

174. Antonakou, E., Lappas, A., Nilsen, M.H., Bouzga, A., Stöcker, M. (2006) Evaluation of various types of Al-MCM-41 materials as catalysts in biomass pyrolysis for the production of bio-fuels and chemicals. *Fuel*, **85** (14-15), 2202–2212.

175. Jimenez-Morales, I., Santamaria-Gonzalez, J., Maireles-Torres, P., Jimenez-Lopez, A. (2011) Aluminum doped SBA-15 silica as acid catalyst for the methanolysis of sunflower oil. *Appl. Catal. B. Environ.*, **105** (1-2), 199–205.

176. Zhang, S.L., Zhang, Y.H., Huang, S.P., Liu, H., Tian, H.P. (2010) First-principles study of structural, electronic and vibrational properties of aluminum-doped silica nanotubes. *Chem. Phys. Lett.*, **498** (1-3), 172–177.

177. Wang, L., Zhang, J., Wang, X.F., Zhang, B.S., Ji, W.J., Meng, X.J., Li, J.X., Su, D.S., Bao, X.H., Xiao, F.S. (2014) Creation of Brønsted acid sites on Sn-based solid catalysts for the conversion of biomass. *J. Mater. Chem. A*, **2** (11), 3725–3729.

178. Nicholas, C.P., Ahn, H., Marks, T.J. (2003) Synthesis, spectroscopy, and catalytic properties of cationic organozirconium adsorbates on "super acidic" sulfated alumina. "single-site" heterogeneous catalysts with virtually 100 active sites. *J. Am. Chem. Soc.*, **125** (14), 4325–4331.

179. Richardson, Y., Motuzas, J., Julbe, A., Volle, G., Blin, J. (2013) Catalytic investigation of in situ-generated Ni metal nanoparticles for tar conversion during biomass pyrolysis. *J. Phys. Chem. C*, **117** (45), 23812–23831.

180. Wang, D.R., Ma, B., Wang, B., Zhao, C., Wu, P. (2015) One-pot synthesized hierarchical zeolite supported metal nanoparticles for highly efficient biomass conversion. *Chem. Commun.*, **51** (82), 15102–15105.

181. Crooks, R.M., Zhao, M.Q., Sun, L., Chechik, V., Yeung, L.K. (2001) Dendrimer-encapsulated metal nanoparticles: Synthesis, characterization, and applications to catalysis. *Acc. Chem. Res.*, **34** (3), 181–190.

182. Murphy, C.J., San, T.K., Gole, A.M., Orendorff, C.J., Gao, J.X., Gou, L., Hunyadi, S.E., Li, T. (2005) Anisotropic metal nanoparticles: Synthesis, assembly, and optical applications. *J. Phys. Chem. B*, **109** (29), 13857–13870.

183. Murray, R.W. (2008) Nanoelectrochemistry: Metal nanoparticles, nanoelectrodes, and nanopores. *Chem. Rev.*, **108** (7), 2688–2720.

184. Fujikawa, T., Idei, K., Ohki, K., Mizuguchi, H., Usui, K. (2001) Kinetic behavior of hydrogenation of aromatics in diesel fuel over silica-alumina-supported bimetallic Pt-Pd catalyst. *Appl. Catal. A*, **205** (1-2), 71–77.

185. Nakagawa, Y., Tomishige, K. (2010) Total hydrogenation of furan derivatives over silica-supported Ni-Pd alloy catalyst. *Catal. Commun.*, **12** (3), 154–156.

186. Lu, Q., Tang, Z., Zhang, Y., Zhu, X.-f. (2010) Catalytic upgrading of biomass fast pyrolysis vapors with Pd/SBA-15 catalysts. *Ind. Eng. Chem. Res.*, **49** (6), 2573–2580.

187. Branca, C., Giudicianni, P., Di Blasi, C. (2003) GC/MS characterization of liquids generated from low-temperature pyrolysis of wood. *Ind. Eng. Chem. Res.*, **42** (14), 3190–3202.

188. Chiaramonti, D., Oasmaa, A., Solantausta, Y. (2007) Power generation using fast pyrolysis liquids from biomass. *Renew. Sust. Energ. Rev.*, **11** (6), 1056–1086.

189. Laresgoiti, M.F., Caballero, B.M., de Marco, I., Torres, A., Cabrero, M.A., Chomon, M.J. (2004) Characterization of the liquid products obtained in tyre pyrolysis. *J. Anal. Appl. Pyrolysis*, **71** (2), 917–934.

190. Yu, W., Tang, Y., Mo, L., Chen, P., Lou, H., Zheng, X. (2011) Bifunctional Pd/Al-SBA-15 catalyzed one-step hydrogenation–esterification of furfural and acetic acid: A model reaction for catalytic upgrading of bio-oil. *Catal. Commun.*, **13** (1), 35–39.

191. Wen, C., Barrow, E., Hattrick-Simpers, J., Lauterbach, J. (2014) One-step production of long-chain hydrocarbons from waste-biomass-derived chemicals using bi-functional heterogeneous catalysts. *PhysChemChemPhys*, **16** (7), 3047–3054.

192. Corma, A., de la Torre, O., Renz, M., Villandier, N. (2011) Production of high-quality diesel from biomass waste products. *Angew. Chem. Int. Ed.*, **50** (10), 2375–2378.

193. Iojoiu, E.E., Domine, M.E., Davidian, T., Guilhaume, N., Mirodatos, C. (2007) Hydrogen production by sequential cracking of biomass-derived pyrolysis oil over noble metal catalysts supported on ceria-zirconia. *Appl. Catal. A*, **323**, 147–161.

194. Tomishige, K., Miyazawa, T., Asadullah, M., Ito, S., Kunimori, K. (2003) Catalyst performance in reforming of tar derived from biomass over noble metal catalysts. *Green Chem.*, **5** (4), 399–403.

195. Zhu, G.L., Li, P.H., Zhao, F., Song, H.L., Xia, C.G. (2016) Selective aromatization of biomass derived diisobutylene to *p*-xylene over supported non-noble metal catalysts. *Catal. Today*, **276**, 105–111.

196. Yan, K., Liao, J.Y., Wu, X., Xie, X.M. (2013) A noble-metal free Cu-catalyst derived from hydrotalcite for highly efficient hydrogenation of biomass-derived furfural and levulinic acid. *RSC Adv.*, **3** (12), 3853–3856.

197. Kang, S.C., Jun, K.W., Lee, Y.J. (2013) Effects of the CO/CO_2 ratio in synthesis gas on the catalytic behavior in Fischer-Tropsch synthesis using K/Fe-Cu-Al catalysts. *Energy Fuels*, **27** (11), 6377–6387.

198. Mirzaei, A.A., Sarani, R., Azizi, H.R., Vahid, S., Torshizi, H.O. (2015) Kinetics modeling of Fischer-Tropsch synthesis on the unsupported Fe-Co-Ni (ternary) catalyst prepared using co-precipitation procedure. *Fuel*, **140**, 701–710.

199. Riedel, T., Claeys, M., Schulz, H., Schaub, G., Nam, S.S., Jun, K.W., Choi, M.J., Kishan, G., Lee, K.W. (1999) Comparative study of Fischer-Tropsch synthesis with H_2/CO and H_2/CO_2 syngas using Fe- and Co-based catalysts. *Appl. Catal. A*, **186** (1-2), 201–213.

200. He, J., Zhao, C., Lercher, J.A. (2012) Ni-catalyzed cleavage of aryl ethers in the aqueous phase. *J. Am. Chem. Soc.*, **134** (51), 20768–20775.

201. Wang, X., Rinaldi, R. (2016) Bifunctional Ni catalysts for the one-pot conversion of Organosolv lignin into cycloalkanes. *Catal. Today*, **269**, 48–55.

6

Porous Polydivinylbenzene-Based Solid Catalysts for Biomass Transformation Reactions

Fujian Liu[1] and Yao Lin[2]

[1]*Department of Chemistry, Shaoxing University, China*
[2]*Department of Chemistry and Institute of Materials Science, University of Connecticut, USA*

6.1 Introduction

The use of fossil fuels has substantially negative impacts on the environment, including air and water pollution, global warming, and wildlife and habitat loss [1–9]. Developing renewable and sustainable energies and technologies is imperative to the global economy and a world that will see its population continue to grow for many decades [1–18]. Among various green and renewable energy sources, biofuels – fuels made from renewable biomass – have the advantages of global abundance, a small carbon footprint, and relatively low cost. Increasing the share of modern biofuels in energy use (i.e., bioenergy) will not only reduce greenhouse gas emissions significantly in the long term, but also stimulate local agricultural progress immediately [1–18]. The transformation of biomass into biofuels relies on the catalytic reactions induced by acids, bases, and a variety of chemical or biological catalysts. Examples of current biomass transformation processes include the transesterification of vegetable oil for biodiesel production, the fermentation of corn sugar for bioethanol production, the depolymerization of crystalline cellulose into sugars, and the transformation of sugars into various fine chemicals [8,9,19]. Compared with base catalysts and enzymes, acid catalysts such as H_2SO_4 and HCl are low-cost, more versatile for different reaction conditions, and capable of transformation of even low-quality

Nanoporous Catalysts for Biomass Conversion, First Edition. Edited by Feng-Shou Xiao and Liang Wang.
© 2018 John Wiley & Sons Ltd. Published 2018 by John Wiley & Sons Ltd.

feedstock into biofuels [8,20–22]. Unfortunately, mineral acids have drawbacks such as toxicity to the environment, corrosion of the reactors, and high costs of separating and recycling, all of which strongly constrain their industrial applications [23–28].

Compared with mineral acids, solid-acid catalysts have shown promising characteristics for biomass transformation, including reduced corrosivity, environmental friendliness, easy regeneration, and good catalytic selectivity [23–27]. Much effort has been made on replacing mineral acids with solid acids to catalyze biomass transformation for green and sustainable biofuel production [8–12,16–20,29–31]. Traditional solid acids such as heteropolyacids, sulfated metal oxides, phosphates and acidic resins [28] have shown good activity and recyclability for catalyzing the transformation reactions of biomass. However, their limited Brunauer–Emmett–Teller (BET) surface area results in poor mass transfer and an inadequate accessibility of catalytically active sites to the reactants in the reaction, which reduces their specific activities [23,24,32]. In order to improve mass transfer and expose more acidic sites for the catalytic reaction, porous structures have been introduced into solid acids. Porous solid acids such as H-form zeolites have large surface areas, uniform and intricate micropores, strong acidity, and excellent stabilities [6,24,33], which shows good catalytic activity in various reactions. However, relatively small pores in zeolites (e.g., 0.3~1.0 nm in ZSM-5, Beta, Y zeolites) severely limit their catalytic performances in breaking down the biomass that contains long polymer chains, or molecules with large molecular weights [24,34–37]. The recent development of mesoporous forms of silica, such as MCM-41 and SBA-15 (e.g., with pore sizes ranging from 2.0 nm to 50 nm) has offered a new way to overcome the limitation in microporous solid acids based on zeolites [38,39]. Various acid sites such as sulfonic groups, Al and Sn (Lewis acids), and heteropolyacids can be incorporated into the hexagonally ordered mesoporous walls of silica [20,26,40–44], and used as catalytic sites for biomass transformation reactions [8,20]. However, the hydrophilic nature of silica usually results in a partial deactivation of the acidic sites and hydrolysis of the framework, as water is the byproduct in many acid-catalyzed reactions. This considerably limits the catalytic performance of mesoporous silica [45]. In addition, the surface wettability of solid acids has also played a very important role in their catalytic activity [32,46,47]. The inorganic framework of silica usually has a poor wettability with organic reactants, which leads to a reduced mass transfer of reactants to the catalytic sites, and may also affect the activation energy of the reaction.

Compared with porous solid acids with inorganic frameworks, porous polymeric solid acids can be rationally designed and functionalized to achieve optimized chemical stability and surface wettability for targeted molecules, for example, by taking advantage of the unique π–π interactions between organic reactants and polymeric networks [46,48,49]. Well-known polymeric solid acids such as Amberlyst 15 and Nafion NR 50 are macroporous acid resins, which show strong and tunable acidity, and are capable of catalyzing a variety of reactions [25,32,50]. The further development of mesoporous polymeric solid acids offers new opportunities for the preparation of porous solid acids with even larger BET surface areas for grafting with acidic groups [25,51–53]. One of the most successful examples is mesoporous phenolic resin, which has ordered mesopores (e.g., packed in a hexagonal and cubic lattice with diameters of between 2 and 50 nm). The resultant mesoporous resin has been successfully grafted with sulfonic groups and has shown excellent activity in the condensation of phenol with acetone [25,51,52].

In general, homogeneous catalysts usually show better catalytic activities than heterogeneous catalysts, because of their higher degree of exposure of catalytically active sites to the reactants. The procedures required to recycle homogeneous catalysts are usually complicated, and therefore much effort has been expended to identify ways of improving the specific catalytic activities of heterogeneous catalysts used in the biomass transformation reactions. It is obvious that increasing the exposure degree of active sites to the reactants, and improving the desorption rates of the products, should be instructive for enhancing the activities of heterogeneous catalysts. In particular, optimized porous structures and surface wettability may increase the exposure degree of active sites in solid catalysts, which then facilitates the rapid diffusion of reactants and the desorption of products. Recently, the engineering of porous structures and surface wettability of solid catalysts have been extensively exploited in different biomass transformation reactions. For example, Xiao and coworkers successfully developed a type of polydivinylbenzene (PDVB)-based porous polymer under unique solvothermal conditions, without the need of using organic templates. The polymers possess large BET surface areas, good stability, and controllable hydrophobicity and wettability [53]. Subsequent functionalization of the porous polymers with sulfonic group and acid ionic liquids groups allowed for the preparation of polymeric solid acids with superior catalytic activity and excellent stability [32,54–59].

The aim of this chapter is to highlight some of the recent developments in PDVB-based solid acids and their catalytic applications, particularly in the transformation of biomass into biofuels. A brief summary of porous polymeric solid acids with controllable acidity and wettability, and their potential applications in converting biomass into fuels, completes the chapter.

6.2 Synthesis of Porous PDVB-Based Solid Acids and Investigation of their Catalytic Performances

6.2.1 Sulfonic Group-Functionalized Porous PDVB

Sulfonic group-functionalized porous PDVB-based solid acids (PDVB-SO_3H) were synthesized via two routes: (i) sulfonation of superhydrophobic porous PDVB by using HSO_3Cl; or (ii) copolymerization of DVB with sodium *p*-styrene sulfonate under solvothermal conditions. The porous PDVB support was synthesized from the polymerization of DVB monomers under template-free solvothermal conditions (Figure 6.1). The resultant porous PDVB possessed abundant nanopores with sizes ranging from 3.0 nm to 25 nm, superhydrophobic surfaces, and very good wettability for various organic solvents, which facilitate their superior swelling property (Figure 6.1). The good swelling property leads to easy sulfonation of the polymer network, which allows for the preparation of PDVB-SO_3H with controllable wettability, large BET surface areas, and very high concentrations of acidic sites (Figure 6.2).

Compared with the sulfonation approach, PDVB-SO_3H synthesized from the copolymerization of DVB with sodium *p*-styrene sulfonate had relatively lower acid

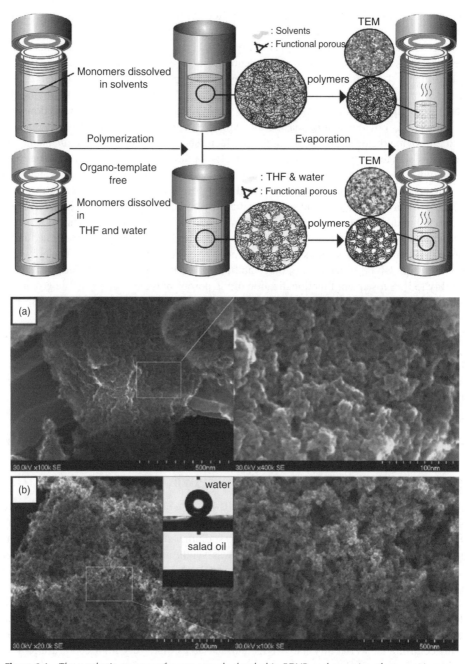

Figure 6.1   *The synthetic process of super superhydrophobic PDVB and scanning electron microscopy images of the resultant PDVB sample.*

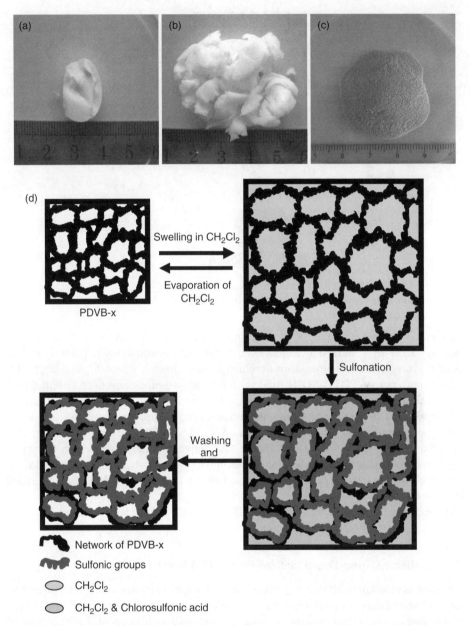

Figure 6.2 *The sulfonation process of PDVB by using HSO₃Cl in CH₂Cl₂ solvent.* (See color plate section for the color representation of this figure.)

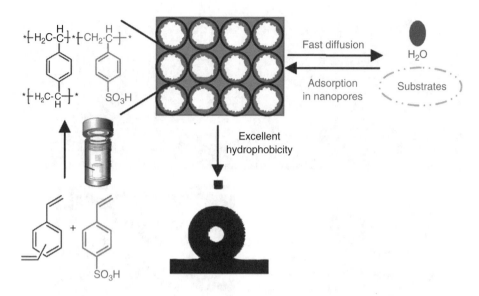

Figure 6.3   *The synthetic procedures of superhydrophobic PDVB-based solid acids.*

concentrations; however, the synthetic route allowed for the preparation of samples with very large BET surface areas and even better surface wettability (Figure 6.3). As a result of synergistic effects from their excellent hydrophobicity, wettability and large BET surface areas, porous PDVB-SO$_3$H solid acids have shown much improved activities than the conventional solid acids (e.g., acid resins, H-form zeolites, sulfonated metal oxides, and mesoporous silicas) for catalyzing the biodiesel production from sunflower oil, waste cooking oil and glycerol tripalmitate (Figures 6.4 and 6.5) [59]. Xiao and coworkers also found that PDVB-SO$_3$H-type solid acids are very active for catalyzing the transformation reactions of other abundant biomass feedstocks into fine chemicals, such as the dehydration of sorbitol into isosorbide, and the dehydration of fructose into 5-hydroxymethylfurfural (HMF) [60,61].

The above results suggest that PDVB-SO$_3$H-type solid acids are very promising catalysts for applications in effective biomass transformations.

6.2.2   Sulfonic Group-Functionalized Porous PDVB-SO$_3$H-SO$_2$CF$_3$

In general, acid strength plays a very important role in the enhancement of catalytic performance of solid acids. To investigate on how modulating the acid strengths in PDVB-type solid acids may affect their catalytic activities in biomass transformations, Liu *et al.* grafted strong electron-withdrawing groups (e.g., –SO$_2$CF$_3$) onto the network of PDVB-SO$_3$H, and found largely enhanced BET surface areas and acid strength in these modified solid acids [55]. Figure 6.6 shows the synthesis of these PDVB-SO$_3$H-SO$_2$CF$_3$ solid acids, which was based on the treatment of porous PDVB-SO$_3$H with super acid of HSO$_3$CF$_3$. The improved acid strength in PDVB-SO$_3$H-SO$_2$CF$_3$ further enhanced the activity of the solid acids, making them very useful in biofuel applications such as biodiesel production, and the depolymerization of crystalline cellulose into sugars (Table 6.1).

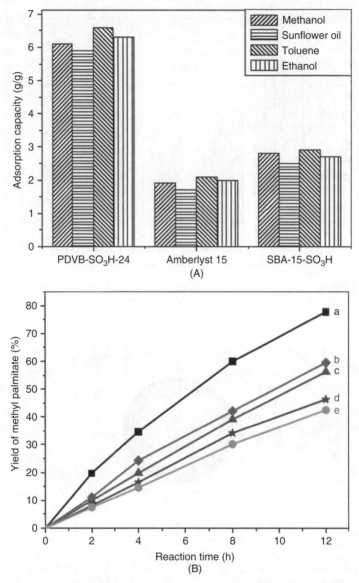

Figure 6.4 *(A) Absorption capacities for methanol, toluene, ethanol, and sunflower oil over PDVB-SO₃H-24, Amberlyst 15 and SBA-15-SO₃H. (B) Catalytic kinetics curves in the transesterification of tripalmitin with methanol over (a) PDVB-SO₃H-24, (b) H₃PO₄₀W₁₂, (c) SBA-15-SO₃H, (d) Amberlyst 15, and (e) S-ZrO₂.*

6.2.3   PDVB-Based Porous Solid Bases for Biomass Transformation

Basic catalysts also show excellent catalytic activities in biomass transformation reactions, including the transesterification of glycerol tripalmitate into biodiesels, and the isomerization of glucose into fructose [61,62]. However, the catalytic activities of solid bases are

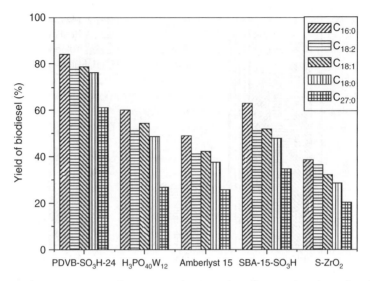

Figure 6.5 *Catalytic activities in the transesterification of sunflower oil with methanol over (a) PDVB-SO₃H-24, (b) H₃PO₄₀W₁₂, (c) SBA-15-SO₃H, (d) Amberlyst 15, and (e) S-ZrO₂.*

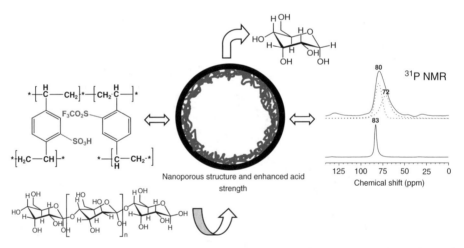

Figure 6.6 *The enhanced acid strength of PDVB-SO₃H-SO₂CF₃ and its applications for catalyzing the depolymerization of crystalline cellulose.*

Table 6.1 *Yields of sugars and dehydration products in the depolymerization of crystalline cellulose catalyzed by various solid acids.*

Samples	Glucose yield (%)[a]	Cellobiose yield (%)[a]	TRS (%)[b]	TTM biodiesel yields (%)[c]
Amberlyst 15	24.5	10.8	50.3	59.8
PDVB-SO₃H	34.0	11.2	60.7	73.8
PDVB-SO₃H-SO₂CF₃	66.2	9.1	86.7	91.4

[a]Monitored by HPLC method.
[b]Total Reducing Sugar (TRS) was monitored using the DNS assay.
[c]TTM: Transesterifications of tripalmitin with methanol.

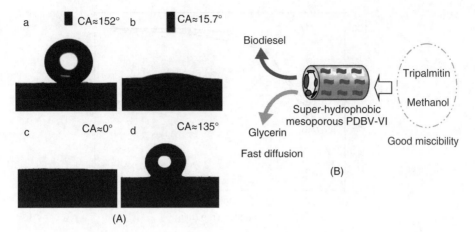

Figure 6.7 *(A) Contact angle of (a) a water droplet, (b) a salad oil droplet, (c) a methanol droplet, and (d) a glycerol droplet on a PDVB-type solid base. (B) Scheme for the application of PDVB solid base for catalyzing biodiesel production.*

very sensitive to water and acidic molecules (e.g., CO_2) in the environment [63]. Therefore, the preparation of solid bases with good stability in normal environmental conditions is imperative to utilizing solid-base catalysts in biomass transformations. Recently, stable PDVB-based solid bases have been developed by Xiao and coworkers, where the porous solid bases were synthesized via the copolymerization of DVB with basic monomers such as 1-vinylimidazole (PDVB-VI) and 4-vinylpyridine (PDVB-Py) under solvothermal conditions without templates (Figure 6.7). The resultant solid bases contained abundant nanopores, had excellent base strength, and showed superhydrophobicity and superwettability for various organic reactants such as methanol and plant oils [63]. The superhydrophobicity found in PDVB-VI and PDVB-Py facilitates their chemical stability against water and CO_2 in the environment, while at the same time their superwettability allowed for their miscibility with organic molecules in different reaction systems. When combined, these desired characteristics make the PDVB-based porous solid bases an excellent candidate to be used as cost-effective, reusable catalysts in biodiesel production processes such as the transformation of plant oil into biodiesel toward transesterification.

For example, the tripalmitin could be effectively transformed into biodiesel within 1 h towards transesterification in the presence of PDVB-VI solid base, which was much better than those of hydrotalcite, CaO, commercial basic resin. Its catalytic activity was even comparable with that of homogeneous KOH (Figure 6.8; Table 6.2).

6.2.4   Strong Acid Ionic Liquid-Functionalized PDVB-Based Catalysts

Moe recently, acid ionic liquids have received much attention in the area of acid catalysis because of their solubility for a wide range of reactants, and strong acid strength. Hence, the use of acid ionic liquids as both solvents and catalysts to achieve effective biomass transformation has also been explored [30,64–66]. Unfortunately, most ionic liquids are very expensive and achieving their cost-effective recycling on an industrial scale remains

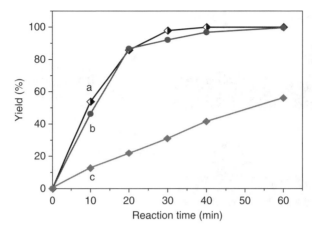

Figure 6.8   *Dependences of catalytic activities on time in transesterification of tripalmitin with methanol over (a) PDVB solid base, (b) KOH, and (c) CaO catalysts. Reaction condition: 0.05 g catalyst; tripalmitin 0.84 g; methanol 3.76 ml; 65 °C.*

a challenge. To overcome these issues, acid ionic liquids have been immobilized on solid, porous material supports, and used as heterogeneous catalysts instead of homogenous catalysts [67]. For applications of these catalysts in biomass transformation, it is clear that their specific surface areas and surface wettability will again play important roles in their catalytic activities.

Xiao and coworkers reported superhydrophobic and porous PDVB-based solid acids that are functionalized by acid ionic liquids. The solid acids were synthesized from the *in-situ* treatment of imidazole or pyridine on a PDVB support with quaternary ammonization reagents such as 1,3-propanesultone and iodomethane, followed by reactions with H_2SO_4 or HSO_3CF_3 (Figure 6.9). The resultant solid acids showed even higher activity than that of homogeneous acid ionic liquids in catalyzing the biodiesel productions (Table 6.3). Presumably, the activity enhancement was due to the superwettability and the abundant nanopores of the materials, which facilitate the effective contacts between the reactants and the catalytically active sites in the solid acids [32]. These studies shed light on the rational design and synthesis of heterogeneous acid ionic liquids catalysts with both high activity and good re-usability. Following these pioneering investigations, Wang and coworkers demonstrated the immobilization of heteropolyacids on porous PDVB supports, for catalyzing the Friedel–Crafts reaction and for biomass transformation (Figure 6.10) [68]. Liu and coworkers successfully grafted both sulfuric acid and ionic liquid groups onto PDVB networks (PDVB-SO_3H-[C_3vim][SO_3CF_3]) with enhanced acid strength and surface polarity. Figure 6.11 shows the synthesis of PDVB-SO_3H-[C_3vim][SO_3CF_3] by the solvothermal copolymerization of divinylbezene (DVB) with 1-vinylimidazolate (vim) and sodium *p*-styrenesulfonate, followed by treatment with 1,3-propanesultone to form quaternary ammonium salts, and ion-exchanging with strong acids such as HSO_3CF_3 and sulfuric acid.

The immobilized ionic liquid groups were capable of destroying the crystalline cellulose, and accelerated the rate of acid-catalyzed depolymerization of cellulose into sugars, much

Table 6.2 The textural parameters and catalytic data in transesterification of tripalmitin with methanol over various solid base catalysts.[a]

Entry	Sample	Alkaline content [mmol g^{-1}]	S_{BET}[b] [m^2 g^{-1}]	V_p[b] [cm^3 g^{-1}]	D_p[c] [nm]	Alcohol	Alkyl palmitate yield [%] $t = 1$ h	$t = 3$ h
1	PDVB-VI-0.2	0.953[d]	670	1.13	24.2	methanol	99.5	99.7
2	PDVB-VI-0.33	2.063[d]	594	1.241	38.1	methanol	99.3	99.9
3	PDVB-VI-0.5	2.822[d]	513	0.815	19.2	methanol	99.7	99.6
4	M-PDVB-VI-0.5[e]	2.822[d]	486	0.915	26.3	methanol	99.8	99.2
5	T-PDVB-VI-0.12[f]	0.953[d]	576	0.454	6.83	methanol	–	–
6	PDVB-VI-0.5-Q	2.063[d]	680	1.0	29.1	methanol	63.1	89.3
7	PDVB-VI-0.5[g]	2.822[d]	521	0.813	24.6	methanol	99.2	99.7
8	PDVB-VI-0.5[h]	2.822[d]	510	0.811	26.8	methanol	98.9	99.3
9	PDVB-VI-0.5	2.822[d]	513	0.815	19.2	ethanol[i]	30.1	64.3
10	PDVB-VI-0.5	2.822[d]	513	0.815	19.2	1-butanol	24.6	55.1
11	Amberlite 400	5.18	0.28	–	–	methanol	57.5	79.8
12	Hydrotalcite [Mg$_6$Al$_2$(OH)$_{16}$CO$_3$]	30.2	103	0.85	33	methanol	61.6	80.7
13	PVI	10.625	–	–	–	methanol	59.3	85.5
14	VI	10.625	–	–	–	methanol	67.2	97.3
15	CaO	17.86	–	–	–	methanol	55.8	82.4
16	KOH	17.86	–	–	–	methanol	99.6	99.9
17	KOH[k]	17.86	–	–	–	methanol	94.6	98.1
18	NaOH	25.0	–	–	–	methanol	38.1	85.2
19	DVB	0	–	–	–	methanol	0	0

[a]For each run, 0.05 g catalyst, 0.84 g tripalmitin, and 3.76 mL alcohol were used.
[b]BET surface area and pore volume estimated from N$_2$ adsorption results.
[c]Pore size distribution estimated from BJH model.
[d]Calculated from the imidazole content of the starting composition.
[e]Mesoporous PDVB-PVI synthesized from using methyl acetate as solvent.
[f]Mesoporous PDVB-PVI synthesized from using THF as solvent.
[g]Catalyst recycled three times.
[h]Catalyst recycled five times.
[i]$T = 78°C$.
[j]$T = 115°C$.
[k]Catalyst exposed to air for 7 days.

higher than those of commercial Amberlyst 15 and homogeneous HCl [57]. In addition, the resultant porous polymers could be used as efficient solid acids for catalyzing the depolymerization of *Gracilaria*, a type of waste seaweed that contains abundant cellulose with a high degree of crystallinity. The activities of the polymers were even higher than those of homogeneous acidic ionic liquid and HCl. The excellent catalytic activity found in PDVB-SO$_3$H-[C$_3$vim][SO$_3$CF$_3$] was attributed to the synergistic effects between the strongly acidic group and the grafted ionic liquid on PDVB support, which could effectively break down the crystalline structures of cellulose and decrease the reaction activation energy (Table 6.4). In addition, Zhang *et al.* found that PDVB-SO$_3$H was very active for catalyzing the transformation of sorbitol into isosorbide. This is an important

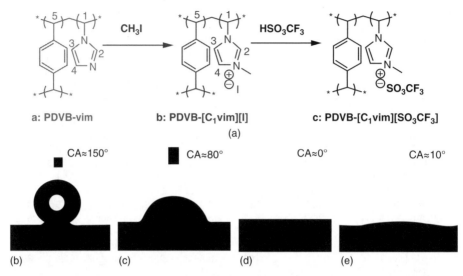

Figure 6.9   *(A) Scheme for the synthesis of PDVB-[C₁vim][SO₃CF₃] from PDVB-vim. (B, C) Contact angle for a water droplet on the surface of (B) PDVB-vim and (C) PDVB-[C₁vim][SO₃CF₃]. (D, E) Contact angle of a droplet of (D) methanol and (E) tripalmitin on the surface of PDVB-[C₁vim][SO₃CF₃]. Figure reproduced with permission from Ref. [32]. Copyright 2012, American Chemical Society.*

Table 6.3   *Activities in transesterification of tripalmitin with methanol over various catalysts.[a]*

Run	Catalyst	Yield of methyl palmitate (%)
1	PDVB-[C₁vim][SO₃CF₃]	96.9
2	PDVB-[C₃vim][SO₃CF₃]	>99.9
3	PDVB-[C₄vim][SO₃CF₃]	97.4
4	PDVB-[C₃vpy][SO₃CF₃]	99.5
5	PDVB-[C₄vpy][SO₃CF₃]	92.9
6	PDVB-[C₃vpr][SO₃CF₃]	99.3
7[b]	[C₁vim][SO₃CF₃]	89.1
8	SBA-15-[C₁vim][SO₃CF₃]	24.1
9	Amberlyst 15	28.6
10[c]	PDVB-[C₁vim][SO₃CF₃]	93.5
11[c]	PDVB-[C₃vim][SO₃CF₃]	98.1
12[c]	Amberlyst 15	20.2

[a]0.05 g catalyst, 1.04 mmol tripalmitin; 92.7 mmol methanol; reaction temperature 65 °C; reaction time 16 h.
[b]The same number of catalytic sites as those in PDVB-[C₁vim][SO₃CF₃].
[c]Fifth run to test the re-usability of the catalyst.

model reaction of biomass dehydration, and PDVB-SO₃H showed much improved catalytic activity compared with commercial Amberlyst 15 and SBA-15-SO₃H (Figure 6.12), due mainly to its large BET surface area, controllable hydrophobic network, and surface wettability. These studies may help in the industrial development of cost-effective and green routes for low-cost biomass transformation into biofuels and fine chemicals.

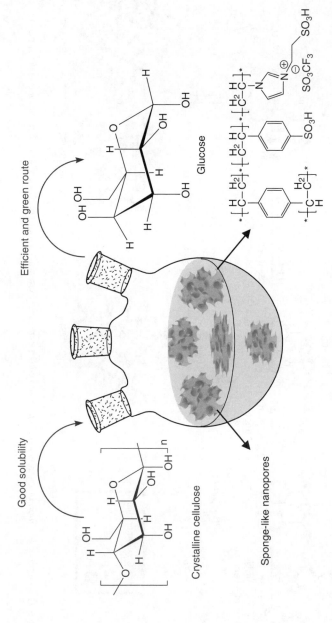

Figure 6.10 *The scheme for the synthesis of the mesoporous ionic copolymers and diagrams of their structures.*

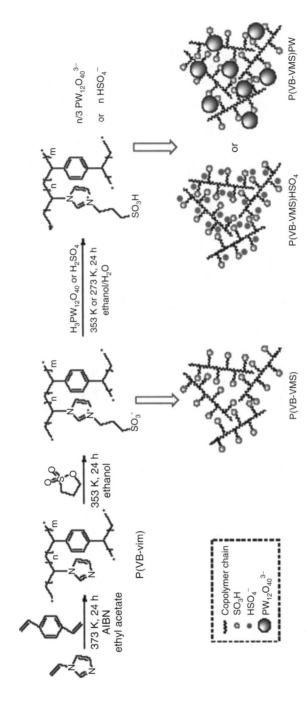

Figure 6.11 Depolymerization of crystalline cellulose into sugars catalyzed by sulfonic group- and acidic ionic liquid-functionalized nanoporous PDVB.

Table 6.4 Yield of sugars and dehydration products in the depolymerization of Avicel catalyzed by various solid acids and mineral acids.

Catalysts	Glucose yield (%)[a]	Cellobiose yield (%)[a]	TRS (%)[b]
Amberlyst 15	25.1	14.8	56.2
HCl	63.4	13.4	94.1
$[C_3vim][SO_3CF_3]^c$	66.2	12.2	93.8
H_2SO_4	59.6	9.8	86.8
$PDVB\text{-}SO_3H$	56.8	10.2	82.6
$PDVB\text{-}SO_3H\text{-}[C_3vim][SO_3CF_3]$	77.0	8.2	99.6
$PDVB\text{-}[C_3vim][SO_3CF_3]$	75.9	6.8	98.1
$PDVB\text{-}SO_3H\text{-}[C_3vim][SO_4H]$	76.8	5.6	98.5
$PDVB\text{-}SO_3H\text{-}[C_3vim][Cl]$	74.1	6.4	96.3

[a]Measured by high-performance liquid chromatography (HPLC) method; the reaction time was 5 h.
[b]Total reducing sugar (TRS) was measured by the DNS method.
[c]The same number of acidic sites as in $PDVB\text{-}SO_3H\text{-}[C_3vim][SO_3CF_3]$.

6.2.5 Cooperative Effects in Applying both PDVB-Based Solid Acids and Solid Bases for Biomass Transformation

The inclusion of both PDVB-based solid acids and bases to catalyze biomass transformed reactions has been recently investigated. For example, the synergistic actions of the solid-acid and -base catalysts were exploited for the transformation of crystalline cellulose into valuable fine chemicals such as 5-hydroxymethylfurfural (HMF) [61]. In this process, the solid acids first catalyzed the depolymerization reaction of cellulose into glucose, which was subsequently transformed into fructose by solid bases. Fructose can then be further transformed into HMF, levulinic acid and formic acid by acid catalysts [6,69]. Xiao and colleagues demonstrated a one-pot process to directly transform glucose into HMF via an integrated catalytic system that included both PDVB-based solid acids and solid bases. The integrated system showed superior catalytic activity and selectivity for HMF (Table 6.5; Figure 6.13). The excellent activities and selectivity found in this system was attributed to the cooperative effects of superhydrophobic solid acids and superhydrophilic solid bases. In the isomerization of glucose, PDVB solid bases showed a good wettability for glucose, and the fructose product could also be quickly transformed into HMF, catalyzed by PDVB solid acids. Moreover, the HMF product in this reaction was quickly dispersed into the organic phase, but could not be transformed into other products by water in the organic phase. More importantly, the solid acids and bases could be easily separated from each other, based on their quite different water wettability after the catalytic reactions [61]. These studies led to the development of novel nanoporous, PDVB-based acid–base catalysts with cooperative actions on the molecules, and may open the way to a cost-effective method of preparing catalysts with a high selectivity for the conversion of biomass into targeted chemicals in the future.

In summary, a series of nanoporous PDVB-based solid catalysts with controllable surface characteristics and catalytically active sites has been synthesized during recent years, via a simple, template-free, solvothermal route. These catalysts are low-cost, re-usable, and show superior activities in catalyzing biomass conversions into fuels. It is envisioned that these types of polymeric solid catalysts and their derivatives will find broad applications in bioenergy-related fields.

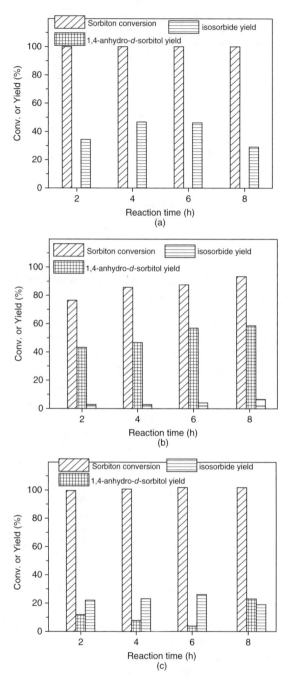

Figure 6.12   *Dehydration of sorbitol over (a and b) P-SO₃H and (c and d) Amberlyst-15 catalysts. Reaction conditions: 100 mg sorbitol; molar ratio of sorbitol to acid sites at 7.2 (a and c) in 5 ml THF or (b and d) in a mixture of 4 ml of THF and 1 ml water; 2 MPa N_2; 175 °C.*

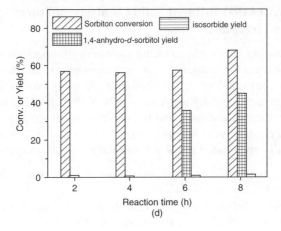

Figure 6.12 *(Continued)*

Table 6.5 *The catalytic yields of HMF from the conversion of glucose over various catalysts.*[a]

	Catalysts		
Entry	Base	Acid	HMF yield (%)
1	P-VI-0	—	Undetectable
2	—	P-SO$_3$H-154	Undetectable
3	P-VI-0	P-SO$_3$H-154	95.4
4	P-VI-108	P-SO$_3$H-154	94.0
5	P-VI-150	P-SO$_3$H-154	76.2
6	1-vinylimidazole	P-SO$_3$H-154	Undetectable
7	P-VI-0	H$_2$SO$_4$	Undetectable
8	P-VI-0	Amberlyst-15	58.1
9[b]	P-VI-0	P-SO$_3$H-154	92.8

[a]Reaction conditions: 100 mg of glucose, 50 mg of solid acid catalyst, 100 mg of solid base catalyst, 5 g of THF-DMSO mixed solvent (weight ratio at 1.5), 100 °C for 10 h. The major by-products were levoglucosan, LA, and FA;
[b]10 mg of solid acid, 20 mg of solid base in 26 h.

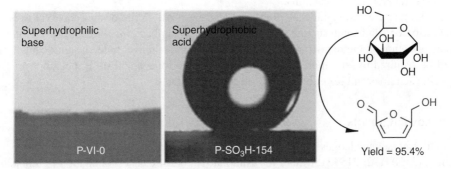

Figure 6.13 *Transformation of glucose into HMF catalyzed by a superhydrophilic PDVB-based solid base and a superhydrophobic PDVB-based solid acid.*

6.3   Perspectives of PDVB-Based Solid Catalysts and their Application for Biomass Transformations

The previous discussions have centered on the findings that nanoporous PDVB-based solid catalysts are highly active for catalyzing biomass transformation, and this ability is strongly related to their characteristics that include good hydrophobicity, controllable wettability for various reactants, and catalytically active sites [47]. Nonetheless, the low concentrations of active sites, low thermal stabilities and re-usability remain as challenges in the use of PDVB-based solid catalysts in industrial biomass transformation. It is recommended that future research in this area should investigate and solve the following issues:

- The relatively high cost of monomers and the complicated procedures involved in the synthesis of PDVB-based solid catalysts.
- That PDVB-based solid catalysts show a relatively low thermal stability due to their polymeric networks, which cannot withstand temperatures above 150 °C for extended periods.
- The limited re-usability of PDVB-based solid catalysts that results from the deactivation of catalytic sites and the destruction of porous structures.
- The still low abundance of active sites relative to homogeneous catalysts.

In order to overcome these 'roadblocks', new developments in the synthesis and mechanistic understanding of the reaction processes involved with PVDB-based solid catalysts will be needed, including:

- The design of functional monomers with high contents of double bonds, which will result in solid catalysts with a very high degree of crosslinking, to enhance the catalysts' thermal stabilities.
- A one-step polymerization of functional monomers without the use of an additional DVB crosslinker, which could considerably increase the densities of active sites in the solid catalysts.
- The introduction of more active sites *in situ* via a one-step polymerization method to simplify the complicated procedures involved in the current synthesis route.

These approaches may ultimately lead to new developments of highly efficient and re-usable PDVB-based solid catalysts and the industrial application of these catalysts in biomass transformation reactions.

In summary, while significant progress has been made in the design and synthesis of PDVB-based solid catalysts, the identification of new applications for catalyzing biomass transformation on the industrial scale remains a major challenge. Indeed, significant effort will be required to seek more cost-effective synthetic routes, and to further improve the specific activities and re-usability of these solid catalysts.

Acknowledgments

Fujian Liu acknowledges support from the National Natural Science Foundation of China (21203122, 21573150) and the National Natural Science Foundation of Zhejiang Province (LY15B030002), while Yao Lin acknowledges support from US DOE BES (DE-SC0005039) and a Senior Visiting Scholarship from Shaoxing University.

References

1. Luque, R., Lovett, J.C., Datta, B., Clancy, J., Campelo, J.M., Romero, A.A. (2010) Biodiesel as feasible petrol fuel replacement: a multidisciplinary overview. *Energy Environ. Sci.*, **3**, 1706–1721.
2. Aslani, A., Wong, K.F.V. (2014) Analysis of renewable energy development to power generation in the United States. *Renew. Energy*, **63**, 153–161.
3. Noshadi, I., Kanjilal, B., Du, S.C., Bollas, G.M., Suib, S.L., Provatas, A., Liu, F.J., Parnas, R.S. (2014) Catalyzed production of biodiesel and bio-chemicals from brown grease using Ionic Liquid functionalized ordered mesoporous polymer. *Appl. Energy*, **129**, 112–122.
4. Corma, A., Iborra, S., Velty, A. (2007) Chemical routes for the transformation of biomass into chemicals. *Chem. Rev.*, **107**, 2411–2502.
5. Román-Leshkov, Y., Barrett, C.J., Liu, Z.Y., Dumesic, J. (2007) Production of dimethylfuran for liquid fuels from biomass-derived carbohydrates. *Nature*, **447**, 982–985.
6. Nikolla, E., Román-Leshkov, Y., Moliner, M., Davis, M.E. (2011) 'One-pot' synthesis of 5-(hydroxymethyl) furfural from carbohydrates using tin-beta zeolite. *ACS Catal.*, **1**, 408–410.
7. Rinaldi, R., Schüth, F. (2009) Acid hydrolysis of cellulose as the entry point into biorefinery schemes. *ChemSusChem*, **2**, 1096–1107.
8. Su, F., Guo, Y.H. (2014) Advancements in solid acid catalysts for biodiesel production. *Green Chem.*, **16**, 2934–2957.
9. Wang, L., Xiao, F.-S. (2015) Nanoporous catalysts for biomass conversion. *Green Chem.*, **17**, 24–39.
10. Huber, G.W., Iborra, S., Corma, A. (2006) Synthesis of transportation fuels from biomass: chemistry, catalysts, and engineering. *Chem. Rev.*, **106**, 4044–4098.
11. Corma, A. (2014) Introduction: chemicals from coal, alkynes, and biofuels. *Chem. Rev.*, **114**, 1545–1546.
12. Xu, L.L., Wang, Y.H., Yang, X., Yu, X.D., Guo, Y.-H., Clark, J.H. (2008) Preparation of mesoporous polyoxometalate-tantalum pentoxide composite catalyst and its application for biodiesel production by esterification and transesterification. *Green Chem.*, **10**, 746–755.
13. Tessonnier, J.P., Villa, A., Majoulet, O., Su, D.S., Schlöl, R. (2009) Defect-mediated functionalization of carbon nanotubes as a route to design single-site basic heterogeneous catalysts for biomass conversion. *Angew. Chem. Int. Ed.*, **48**, 6543–6546.
14. de Lasa, H., Salaices, E., Mazumder, J., Lucky, R. (2011) Catalytic steam gasification of biomass: catalysts, thermodynamics and kinetics. *Chem. Rev.*, **111**, 5404–5433.
15. Jarvis, M. (2003) Chemistry: cellulose stacks up. *Nature*, **426**, 611–612.
16. Binder, J.B., Raines, R.T. (2010) Fermentable sugars by chemical hydrolysis of biomass. *Proc. Natl Acad. Sci. USA*, **107**, 4516–4521.
17. Zhang, H., Holladay, J.E., Brown, H., Zhang, Z.C. (2007) Metal chlorides in ionic liquid solvents convert sugars to 5-hydroxymethylfurfural. *Science*, **316**, 1597–1600.
18. Hu, S., Zhang, Z., Song, J., Zhou, Y., Han, B.X. (2009) Efficient conversion of glucose into 5-hydroxymethylfurfural catalyzed by a common Lewis acid $SnCl_4$ in an ionic liquid. *Green Chem.*, **11**, 1746–1749.
19. Huber, G.W., Chheda, J.N., Barrett, C.J., Dumesic, J.A. (2005) Production of liquid alkanes by aqueous-phase processing of biomass-derived carbohydrates. *Science*, **308**, 1446–1450.
20. Li, W., Jiang, Z.J., Ma, F.Y., Su, F., Chen, L., Zhang, S.Q., Guo, Y.H. (2010) Design of mesoporous SO_4^-/ZrO_2-SiO_2 (Et) hybrid material as an efficient and reusable heterogeneous acid catalyst for biodiesel production. *Green Chem.*, **12**, 2135–2138.
21. Rinaldi, R., Schüth, F. (2009) Design of solid catalysts for the conversion of biomass. *Energy Environ. Sci.*, **2**, 610–626.
22. Li, C.Z., Zhao, X.C., Wang, A.Q., Huber, G.W., Zhang, T. (2015) Catalytic transformation of lignin for the production of chemicals and fuels. *Chem. Rev.*, **115**, 11559–11624.

23. Corma, A. (1997) From microporous to mesoporous molecular sieve materials and their use in catalysis. *Chem. Rev.*, **97**, 2373–2420.
24. Davis, M.E. (2002) Ordered porous materials for emerging applications. *Nature*, **417**, 813–821.
25. Xing, R., Liu, N., Liu, Y.M., Wu, H.H., Jiang, Y.W., Chen, L., He, M.Y., Wu, P. (2007) Novel solid acid catalysts: sulfonic acid group-functionalized mesostructured polymers. *Adv. Funct. Mater.*, **17**, 2455–2461.
26. De Vos, D.E., Dams, M., Sels, B.F., Jacobs, P.A. (2002) Ordered mesoporous and microporous molecular sieves functionalized with transition metal complexes as catalysts for selective organic transformations. *Chem. Rev.*, **102**, 3615–3640.
27. Zhang, X.M., Zhao, Y.P., Xu, S.T., Yang, Y., Liu, J., Wei, Y.X., Yang, Q.H. (2014) Polystyrene sulphonic acid resins with enhanced acid strength via macromolecular self-assembly within confined nanospace. *Nat. Comm.*, **5**, 3170–3178.
28. Corma, A. (1995) Inorganic solid acids and their use in acid-catalyzed hydrocarbon reactions. *Chem. Rev.*, **95**, 559–614.
29. Liu, F.J., Kong, W.P., Wang, L., Yi, X.F., Noshadi, I., Zheng, A.M., Qi, C.Z. (2015) Efficient biomass transformations catalyzed by graphene-like nanoporous carbons functionalized with strong acid ionic liquids and sulfonic groups. *Green Chem.*, **17**, 480–489.
30. Rinaldi, R., Palkovits, R., Schüth, F. (2008) Depolymerization of cellulose using solid catalysts in ionic liquids. *Angew. Chem. Int. Ed.*, **47**, 8047–8050.
31. Liu, F.J., Zuo, S.F., Kong, W.P., Qi, C.Z. (2012) High-temperature synthesis of strong acidic ionic liquids functionalized, ordered and stable mesoporous polymers with excellent catalytic activities. *Green Chem.*, **14**, 1342–1349.
32. Liu, F.J., Wang, L., Sun, Q., Zhu, L.F., Meng, X.J., Xiao, F.-S. (2012) Transesterification catalyzed by ionic liquids on superhydrophobic mesoporous polymers: heterogeneous catalysts that are faster than homogeneous catalysts. *J. Am. Chem. Soc.*, **134**, 16948–16950.
33. Corma, A., Nemeth, L.T., Renz, M., Valencia, S. (2001) Sn-zeolite beta as a heterogeneous chemoselective catalyst for Baeyer–Villiger oxidations. *Nature*, **412**, 423–425.
34. Hartmann, M. (2004) Hierarchical zeolites: A proven strategy to combine shape selectivity with efficient mass transport. *Angew. Chem. Int. Ed.*, **43**, 5880–5882.
35. Meng, X.J., Nawaz, F., Xiao, F.-S. (2009) Templating route for synthesizing mesoporous zeolites with improved catalytic properties. *Nano Today*, **4**, 292–301.
36. Tao, Y., Kanoh, H., Abrams, L., Kaneko, K. (2006) Mesopore-modified zeolites: preparation, characterization, and applications. *Chem. Rev.*, **106**, 896–910.
37. Choi, M., Cho, H.S., Srivastava, R., Venkatesan, C., Choi, D.-H., Ryoo, R. (2006) Amphiphilic organosilane-directed synthesis of crystalline zeolite with tunable mesoporosity. *Nat. Mater.*, **5**, 718–723.
38. Kresge, C.T., Leonowicz, M.E., Roth, W.J., Vartuli, J.C., Beck, J.S. (1992) Ordered mesoporous molecular sieves synthesized by a liquid-crystal template mechanism. *Nature*, **359**, 710–712.
39. Zhao, D.Y., Feng, J.L., Huo, Q.S., Melosh, N., Fredrickson, G.H., Chmelka, B.F., Stucky, G.D. (1998) Triblock copolymer syntheses of mesoporous silica with periodic 50 to 300 angstrom pores. *Science*, **279**, 548–552.
40. Margolese, D., Melero, J.A., Christiansen, S.C., Chmelka, B.F., Stucky, G.D. (2000) Direct syntheses of ordered SBA-15 mesoporous silica containing sulfonic acid groups. *Chem. Mater.*, **12**, 2448–2459.
41. Melero, J.A., Stucky, G.D., Griekena, R.V., Morales, G. (2002) Direct syntheses of ordered SBA-15 mesoporous materials containing arenesulfonic acid groups. *J. Mater. Chem.*, **12**, 1664–1670.
42. Yang, Q.H., Liu, J., Yang, J., Kapoor, M.P., Inagaki, S., Li, C. (2004) Synthesis, characterization, and catalytic activity of sulfonic acid-functionalized periodic mesoporous organosilicas. *J. Catal.*, **228**, 265–272.
43. Wu, S., Han, Y., Zou, Y.C., Song, J.W., Zhao, L., Di, Y., Liu, S.Z., Xiao, F.-S. (2004) Synthesis of heteroatom substituted SBA-15 by the 'pH-adjusting' method. *Chem. Mater.*, **16**, 486–492.

44. Yang, X.Y., Vantomme, A., Lemaire, A., Xiao, F.-S., Su, B.-L. (2006) A highly ordered mesoporous aluminosilicate, CMI-10, with a Si/Al ratio of one. *Adv. Mater.*, **18**, 2117–2122.

45. Morales, G., Athens, G., Chmelka, B.F., van Grieken, R., Melero, J.A. (2008) Aqueous-sensitive reaction sites in sulfonic acid-functionalized mesoporous silicas. *J. Catal.*, **254**, 205–217.

46. Liu, F.J., Feng, G.F., Lin, M.Y., Wang, C., Hu, B.W., Qi, C.Z. (2014) Superoleophilic nanoporous polymeric ionic liquids loaded with palladium acetate: Reactants enrichment and efficient heterogeneous catalysts for Suzuki–Miyaura coupling reaction. *J. Coll. Inter. Sci.*, **435**, 83–90.

47. Wang, L., Xiao, F.-S. (2014) The importance of catalyst wettability. *ChemCatChem*, **6**, 3048–3052.

48. Zhao, Y.J., Domoto, Y., Orentas, E., Beuchat, C., Emery, D., Mareda, J., Sakai, N., Matile, S. (2013) Catalysis with anion-pi interactions. *Angew. Chem. Int. Ed.*, **52**, 9940–9943.

49. Wheeler, S.E., McNeil, A.J., Müller, P., Swager, T.M., Houk, K.N. (2010) Probing substituent effects in aryl–aryl interactions using stereoselective Diels–Alder cycloadditions. *J. Am. Chem. Soc.*, **132**, 3304–3311.

50. Barbaro, P., Liguori, F. (2009) Ion exchange resins: catalyst recovery and recycle. *Chem. Rev.*, **109**, 515–529.

51. Meng, Y., Gu, D., Zhang, F.Q., Shi, Y.F., Yang, H.F., Li, Z., Yu, C.Z., Tu, B., Zhao, D.Y. (2005) Ordered mesoporous polymers and homologous carbon frameworks: Amphiphilic surfactant templating and direct transformation. *Angew. Chem. Int. Ed.*, **44**, 7053–7059.

52. Zhang, F.Q., Meng, Y., Gu, D., Yan, Y., Yu, C.Z., Tu, B., Zhao, D.Y. (2005) A facile aqueous route to synthesize highly ordered mesoporous polymers and carbon frameworks with Ia3d bicontinuous cubic structure. *J. Am. Chem. Soc.*, **127**, 13508–13509.

53. Zhang, Y.L., Wei, S., Liu, F.J., Du, Y.C., Liu, S., Ji, Y.Y., Yokoi, T., Tatsumi, T., Xiao, F.-S. (2009) Superhydrophobic nanoporous polymers as efficient adsorbents for organic compounds. *Nano Today*, **4**, 135–142.

54. Liu, F.J., Meng, X.J., Zhang, Y.L., Ren, L.M., Nawaz, F., Xiao, F.-S. (2010) Efficient and stable solid acid catalysts synthesized from sulfonation of swelling mesoporous polydivinylbenzenes. *J. Catal.*, **271**, 52–58.

55. Liu, F.J., Zheng, A.M., Noshadi, I., Xiao, F.-S. (2013) Design and synthesis of hydrophobic and stable mesoporous polymeric solid acid with ultra strong acid strength and excellent catalytic activities for biomass transformation. *Appl. Catal. B. Environ.*, **136-137**, 193–201.

56. Liu, F.J., Kong, W.P., Qi, C.Z., Zhu, L.F., Xiao, F.-S. (2012) Design and synthesis of mesoporous polymer-based solid acid catalysts with excellent hydrophobicity and extraordinary catalytic activity. *ACS Catal.*, **2**, 565–572.

57. Liu, F.J., Kamat, R.K., Noshadi, I., Peck, D., Parnas, R.S., Zheng, A., Qi, C.Z., Lin, Y. (2013) Depolymerization of crystalline cellulose catalyzed by acidic ionic liquids grafted onto sponge-like nanoporous polymers. *Chem. Commun.*, **49**, 8456–8458.

58. Noshadi, I., Kumar, R.K., Kanjilal, B., Parnas, R., Liu, H., Li, J.T., Liu, F.J. (2013) Transesterification catalyzed by superhydrophobic–oleophilic mesoporous polymeric solid acids: an efficient route for production of biodiesel. *Catal. Lett.*, **143**, 792–797.

59. Xia, P., Liu, F.J., Wang, C., Zuo, S.F., Qi, C.Z. (2012) Efficient mesoporous polymer-based solid acid with superior catalytic activities towards transesterification to biodiesel. *Catal. Commun.*, **26**, 140–143.

60. Zhang, J., Wang, L., Liu, F.J., Meng, X.J., Mao, J.X., Xiao, F.-S. (2015) Enhanced catalytic performance in dehydration of sorbitol to isosorbide over a superhydrophobic mesoporous acid catalyst. *Catal. Today*, **242**, 249–254.

61. Wang, L., Wang, H., Liu, F.J., Zheng, A.M., Zhang, J., Sun, Q., Lewis, J.P., Zhu, L.F., Meng, X.J., Xiao, F.-S. (2014) Selective catalytic production of 5-hydroxymethylfurfural from glucose by adjusting catalyst wettability. *ChemSusChem*, **7**, 402–406.

62. Liu, F.J., Li, W., Sun, Q., Zhu, L.F., Meng, X.J., Guo, Y.H., Xiao, F.-S. (2011) Transesterification to biodiesel with superhydrophobic porous solid base catalysts. *ChemSusChem*, **4**, 1059–1062.

63. Hattori, H. (1995) Heterogeneous basic catalysis. *Chem. Rev.*, **95**, 537–558.

64. Liang, X.Z., Yang, J.G. (2010) Synthesis of a novel multi-SO$_3$H functionalized ionic liquid and its catalytic activities for biodiesel synthesis. *Green Chem.*, **12**, 201–204.
65. Li, C.Z., Zhao, Z.B. (2007) Efficient acid-catalyzed hydrolysis of cellulose in ionic liquid. *Adv. Synth. Catal.*, **349**, 1847–1850.
66. Cai, H.L., Li, C.Z., Wang, A.Q., Xu, G.L., Zhang, T. (2012) Zeolite-promoted hydrolysis of cellulose in ionic liquid, insight into the mutual behavior of zeolite, cellulose and ionic liquid. *Appl. Catal. B. Environ.*, **123-124**, 333–338.
67. Mehnert, C.P., Mozeleski, E.J., Cook, R.A. (2002) Supported ionic liquid catalysis investigated for hydrogenation reactions. *Chem. Commun.*, **38**, 3010–3011.
68. Li, J., Zhou, Y., Mao, D., Chen, G.J., Wang, X.C., Yang, X.N., Wang, M., Peng, L.M., Wang, J. (2014) Heteropolyanion-based ionic liquid-functionalized mesoporous copolymer catalyst for Friedel–Crafts benzylation of arenes with benzyl alcohol. *Chem. Eng. J.*, **254**, 54–62.
69. Pagán-Torres, Y.J., Wang, T.F., Gallo, J.M.R., Shanks, B.H., Dumesic, J.A. (2012) Production of 5-hydroxymethylfurfural from glucose using a combination of Lewis and Brønsted acid catalysts in water in a biphasic reactor with an alkylphenol solvent. *ACS Catal.*, **2**, 930–934.

7

Designing Zeolite Catalysts to Convert Glycerol, Rice Straw, and Bio-Syngas

Chuang Xing[1], Guohui Yang[2], Ruiqin Yang[1], and Noritatsu Tsubaki[2]

[1]*School of Biological and Chemical Engineering, Zhejiang University of Science and Technology, China*
[2]*Department of Applied Chemistry, Graduate School of Engineering, University of Toyama, Japan*

7.1 Glycerol Conversion to Propanediols

7.1.1 Introduction

Biodiesel as an alternative fuel has received much attention due to its efficient, re-usable, sustainable and environment-friendly nature. It is usually obtained via the transesterification of triglycerides with methanol in the presence of basic or acidic catalysts (see Scheme 7.1), where glycerol amounts to about 10 wt% of the total biodiesel produced [1]. However, the rapidly increasing production of biodiesel has led to a surplus of glycerol as a byproduct. Glycerol (1,2,3-propanetriol), as a renewable biomass resource and a highly polyfunctionalized platform chemical, is converted into many value-added chemicals via a chemoselective catalytic process [2], which is proverbially applied in the production of cosmetic products, pharmaceuticals, foods, and so on [3].

With the anticipated depletion of petrochemical resources and an escalating production of biomass, the use of glycerol for the synthesis of value-added chemicals has become more readily available as a renewable resource. Therefore, great interest has been taken in the sustainable development of the catalytic transformation of glycerol to value-added chemicals (Scheme 7.2) such as glyceric acid [4–6], dihydroxyacetone [7–13], acrolein [14–18], hydrogen [19] and propanediols [20–25]. In particular, the hydrogenolysis of glycerol to form propanediols has attracted increasing attention as it provides an

Nanoporous Catalysts for Biomass Conversion, First Edition. Edited by Feng-Shou Xiao and Liang Wang.
© 2018 John Wiley & Sons Ltd. Published 2018 by John Wiley & Sons Ltd.

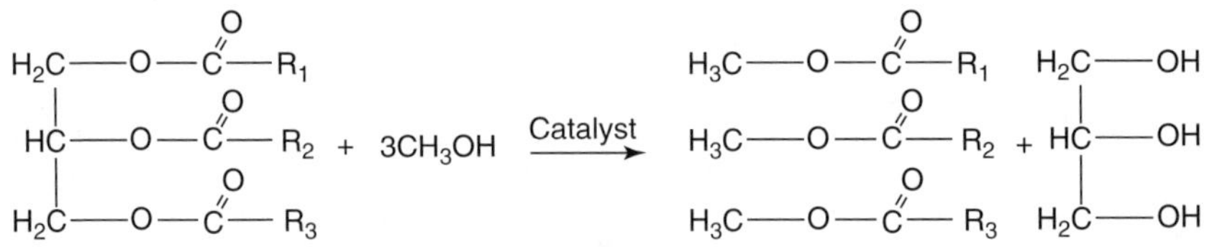

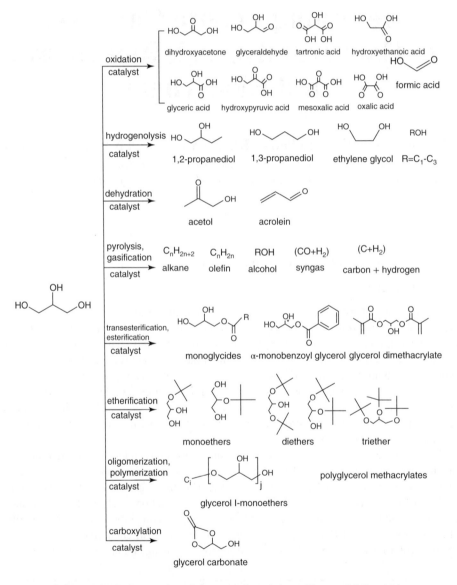

Scheme 7.1 *Transesterification of triglycerides to create glycerol.*

Scheme 7.2 *Processes of catalytic conversion of glycerol into useful chemicals.*

economical and environment-friendly alternative route to the depletion of petroleum [4,26]. 1,2-Propanediol is a valuable chemical intermediate in the manufacture of paints, antifreeze coolants, liquid detergents, cosmetics, foods, tobacco and polymer–monomers [1]. 1,3-Propanediol is mainly used as a feedstock in the production of food, cosmetics, in the pharmaceutical industry, and in fiber manufacture. 1,3-Propanediol is also an important intermediate in the fabrication of polyethers, polyurethanes, and polyesters [3].

In view of the rapid progress and increasing achievements of propanediol synthesis during recent years, the selective conversion of glycerol to 1,2-propanediol and 1,3-propanediol over zeolite-based catalysts will be discussed in greater detail.

7.1.2 Mechanisms of Propanediol Synthesis

Three typical reaction mechanisms have been identified in relation to the selective hydrogenation of glycerol to propanediol, including dehydration–hydrogenation, dehydrogenation–dehydration–hydrogenation, and a recently proposed direct-hydrogenolysis mechanism [1] (see Scheme 7.3). For the dehydration–hydrogenation mechanism, glycerol is first dehydrated through an acidic catalyst reaction to generate acetol or 3-hydroxypropionaldehyde intermediates (Scheme 7.3; Route 1), after which the intermediate is hydrogenated to form the final products [27,28]. The dehydrogenation–dehydration–hydrogenation mechanism (see Scheme 7.3; Route 2) usually occurs in neutral water and alkaline environments, with the glycerol being initially dehydrogenated to form glyceraldehyde on the metal sites of the catalysts. Dehydration of the glyceraldehyde intermediate then occurs on the base sites to form 2-hydroxyacrylaldehyde. Finally, 2-hydroxyacrylaldehyde is hydrogenated sequentially on the metal sites to obtain propanediol [29]. The mechanism of direct glycerol

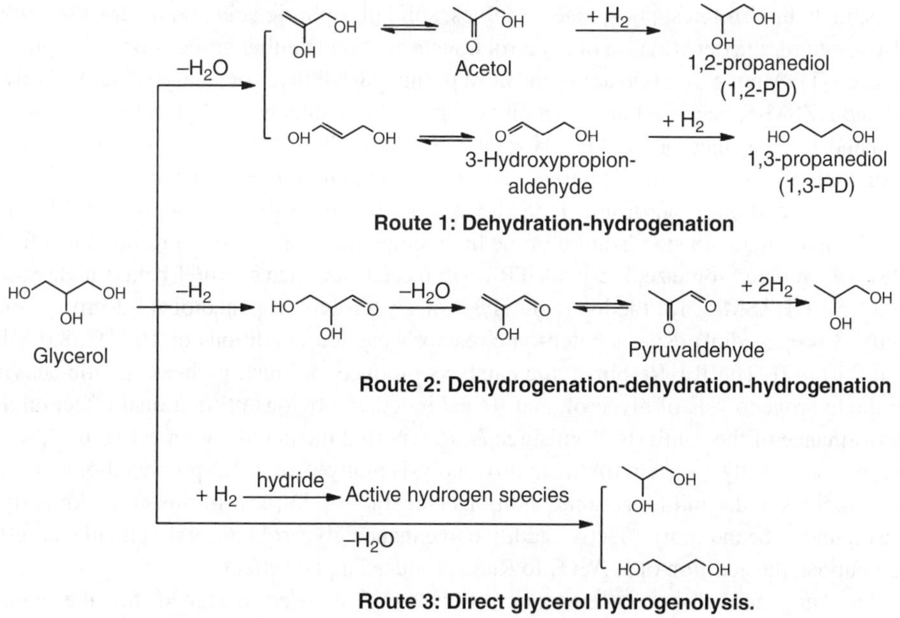

Scheme 7.3 *Proposed mechanisms of glycerol hydrogenolysis to produce propanediols.*

hydrogenolysis was reported recently by Tomishige *et al.* [30,31], and is illustrated in Scheme 7.3 (Route 3). These authors proposed that glycerol is adsorbed onto the interface between the Ir metal surface and the ReO_x cluster to form 2,3-dihydroxypropoxide and 1,3-dihydroxyisopropoxide, after which the products are created by hydrolysis of the reduced alkoxide.

7.1.3   Zeolite Catalysts for Propanediol Synthesis

Currently, glycerol hydrogenolysis to propanediols is a catalytic chemical process in which carbon–carbon or carbon–oxygen bonds are broken. The selective catalytic hydrogenolysis pathways involve a two-step dehydration–hydrogenation reaction, whereby glycerol is initially dehydrated to acetol or 3-hydroxypropanal via an acidic catalyst, and the intermediate is subsequently hydrogenolyzed to 1,2-propanediol or 1,3-propanediol [32,33] (see Scheme 7.3; Route 1). The selectivity towards 1,3-propanediol or 1,2-propanediol is highly dependent on the catalyst used and the reaction conditions. Numerous studies have been conducted to optimize the selectivity of the propanediols production.

During recent years, significant efforts have been made to investigate the heterogeneous catalysts involved in the catalytic hydrogenolysis of glycerol to propanediols. Currently, the catalysts can be classified as two groups, where one group contains noble-metal-based catalysts such Ru, Pt, Pd, Rh, and Ag, and the other group contains transition metals such as Cu, Ni, and Co. The non-noble metals are more competitive because of their lower prices and higher resistance to poisoning by trace impurities, even if their activity is lower than that of the noble metals [34]. The different supports, such as active carbon, SiO_2, Al_2O_3, TiO_2 and various zeolites, when loaded onto metal catalysts, have a great influence on the conversion of glycerol, and also on propanediol selectivity [35]. Various solid-acid catalysts, including sulfates, phosphates, zeolites, solid phosphoric acid and others, have been also tested for the dehydration of glycerol transformation in either gaseous or liquid phases [16,36–41]. Among the solid acids, the most promising additives or catalysts are Amberlyst 70 and HZSM-5, because their reaction systems exhibit an enhanced glycerol transformation and 1,3-propanediol yield [1]. However, the Amberlyst 70 system as an ion-exchange resin with poor re-usability, is burned during the recycling process. When considering the re-usability and duration of use, HZSM-5 zeolite is a more suitable catalyst for glycerol transformation due to stabilization of the high temperature and a good promotional effect. He *et al.* prepared some zeolite-loaded Ru-based catalysts, such as Ru/H-beta, Ru/HZSM-5 and Ru-Re/HZSM-5, for the hydrogenolysis of glycerol to propanediols (1,2-propanediol and 1,3-propanediol) using a batch-type reactor (reaction conditions of 160 °C, 8.0 MPa and 8 h) [42]. The Ru–Re bimetallic catalysts produced a much higher catalytic activity in the hydrogenolysis of glycerol, and Re exhibited an obvious promotional effect on the performance of the catalysts. Tomishige *et al.* reported the details of an active metal-acid bifunctional catalyst system for the hydrogenolysis of glycerol to 1,2-propanediol with various zeolites under mild reaction conditions (see Table 7.1) [43]. In this case, Amberlyst was found to be the most effective additive to enhance glycerol conversion in all catalysts. In contrast, the addition of H_2WO_4 to Ru/C produced a poor effect.

An autogenous reducing environment system has been designed for the in-situ transformation of glycerol into 1,2-propanediol with 2.7 wt% Pt/NaY catalyst, where

Table 7.1 *Results of glycerol hydrogenolysis over Ru/C + acid catalyst at 180 °C.[a]*

Catalyst	Conversion (%)	Selectivity (%)[b]				
		1,2-PDO	1,3-PDO	1-PO	2-PO	Others
Ru/C + SO$_4^{2-}$/ZrO$_2$	8.5	58.9	1.1	0.7	0.1	39.2
Ru/C + BEA	9.8	51.4	2.5	3.4	0.2	42.5
Ru/C + USY	6.7	82.0	0.4	0.9	0.4	16.3
Ru/C + MFI	7.4	44.2	2.6	17.5	0.3	35.4
Ru/C + H$_2$WO$_4$	5.9	79.4	1.4	1.2	0.2	17.8
Ru/C + Amberlyst	15.0	53.4	1.6	5.4	1.0	38.6
Rh/C + H$_2$WO$_4$	1.3	56.7	20.9	10.4	0.7	11.3

[a]Reaction conditions: 20 mass % glycerol aqueous solution 20 mL, the initial H$_2$ pressure 8.0 MPa, reaction time 10 h, catalyst weight: 150 mg M/C + 300 mg solid acid.
[b]C-based selectivity.

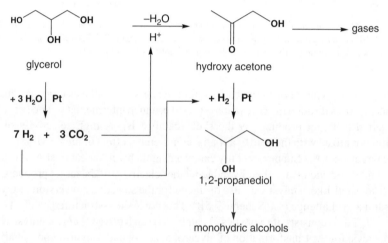

Scheme 7.4 *The main pathways for the conversion of glycerol into 1,2-propanediol via bifunctional catalysis with Pt/NaY.*

in-situ H$_2$ was generated via liquid reforming [44]. The selectivity of 1,2-propanediol was 91.3%, and the glycerol conversion was up to 95.9%. The authors proposed the existence of mechanistic pathways (Scheme 7.4) for the conversion of glycerol into mainly 1,2-propanediol that involved two steps: (i) an acid-catalyzed dehydration to hydroxyacetone; and (ii) subsequent hydrogenation to 1,2-propanediol. There was an optimal balance between the initially formed CO$_2$ and the origin of solution acidity, with subsequent catalysis of the dehydration of glycerol into hydroxyacetone with suitable zeolite acidity.

Indeed, the transformation of glycerol to propanediols occurs as a tandem reaction, with an initial dehydration to acetol or 3-hydroxypropanal via an acidic catalyst, and the intermediate subsequently being hydrogenolyzed to 1,2-propanediol or 1,3-propanediol on a metal catalyst. To increase the efficiency of the tandem reactions, Tsubaki and coworkers designed a type of capsule catalyst with a core–shell structure, and applied this successfully in syngas to a low olefin [45], isoparaffin [46–51] and dimethyl ether (DME) [52–56] by a one-pot process. Scheme 7.5 illustrates one of the most well-used tandem reaction processes,

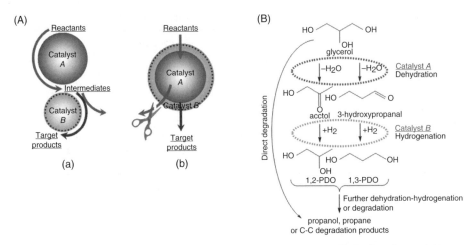

Scheme 7.5 *(A) Illustration of the tandem reaction process on (a) a general hybrid catalyst and (b) a capsule catalyst. (B) General pathway of glycerol conversion to 1,2-propanediol (1,2-PDO) or 1,3-propanediol (1,3-PDO). (See color plate section for the color representation of this figure.)*

proceeding on two types of different catalyst for the one-step synthesis target products [57]. Reactants first pass through the core (catalyst A) to produce intermediates, and subsequently are converted to target products by the shell (catalyst B). A high selectivity of middle isoparaffin is realized with the help of a spatial confinement effect of the ZSM-5 shell. Moreover, the target products can be generally catalyzed again by either catalyst A or B to generate other undesired chemicals, thereby reducing the selectivity of the target products. Hence, a novel core–shell-like catalyst was also designed for the selective conversion of glycerol to propanediol as a challenge (see Scheme 7.5B). This core–shell-structured catalyst (denoted as Ru/Al_2O_3-Pd/S) consists of two types of catalyst: (i) an Ru/Al_2O_3 core catalyst as a dehydration catalyst for the transformation of glycerol to acetol and 3-hydroxypropanal; and (ii) a Pd-doped microporous silicalite-1 zeolite shell (denoted as Pd/S) for the hydrogenation of acetol and 3-hydroxypropanal to form 1,2-propanediol or 1,3-propanediol.

The scanning electron microscopy (SEM) and energy-dispersive X-ray spectroscopy (EDS) analysis of the core–shell-structured catalyst is shown in Figure 7.1. For the naked core, the catalyst Ru/Al_2O_3 has a weight ratio of Ru of 5.1% (Figure 7.1a), while a zero content of Ru is found in the coated microporous silicalite-1 zeolite shell [57]. Notably, the SEM image (Figure 7.1b) of the Ru/Al_2O_3–Pd/S capsule catalyst shows an homogeneous, uniform zeolite shell without defects, even when undergoing harsh calcination before characterization. This further proves the formation of a defect-free zeolite shell that enwraps the Ru/Al_2O_3 core catalyst in compact fashion. The cross-section of Ru/Al_2O_3-Pd/S, using SEM and EDS line analysis methods, suggests the presence of a compact zeolite shell with a thickness of 6 μm on the surface of the Ru/Al_2O_3 core catalyst (Figure 7.1c). These findings confirmed that the capsule catalyst was synthesized successfully.

The catalytic performance of glycerol transformation to propanediol with a pure core catalyst and a hybrid catalyst, capsule catalyst Ru/Al_2O_3-Pd/S, is listed in Table 7.2. The slightly lower reaction activity of the capsule catalyst is probably due to the formation of some zeolite crystals covering part of metallic Ru particles of the core catalyst [58].

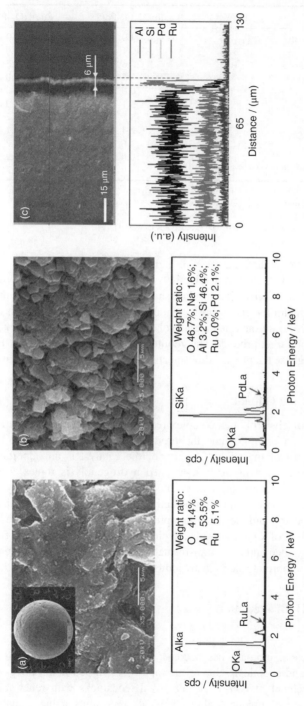

Figure 7.1 (a) Surface scanning electron microscopy (SEM) image (inset: complete morphology under lower magnification) and energy-dispersive X-ray spectroscopy (EDS) analysis of the Ru/Al_2O_3 core catalyst. (b) Surface SEM image and EDS analysis of the zeolite capsule catalyst Ru/Al_2O_3-Pd/S. (c) Cross-sectional SEM image and EDS line analysis of the zeolite capsule catalyst Ru/Al_2O_3-Pd/S. (See color plate section for the color representation of this figure.)

Table 7.2 Reaction properties of core catalyst, zeolite capsule catalyst and hybrid catalyst.[a]

	Catalyst	Ru/Al_2O_3	Ru/Al_2O_3-Pd/S [b]	Ru/Al_2O_3-Pd/S-M [c]
	Conversion	5.7	4.9	6.2
Selectivity/%	Acetol	57.1	40.1	22.0
	1,2-PDO	10.2	27.6	9.2
	1,3-PDO	0.0	2.2	0.0
	1-PO	0.7	0.0	1.0
	2-PO	3.3	5.1	3.9
	EG	2.6	5.1	3.2
	EtOH	3.1	2.4	18.1
	MeOH	3.9	3.6	9.7
	CO_2	14.7	10.0	28.2
	Gas	4.4	3.9	4.7

[a]Reaction conditions: 200 °C; H_2 0.3 MPa; 10 h; 20 ml 20% glycerol; 1.0 g core catalyst.
[b]Pd/S = Pd-doped Silicalite-1 zeolite shell enwrapping the Ru/Al_2O_3 core catalyst.
[c]M = the physical mixture of Ru/Al_2O_3 and Pd-doped Silicalite-1 zeolite.
PDO: Propanediol; PO: Propanol; EG: ethylene glycol; Gas: CH_4, C_2H_6.

In this reaction, an intermediate acetol with higher selectivity up to 57.1% was formed over the Ru/Al_2O_3 catalyst, but lower selectivity of secondary degradation products was consistent with a glycerol conversion mechanism through a tandem reaction process that included an initial dehydration and a subsequent hydrogenation [32,59]. In this case, glycerol is first dehydrated to an intermediate acetol and subsequently hydrogenated to form 1,2-propanediol as the target product. The zeolite capsule catalyst Ru/Al_2O_3-Pd/S, however, exhibited the highest 1,2-propanediol selectivity of 27.6%, via the tandem reaction.

7.1.4 Conclusions and Outlook

The hydrogenolysis of glycerol, as a complex reaction, involves many potential reaction routes and products. In order to tune the product distribution and obtain the correct target products, control of the acidity/basicity of the catalyst is an important strategy to exclude any unexpected acid- or base-assisted hydrogenolysis routes. Zeolite, with its unique micropore structure, high thermal stability and specific acidity, is widely applied in glycerol conversion catalysts. Selectivity control is a major challenge in the transformation of glycerol, and therefore the optimization of catalyst structure and zeolite acidity, and the building of kinetic models are clear requirements in the near future. With ever-increasing energy demands, the transformation of bio-glycerol to propanediols will play an increasingly important role in many industrial applications.

7.2 Rice Straw Hydrogenation

7.2.1 Introduction

During the past decade, biomass has become a major concern for the production of transportation fuels and chemicals, due to the rapid rises in petroleum prices worldwide [60–62]. The replacement of non-renewable fossil feedstocks with renewable biomass feedstocks to generate high-added-value chemicals represents a huge challenge with regards to the creation of greenhouse gases and environment and climate changes [63].

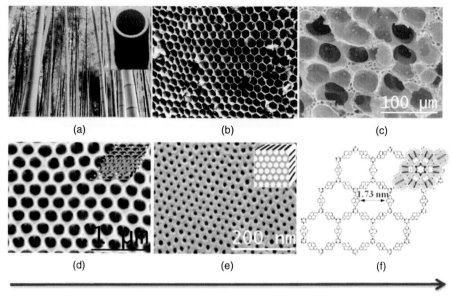

(a)

(b)

(c)

(d)

(e)

(f)

Decreasing the pore size

Figure 1.1 *Illustration of porosity existing in Nature and synthesized frameworks with a decreasing pore size. (a) Bamboo; (b) honeycomb; (c) scanning electron microscopy (SEM) image of alveolar tissue in mouse lung; (d) SEM image of an ordered macroporous polymer; (e) SEM image of an ordered mesoporous polymer from self-assembly of block copolymers; (f) structural representation of the COF structure.*

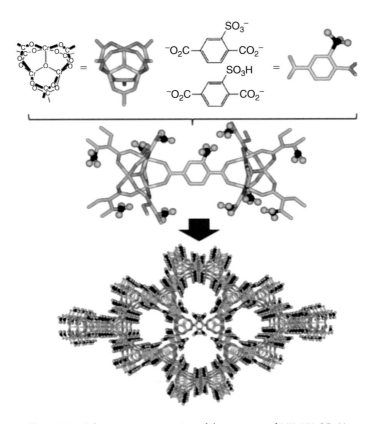

Figure 1.2 Schematic representation of the structure of MIL-101-SO$_3$H.

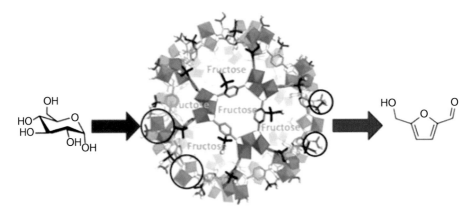

Figure 1.3 Bifunctional catalyst MIL-101(Cr)-SO$_3$H used for glucose conversion to HMF. Reproduced with permission from Ref. [38]. Copyright 2016, American Institute of Chemical Engineers.

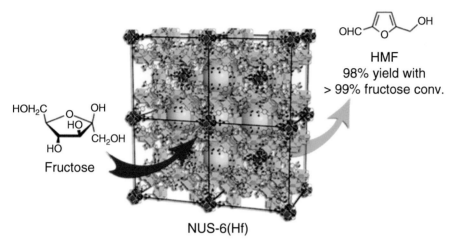

HMF
98% yield with
> 99% fructose conv.

Fructose

NUS-6(Hf)

Figure 1.4 *NUS-6(Hf) used as a heterogeneous catalyst for fructose conversion to HMF. Reproduced by permission of Ref. [41]. Copyright 2012 American Chemical Society.*

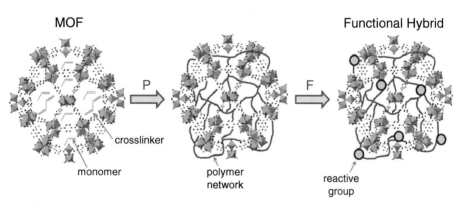

MOF   Functional Hybrid

crosslinker

monomer   polymer   reactive
                          network   group

Figure 1.7 *Scheme of fabrication of functional hybrids of MOFs and polymer networks. Reprinted with permission from Ref. [45]. Copyright 2014, American Chemical Society.*

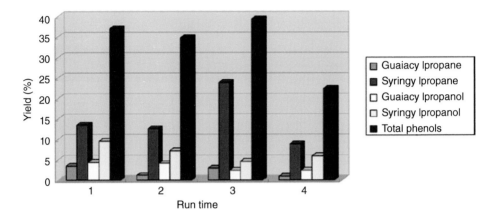

Figure 2.4   *The recycling results of Ni-W$_2$C/AC catalyst for lignin hydrocracking reaction [76b].*

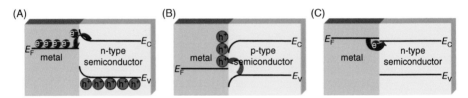

Figure 3.1   *Schematic view of typical (a) rectifying metal–n-type semiconductor contact. (b) Rectifying metal–p-type semiconductor. (c) Metal–semiconductor ohmic contact. Reproduced by permission from Ref. [16]. Copyright 2013, Royal Society of Chemistry.*

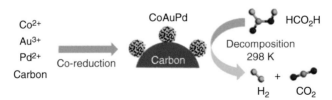

Figure 3.4   *Preparation and application of CoAuPd/C nanocatalyst for formic acid decomposition at 298K. Reproduced with permission from Ref. [40]. Copyright 2013, Wiley.*

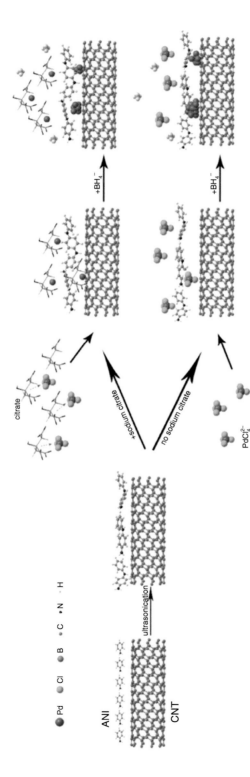

Figure 3.7 Proposed mechanisms for the synthesis of Pd-PANI/CNT catalysts. Reproduced with permission from Ref. [48]. Copyright 2015, Wiley.

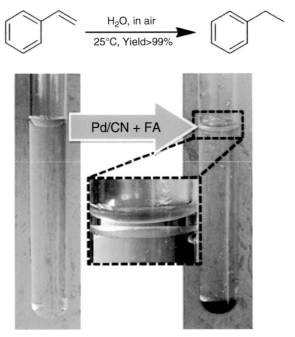

Figure 3.10 *The rapid separation of ethylbenzene after the hydrogenation reaction of styrene. After the reaction, the as-formed ethylbenzene was separated automatically from the water phase, within 10 min. The solid catalyst was precipitated automatically at the bottom and easily separated from the oil phase by filtration or decantation. The reaction conditions were: 10 mmol styrene, 250 ml H_2O, 30 mmol formic acid (FA), and 500 mg Pd/CN, at 298K.*

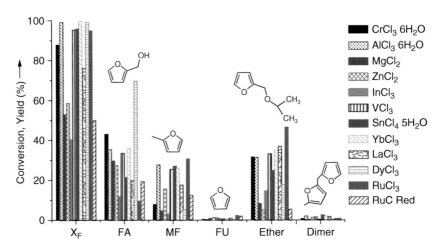

Figure 3.17 *Effect of the addition of Ru/C catalyst on furfural conversion and product yield for the indicated homogeneous Lewis acid catalysts. Reproduced with permission from Ref. [65]. Copyright 2014, Wiley.*

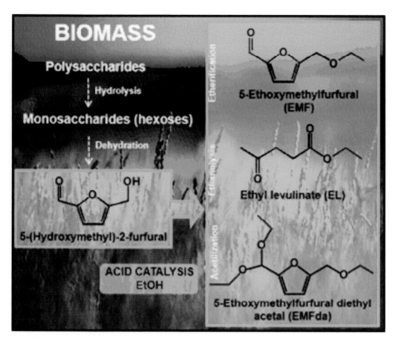

Figure 3.23 *Conversion of HMF into products for biofuels applications. Reproduced with permission from Ref. [77]. Copyright 2014, Wiley.*

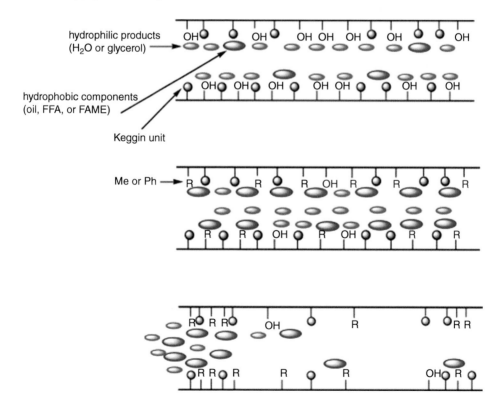

Figure 5.6 *Schematic representation of the influence of surface hydrophobicity on catalytic activity. Top: HPAs/SiO$_2$-Ta$_2$O$_5$. Alkyl-functionalized HPAs/SiO$_2$-Ta$_2$O$_5$ sample synthesized by a co-condensation technique (middle) and a post-synthesis grafting method (bottom).*

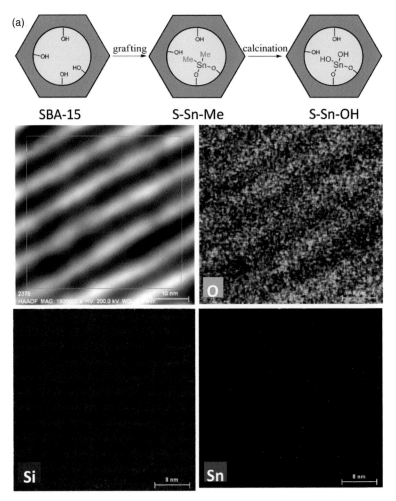

Figure 5.8 (a) The synthesis of S-Sn-OH; (b) Scanning transmission electron microscopy image; (c–e) O, Si, and Sn energy-dispersive X-ray spectroscopy (EDS) elemental maps of S-Sn-OH, respectively. The scale bars in images (b) to (e) are 10, 8, 8, and 8 nm, respectively.

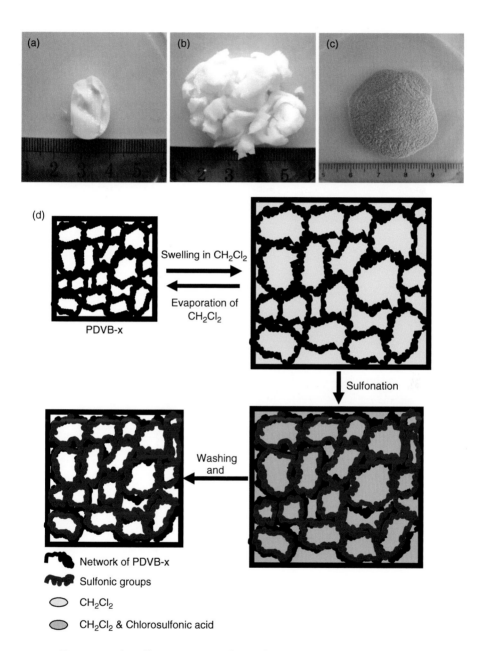

Figure 6.2 The sulfonation process of PDVB by using HSO_3Cl in CH_2Cl_2 solvent.

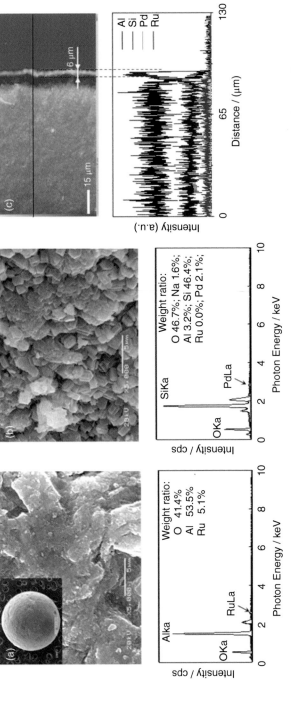

Figure 7.1 (a) Surface scanning electron microscopy (SEM) image (inset: complete morphology under lower magnification) and energy-dispersive X-ray spectroscopy (EDS) analysis of the Ru/Al$_2$O$_3$ core catalyst. (b) Surface SEM image and EDS analysis of the zeolite capsule catalyst Ru/Al$_2$O$_3$-Pd/S. (C) Cross-sectional SEM image and EDS line analysis of the zeolite capsule catalyst Ru/Al$_2$O$_3$-Pd/S.

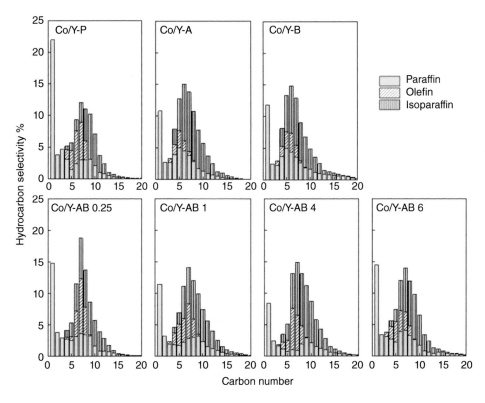

Figure 7.5 *Product distribution for the prepared catalysts.*

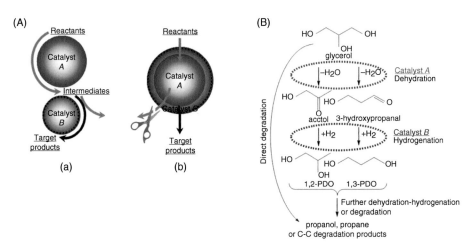

Scheme 7.5 *(A) Illustration of the tandem reaction process on (a) a general hybrid catalyst and (b) a capsule catalyst. (B) General pathway of glycerol conversion to 1,2-propanediol (1,2-PDO) or 1,3-propanediol (1,3-PDO).*

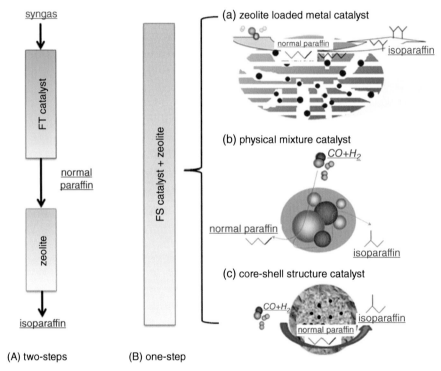

Scheme 7.7 *Representative isoparaffin synthesis through FT catalytic systems. (A) Two-step reactor configuration. (B) Hybrid catalyst for the one-step synthesis: (a) zeolite-loaded metal catalyst; (b) physical mixture catalyst; (c) core–shell structure catalyst.*

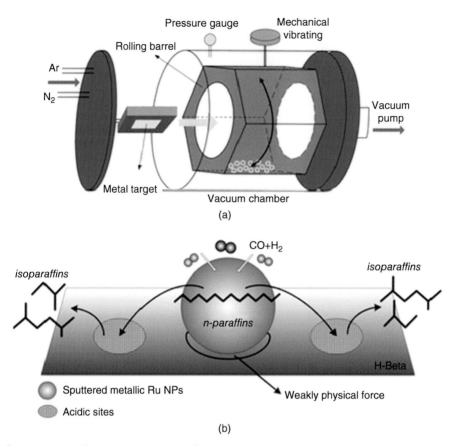

Scheme 7.8 *(a) Schematic representation of the sputtering apparatus. (b) One-step synthesis of isoparaffin from syngas on Ru/H-Beta catalyst.*

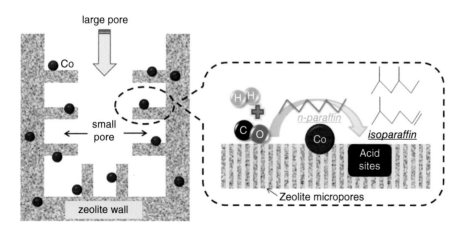

Scheme 7.9 *Representation of the hierarchical zeolite catalyst for one-step isoparaffin synthesis.*

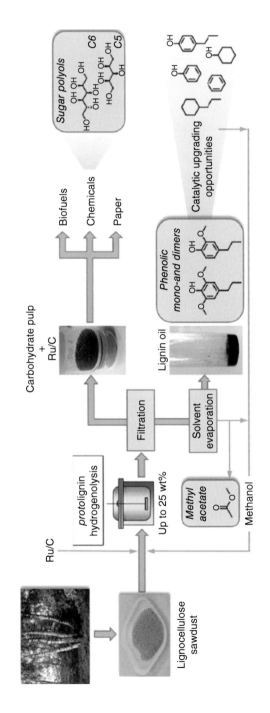

Figure 8.2 *Schematic representation of an integrated bio-refinery process for the full utilization of lignocellulose.*

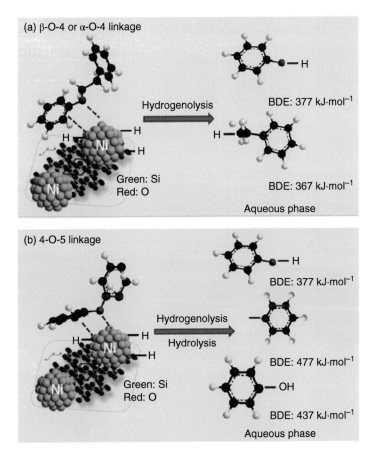

Figure 8.4 *The proposed different mechanisms of cleavage of aryl ether C–O bonds in the α-O-4, β-O-4, and 4-O-5 model compounds of lignin over Ni/SiO₂ in the aqueous phase.*

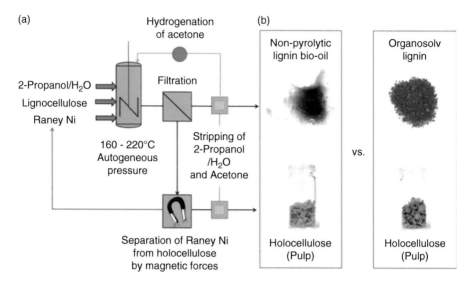

Figure 8.7 (a) Schematic representation of the catalytic bio-refining method over Raney Ni in the mixed solvent of 2-propanol and H_2O. (b) Comparison of lignin-derived bio-oil and the traditional organosolv process.

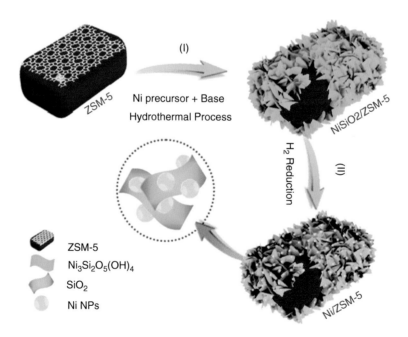

Figure 9.7 Preparation procedures for Ni nanoparticles supported on mesoporous ZSM-5 zeolite crystals.

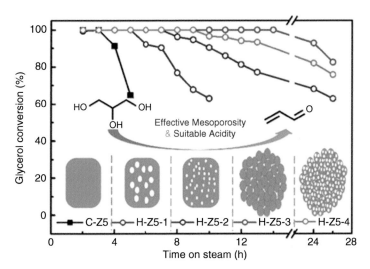

Figure 9.12 *Dependences of glycerol conversion on reaction time over different mesoporous zeolite catalysts.*

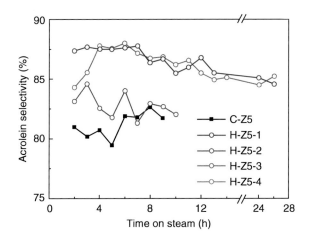

Figure 9.13 *Acrolein selectivity in glycerol conversion over different mesoporous zeolite catalysts.*

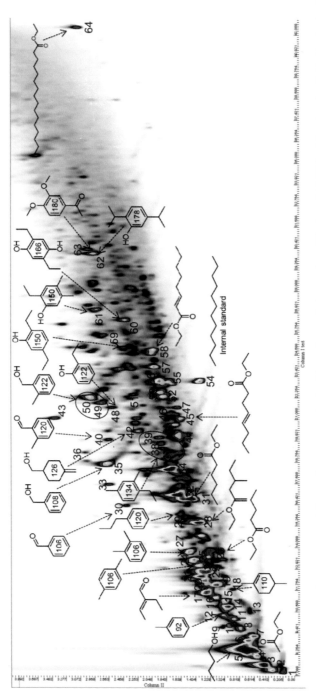

Figure 10.2 GCxGC-MS chromatogram of the product mixture obtained from the catalytic reaction of lignin at 300 °C for 8 h using the CuMgAl mixed-oxide catalyst.

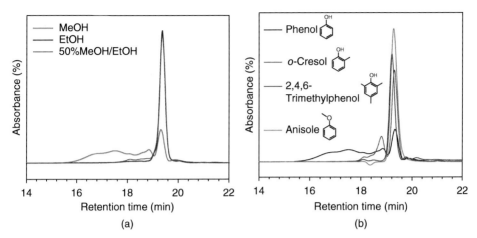

Figure 10.4 *GPC chromatograms of reaction mixtures obtained from the reaction at 300 °C for 1 h over the $Cu_{20}MgAl(2)$ catalyst. (a) Using phenol in different solvents; (b) Using different reactants in methanol solvent (the depicted chromatograms have been normalized by the sum of the peak area) [59].*

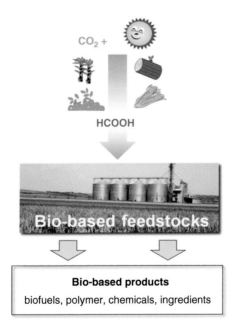

Figure 12.1 *The concept of formic acid-based biorefinery.*

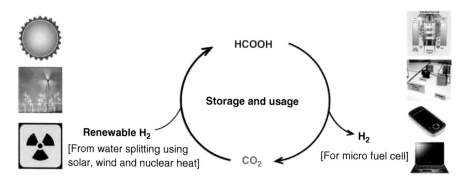

Figure 12.2 *Carbon-neutral H$_2$ store using biorenewable formic acid as an energy carrier.*

Currently, conventional acid catalysts use mineral acid technologies for biomass utilization and rapid catalytic pyrolyses, with concentrated sulfuric acid, dilute sulfuric acid, cellulase, and supercritical water being applied widely [64]. These materials invariably cause a variety of environmental issues and other problems, including the corrosion of equipment, as well as toxicity, recovery, and the generation of neutralization waste. In order to avoid these defects, the exploitation of solid, nontoxic, re-use and recovery has been reported for biomass conversions [65].

The development of stable and recyclable solid-acid catalysts to replace conventional liquid acids continues to show promise for biomass conversion and utilization. Zeolite, as a typical solid-acid catalyst, with a high thermal stability and acidity, and a unique micro-pore structure, is used worldwide as adsorbent and also as a catalyst in many fields such as basic petrochemistry, oil refining and fine-chemical synthesis. Biomass conversion based on zeolite catalysis as an economical and environment-friendly alternative approach, and can be applied to many situations. Catalytic fast pyrolysis (CFP) shows great promise as a process for solid biomass direct conversion into desirable hydrocarbon products [66]. Lappas and coworkers have used FCC catalysts (Y zeolite) for the conversion of beech wood into pyrolysis oil at moderate temperatures [67]. Although the pyrolysis oil yields decreased from 75% to 45–50%, the quality of the oils was greatly improved. Huber *et al.* have reported that aromatics were produced from solid biomass with a ZSM-5 catalyst in a one-step process using CFP [68]. The same authors examined the use of glucose as a model compound for conversion to aromatics over a ZSM-5 zeolite through isotopic tracing [69]. A direct conversion of cellulose to C_3–C_4 hydrocarbons was performed via a decarbona-tion reaction using H-beta zeolite at 170 °C, with water as a medium [64]. T.J. Wang and colleagues focused on the synthesis of C_5–C_6 isoparaffins from sorbitol, using a Ni-ZSM-5 catalyst, by aqueous phase reforming, to achieve a maximal C_5–C_6 isoparaffin selectivity of 32.3% at a calcination temperature of 500 °C [70]. Huber *et al.* reported that C_5–C_6 alka-nes could be produced from sorbitol over the bifunctional Pt/SiO_2–Al_2O_3 by employing an aqueous phase-processing catalyst [71].

Currently, the catalytic conversion of cellulose into sugar alcohol has been achieved using effective bifunctional catalysts – an efficient approach that involves a combination of hydrolysis and the use of hydrogen in a hot-water environment. Usually, cellulose is derived from lignocellulosic rice straw, but this is a complicated and high-cost process due to the dense packing structure of the lignin layers [72]. Thus, direct utilization of the rice straw rather than pure cellulose would be economic, and also challenging.

Rice straw as an attractive lignocellulosic material, is one of the most abundant renewable resources worldwide, with low cost and great abundance [72]. Typically, the composition of rice straw contains 32–47% cellulose, 19–27% hemicellulose, 5–24% lignin and others [65,73,74], which can be converted to fermentable sugars, bioethanol, biochar, hydrocar-bons, and so on [75,76]. The conversion of rice straw, based on acidic zeolite, represents an alternative approach for the generation of transportation fuels and chemicals.

7.2.2 Direct Conversion of Rice Straw into Sugar Alcohol Through In-Situ Hydrogen

The catalytic conversion of rice straw to sugar alcohol involves a combination of rice straw hydrolysis and glucose hydrogenation, where the acquisition of compressed H_2 is indispensable for the hydrogenation process. Taking into consideration the wide availability

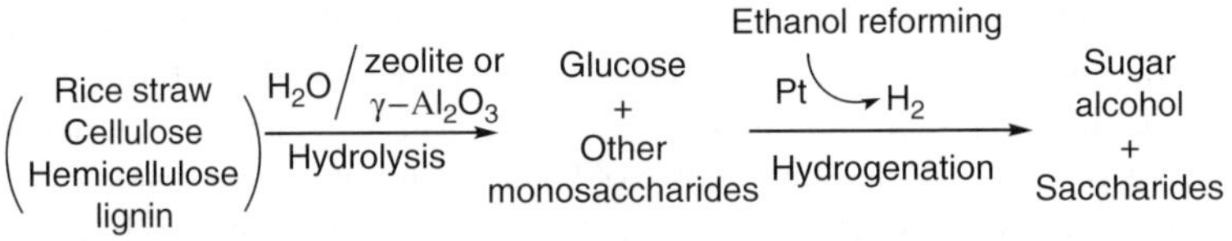

Scheme (flow diagram):

$$\left(\begin{array}{l}\text{Rice straw}\\\text{Cellulose}\\\text{Hemicellulose}\\\text{lignin}\end{array}\right)\xrightarrow[\text{Hydrolysis}]{H_2O\Big/\begin{array}{c}\text{zeolite or}\\\gamma-Al_2O_3\end{array}}\begin{array}{c}\text{Glucose}\\+\\\text{Other}\\\text{monosaccharides}\end{array}\xrightarrow[\text{Hydrogenation}]{\substack{\text{Ethanol reforming}\\Pt\searrow H_2}}\begin{array}{c}\text{Sugar}\\\text{alcohol}\\+\\\text{Saccharides}\end{array}$$

Scheme 7.6 *Direct conversion of rice straw via a solid acid-supported Pt catalyst.*

of rice straw and H_2 transportation problems, a new 'green' catalytic process has been designed and applied by the Tsubaki laboratory for the direct conversion of rice straw to sugar alcohol [77]. The demand of H_2 for the hydrogenation of glucose is achieved by employing an ethanol steam reforming ($C_2H_5OH + H_2O \rightarrow 2CO + 4H_2$), followed by a water gas shift reaction ($CO + H_2O \rightarrow CO_2 + H_2$) to form H_2 in-situ over a relatively wide temperature range [22]. Thus, a direct hydrodegradation route of rice straw is realized in aqueous phase over ethanol addition, achieved via a solid acid-supported Pt bifunctional catalyst (see Scheme 7.6).

Details of the H_2 produced in-situ in the water phase by the ethanol steam reforming reaction using a solid acid-supported Pt catalyst over a wide temperature range (190–230 °C) are listed in Table 7.3 [77]. The amounts of H_2 are clearly increased in line with increases in the reaction temperature, from 190 to 230 °C due to the accelerated reforming reaction. In company with the biomass addition (cellulose, lignin or rice straw), H_2 productivity is decreased significantly due to the consumption of the hydrogenation reaction. A small number of CO, CO_2, C_1–C_6 hydrocarbons is also detected in the gas products.

The reaction performance of the solid acid-supported Pt catalysts is listed in Table 7.4. A 64.3% conversion of the rice straw was observed even in the absence of H_2 or a catalyst, but this derived from hydrothermal decomposition at 190 °C. The main products were water-soluble non-hexitol compounds such as glucose, xylose, arabinose, and mannose. A similar rice straw conversion (Table 7.4; entry 2) was found via a Pt/γ-Al$_2$O$_3$ catalyst as well as ethanol addition, while sugar alcohols were not detected due to the hydrolyzed rice straw. Comparatively, with a further increase in reaction temperature (Table 7.4; entry 3), the rice straw gave the highest conversion of 73.1%, producing a combination of hydrolyzed and subsequently hydrogenated processes, sugar alcohols in 6.3% yield (mannitol 5.5%; sorbitol 0.8%). It should be noted that the yield of hexitols for entry 3 was lower than when using H_2 (entry 8). Thus, a tandem reaction can be successfully designed and operated for

Table 7.3 H_2 productivity via ethanol steam reforming.[a]

Entry	Reactants [g]	Temp. [°C]	H_2 [mmol l^{-1}][b]	H_2 [%][c]	H_2 [mmol]/EtOH [mol]
1	0	190	0.24	32.0	0.15
2	0	230	0.46	61.1	0.30
3	Cellulose, 0.5	230	0.42	56.0	0.27
4	Lignin, 0.15	230	0.19	9.3	0.03
5	Rice straw, 0.5	230	0.13	6.3	0.02

[a]Catalyst Pt/γ-Al$_2$O$_3$: Pt/HZSM-5 (1:1) 0.2 g; H$_2$O 50 ml; ethanol 10 ml; N$_2$ 5 MPa (room temperature); 24 h.
[b]Volume fraction of H$_2$ in the autoclave.
[c]Molar fraction of H$_2$ in the effluent gases.

Table 7.4 Rice straw conversion and yield of sugar alcohols under different conditions.[a]

| | | | | | Mannitol | | Sorbitol | |
| | | | | | Yield [%] | Sel. [%] | Yield [%] | Sel. [%] |
Entry	Catalyst	Temp. [°C]	H$_2$ (1) or (2)	Conv. [%]				
1	No catalyst	190	–	64.3	0	0	0	0
2	Pt/γ-Al$_2$O$_3$	190	Self-supplied	66.4	0	0	0	0
3	Pt/γ-Al$_2$O$_3$	230	Self-supplied	73.1	5.5	3.1	0.8	0.2
4	Pt/γ-Al$_2$O$_3$-H-ZSM-5 (1:1)	190	Self-supplied	64.8	4.5	2.9	2.1	1.3
5	Pt/γ-Al$_2$O$_3$:Pt/H-ZSM-5 (1:1)	190	Self-supplied	64.5	4.3	2.7	1.8	1.1
6	No catalyst	190	Employed	76.1	0	0	0	0
7	Pt/γ-Al$_2$O$_3$	190	Employed	66.8	10.8	7.2	1.9	1.2
8	Pt/γ-Al$_2$O$_3$	230	Employed	68.7	14.4	9.3	2.2	1.3
9	Pt/γ-Al$_2$O$_3$-H-ZSM-5 (1:1)	190	Employed	68.5	6.8	4.7	8.5	5.9
10	Pt/γ-Al$_2$O$_3$:Pt/H-ZSM-5 (1:1)	190	Employed	67.2	4.7	3.1	7.3	4.9

[a]24 h reaction time; 0.5 g rice straw; 0.2 g catalyst. Rice straw: α-cellulose, 34.6%; β-cellulose, 19.4%; lignin, 13.4%; ash, 17.1%; soluble component in toluene and ethanol, 3.5%; others, 12.0%.
(1) 60 ml H$_2$O; 5 MPa employed H$_2$ at room temperatures; or (2) 50 ml H$_2$O + 10 ml C$_2$H$_5$OH; 5 MPa N$_2$ at room temperature were used in the case of using self-supplied H$_2$.

a direct conversion of rice straw into sugar alcohol, achieved by in-situ hydrogen produced via an ethanol steam reforming reaction. Bifunctional catalysts for one-step sugar alcohol direct synthesis, as a hybrid catalyst, containing a zeolite catalyst and a hydrogenation catalyst, may provide novel inspiration to the field of biomass conversion.

7.2.3 Conclusions and Outlook

The conversion of rice straw to produce sugar alcohol has been performed using in-situ H$_2$ through the ethanol steam reforming reaction. However, as the yield of sugar alcohol is lower than when using H$_2$, the way in which the alcohol steam reforming reaction can be used to generate more in-situ H$_2$ has become an urgent issue. Ethanol created by fermentation processes or from the catalytic conversion of syngas represents an important alternative raw material for the production of gasoline, albeit at a high price. Thus, it must be considered whether ethanol could be replaced by methanol or other inexpensive alcohols. Rice straw has three main components, namely cellulose, hemicelluloses, and lignin. The production of bioethanol from rice straw has resulted in a direct competition with food supplies, which includes two routes. In the 'sugar route', cellulose/hemicellulose is first converted to fermentable sugars, which then are fermented to produce ethanol. In the 'syngas route', the rice straw is first gasified and then purified to obtain carbon monoxide and hydrogen, with these gaseous products subsequently being converted catalytically to ethanol. With ever-increasing energy demands, the use of rice straw may prove to be a feasible technology in the very near future.

7.3 Bio-Gasoline Direct Synthesis from Bio-Syngas

7.3.1 Introduction

With the depletion of crude oil and a rapidly growing demand for liquid fuels, the use of non-petroleum carbon resources for the sustainable production of liquid fuels has attracted great interest for the synthesis of alternative fuels [78]. Syngas (CO, H_2), as an important role intermediate medium, is generally converted into clean liquid fuels through the Fischer–Tropsch (FT) synthesis reaction [79], which usually involves the following reactions:

$$(2n + 1)H_2 + nCO \rightarrow C_nH_{2n+2} + nH_2O \tag{7.1}$$

$$2nH_2 + nCO \rightarrow C_nH_{2n} + nH_2O \tag{7.2}$$

The liquid hydrocarbons produced by FT synthesis, with their advantageous sulfur-free, nitrogen-free and aromatics-free properties, are ideal candidates for the production of clean transportation fuels and chemicals [50]. In general, the FT products are normal aliphatic hydrocarbons with small amounts of α-olefin, with their distribution strictly following the Anderson–Schulz–Flory (ASF) law, such that they are suitable only as synthetic diesel fuels [80–82]. In order to resolve these problems, great attention has been focused on the isoparaffin synthesis of gasoline with high octane values. The catalysts for direct isoparaffin synthesis via FT synthesis route (syngas to gasoline) are of two main forms – conventional FT catalysts and acidic zeolites.

Currently, syngas is produced from fossil fuels, mainly coal, natural gas, and naphtha. Syngas from biomass is named bio-syngas or biomass-derived syngas, and shows great promise due to its efficient, re-usable, sustainable and environmentally benign properties [83–86]. Bio-syngas can also be further refined and synthesized to create clean fuels and chemicals such as gasoline, diesel, alcohols, olefins, oxychemicals, and synthetic natural gas [87].

The direct synthesis of bio-gasoline from bio-syngas will be discussed in the following sections.

7.3.2   Biomass Gasification to Bio-Syngas

Biomass gasification is a thermochemical process, in which biomass reacts with oxygen (or air) and steam to generate bio-syngas, a mixture that consists primarily of CO, CO_2, H_2, and H_2O. The generic biomass gasification process is illustrated schematically in Figure 7.2 [88]. Common oxidants are air or oxygen, which can decompose the large polymeric molecules of biomass into permanent gases (CO, H_2, CH_4, and lighter hydrocarbons), ash, char, and byproducts. The ash and char formed represent the incomplete biomass.

Heat is supplied to the gasifier either directly or indirectly, with the gasification temperature between 600 °C and 1000 °C. The gasification temperature is one of the most influential factors affecting the product's gas composition and its properties [89]. The biomass conversion and gas yield are accompanied by an increase in reaction temperature. The most interesting problem is the H_2/CO ratio, which is an important factor of feedstock for the subsequent FT synthesis to liquid fuel. Lv *et al.* have studied the H_2/CO ratio with effects of different fixed-bed reactor temperatures [79], as shown in Figure 7.3. The ratio of H_2/CO first increases and then decreases with the increase in temperature, but reaches a maximum value of 4.45 at a temperature of 750 °C. The flow rate of air, steam, and the type and properties of the biomass are also influenced effectively by the H_2/CO ratio.

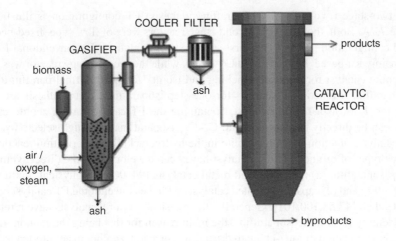

Figure 7.2   *Generic biomass gasification process.*

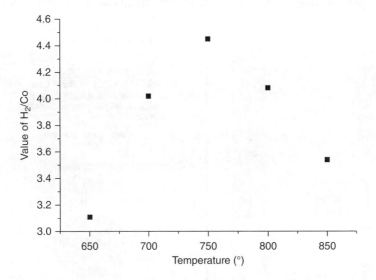

Figure 7.3   *Effect of temperature on the H_2/CO ratio.*

7.3.3   Representative FT Gasoline Synthesis System

Generally, the FT reaction produces mainly normal hydrocarbons and some building-block chemicals such as lower olefins, which cannot be used directly as gasoline. Additional hydrocracking and isomerization steps are usually employed for the production of gasoline, especially isoparaffin in industry [90]. Acidic zeolites have been exploited to construct the bifunctional FT catalyst for isoparaffin direct synthesis [78].

The two-stage (two-steps for isoparaffin synthesis) FT configuration is illustrated in Scheme 7.7A. Both, the first- and second-stage reactors were of flow-type, fixed-bed pressurized configuration [91–93]. The first-stage reactor contained the conventional FT catalysts, mainly Ru-based, Fe-based, and Co-based, while the second-stage reactor was loaded with zeolite catalyst for the hydrocracking and isomerization of normal paraffin to product isoparaffin. Bifunctional catalysts for one-step isoparaffin direct synthesis are shown in Scheme 7.7B, as a hybrid catalyst, containing the FT catalyst and a zeolite catalyst. Syngas can be directly transformed into C_5–C_{11} isoparaffins with high selectivity, owing to the catalytic function of acidic zeolite in the hydrocracking/isomerization reactions. So far, three types of bifunctional FT catalyst have been developed for the direct synthesis of C_5–C_{11} isoparaffin: (a) zeolite-loaded metal catalysts [94–96]; (b) physical mixture catalysts [97–99]; and (c) capsule or coated catalysts with the conventional FT catalyst core and a zeolite shell [47,58,100,101]. Generally, the physical mixture catalysts have a relatively low selectivity for middle isoparaffins, the main reason for this being the random hydrocracking/isomerization of long-chain hydrocarbons in acidic zeolite. In zeolite-loaded metal catalysts, the active metal is dispersed on acidic zeolite supports, exhibiting a low reduction degree due to the strong metal support interaction (SMSI). Capsule or coated catalysts are more effective for isoparaffin synthesis because of the spatial confinement effect and the

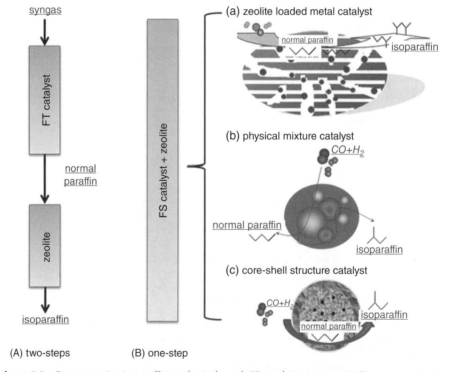

Scheme 7.7 *Representative isoparaffin synthesis through FT catalytic systems. (A) Two-step reactor configuration. (B) Hybrid catalyst for the one-step synthesis: (a) zeolite-loaded metal catalyst; (b) physical mixture catalyst; (c) core–shell structure catalyst. (See color plate section for the color representation of this figure.)*

unique pore structure of the zeolite shell, which can enforce effectively the hydrocracking/ isomerization of long-chain hydrocarbons.

7.3.4 FT Gasoline Synthesis Catalysts

Whilst all Group VIII metals as FT catalysts have noticeable activity, only Fe, Ni, Co, and Ru metals have the required FT activity for commercial application. On a metal basis, the approximate relative costs of Fe:Ni:Co:Ru are 1:250:1000:50000. Nickel-based catalysts produce too much undesirable methane, and ruthenium-based catalysts are very expensive compared to iron- and cobalt-based catalysts [93]. Hence, iron- and cobalt-based catalysts are applied broadly as activity components in FT synthesis.

7.3.4.1 Ru-Based Zeolite Catalyst

Despite the high cost of Ru-based catalysts as compared to conventional Fe- and Co-based catalysts, they are suitable for fundamental research due to their higher FT reaction activities, higher degrees of reduction, chain growth probability, and attractive stability under higher partial pressures of steam or other oxygenated atmospheres [82,102,103]. Sun *et al.* have successfully designed H-Beta zeolite-supported Ru nanoparticles without any reduction, by using a self-made polygonal barrel-sputtering process (Scheme 7.8a). The treated zeolite support was placed in a vacuum chamber, which as a sputtering target was sputtered by a metallic ruthenium plate under an accelerating voltage of 200 kV. The sputtered Ru/H-Beat catalyst exhibited uniform metal particles and maintained a diameter range of 2–4 nm. The sputtered Ru/H-Beta gives a CO conversion that is 1.6-fold as high as the conventional impregnation catalyst under the same FT reaction conditions. Furthermore, a satisfying $C_5–C_{11}$ selectivity of 71.7% was obtained, and the C_{iso}/C_n ratio reached 4.6. According to structure characterization and performance tests, the schematic diagram for direct middle isoparaffin synthesis over the sputtered Ru/H-Beta catalyst is illustrated in Scheme 7.8b. Normal aliphatic (*n*-paraffin) is first produced through the metallic Ru, and then hydrocracked and isomerized to convert the isoparaffin via catalysis with the acidic sites of H-Beta zeolite. Furthermore, the suppression of heavy hydrocarbon formation is conducted more intensely on the sputtered catalyst, resulting from the shorter distance between the highly dispersed Ru and acidic sites.

A representative innovation of mesoporous zeolite-supported Ru-based catalysts was first exploited by Wang and coworkers [82,104]. The designed zeolites contained both micro- and mesopore structures, which have attracted much attention as catalyst supports because of their efficient mass-transport properties. The mesoporous zeolite was prepared by a simple base leaching on the HZSM-5 matrix, supporting the Ru catalyst through a conventional impregnation for one-step isoparaffin synthesis. The pristine beta has a unique pore distribution of around 0.67 nm, which presents a typical beta zeolite. The sample of Y-A was treated by single base leaching with various NaOH concentrations, which resulted in obvious mesopores with a range of 3–10 nm. In addition, the intensity and width of the peaks increased linearly with the increasing base concentration, which indicated that a larger pore volume would be created with a higher NaOH concentration. Further observations of the mesopore structure produced involved transmission electron microscopy (TEM) images of the

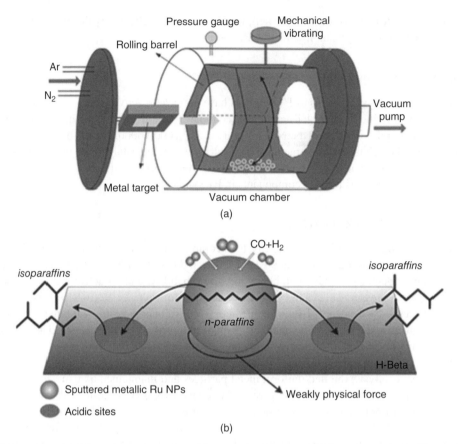

Scheme 7.8   *(a) Schematic representation of the sputtering apparatus. (b) One-step synthesis of isoparaffin from syngas on Ru/H-Beta catalyst. (See color plate section for the color representation of this figure.)*

mesoporous beta zeolites, prepared using different concentrations of NaOH (Figure 7.4A). A high concentration of NaOH, however, led to a partial collapse of the zeolite structure (Figure 7.4A, panel f). The catalytic activity of the beta zeolites is shown in Figure 7.4B. The highest selectivity to C_5–C_{11} hydrocarbons (67% for this series of catalysts) was attained over a 0.15 M NaOH solution-treated beta zeolite-supported Ru catalyst. The selectivity of CH_4 and C_{12+} hydrocarbons was also decreased, due to the mesoporosity and unique acidity of the meso-beta sample in providing rapid reactant and product diffusion, thereby contributing to the selective hydrocracking of heavier C_5–C_{11} hydrocarbons.

7.3.4.2   Co-Based Zeolite Catalyst

Compared to iron catalysts, cobalt-based catalysts are also very active in FT reactions. Notably, they demonstrate lower water-gas-shift reaction (WGSR) activities and superior reaction conditions, higher yields, and longer lifetimes, mainly producing linear and long-chain paraffins [93,105]. However, in order to obtain a controlled product distribution

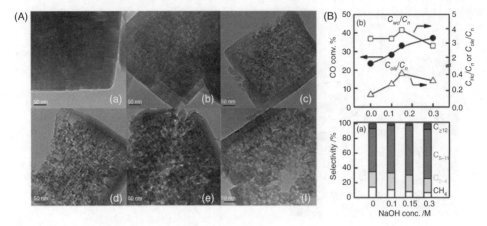

Figure 7.4 *(A) Transmission electron microscopy images of beta zeolite with various NaOH concentrations. (a) H-beta; (b) meso-beta-0.05 M NaOH; (c) meso-beta-0.15 M NaOH; (d) meso-beta-0.3 M NaOH; (e) meso-beta-0.5 M NaOH; (f) meso-beta-0.7 M NaOH. (B) Product selectivity for different catalysts.*

and create a high gasoline selectivity, bifunctional catalysts are required. Usually, the catalytic system contains both an FT metal for the CO hydrogenation reaction to produce heavier hydrocarbons (the primary reaction), and an acidic zeolite to hydrocrack and isomerize these heavier hydrocarbons (the secondary reaction). To overcome the severe transport limitations for the reactants and products of microporous zeolites, a new strategy of hierarchical zeolites with combined mesopores and zeolitic microporous walls, as highly active catalysts, has been reported in recent years [106–112]. Sasaki *et al.* first reported a dealumination method to generate mesoporous channels [113], but more recently the hierarchical zeolite has also been synthesized via a templating method, using various carbon templates [114–118]. Sartipi *et al.* recently investigated a mesoporous HZSM-5 zeolite-supported Co catalyst for the direct synthesis of C_5–C_9 isoparaffins, which exhibited a higher isoparaffin selectivity, as well as FTS activity that compared well with that of a conventional catalyst [95,119–121]. The present authors' group designed a hierarchical zeolite Y-supported cobalt bifunctional catalyst for the simple production of gasoline from syngas [80]. In this case, a two-step method consisting of acid leaching and base leaching was developed and applied to create hierarchical pores on a general microporous Y zeolite.

In this hierarchical zeolite catalyst (Scheme 7.9), with cobalt loaded onto the hierarchical zeolite Y catalyst, the small pores of zeolite could provide a large surface area and a high cobalt dispersion as well as rich acid sites, while the large pore structure would facilitate the diffusion of reactants and products in a direct isoparaffin synthesis reaction from syngas. Syngas first passes through the metallic Co particles to form long-chain *n*-paraffins, which subsequently are convert to isoparaffin through hydrocracking and isomerization reactions at the strong acidic sites of the zeolite.

In the FT synthesis reaction, zeolite Y acts not only as a support but also as an excellent hydrocracking and isomerization catalyst, owing to its acid sites, special pores, cavities, and regular channels. The isoparaffin selectivity is clearly enhanced in line with the base

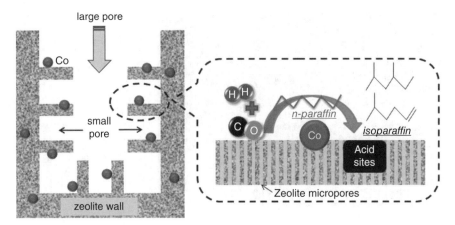

Scheme 7.9 *Representation of the hierarchical zeolite catalyst for one-step isoparaffin synthesis.* (See color plate section for the color representation of this figure.)

leaching time on an acid-treated zeolite support (Y-AB*x*, A: acid leaching, B: base leaching, where '*x*' indicates the base leaching time). Among the tested catalysts, Co/Y-AB4 exhibited the highest isoparaffin selectivity of 52.3%, because of a high Brønsted/Lewis acid ratio (3.14) and hierarchical structure. Compared with Y-AB4, both Y-A (only acid leaching) and Y-B (only base leaching) had a higher B/L ratio, but the surface and volume of mesopores were lower than that of Y-AB4. The results indicated that a strong B/L ratio and textural properties are effective for improved FT synthetic activity and isoparaffin selectivity. The product distribution of the FT reaction over Co/Y-P, Co/Y-A, Co/Y-B and Co/Y-AB*x* series catalysts is presented in Figure 7.5. Generally, FT synthesis products are normal aliphatic hydrocarbons with few olefins and isoparaffins. However, when using Co/Y-AB*x* catalysts with varied hierarchical zeolite Y as supports, the product selectivity – especially isoparaffin selectivity – can be easily tuned with the restrained but simultaneous formation of light hydrocarbons of C_{1-4}.

7.3.4.3   Fe-Based Zeolite Catalyst

Fe-based catalysts have much lower methane selectivity and cost than Ru Co-based catalysts for FT synthesis, which are true candidates for direct isoparaffin synthesis [90,122–124]. Yoneyama *et al.* reported a hybrid catalyst with a combination of conventional Fe-based FT catalyst and HZSM-5 zeolite to achieve a highly selective isoparaffin synthesis via a one-step process [97]. A novel Raney Fe@ZSM-5 catalyst was prepared via a simple one-pot strategy using an FeAl alloy [122]. The C_{12} hydrocarbons were completely suppressed, and CH_4 selectivity was substantially decreased, while the gasoline-range hydrocarbons were the dominant products.

Recently, the present authors' group developed a fused iron-based capsule catalyst without any organic template [51], which provided a remarkable increase in the isoparaffin/n-paraffin ratio compared to the core catalyst (see Scheme 7.10). In the case of this core–shell structure, fused iron (FI) as core can be easily modified via small amounts of 3-aminopropyltrimethoxysilane (AP-TMS) due to its surface hydroxyl groups, leading the

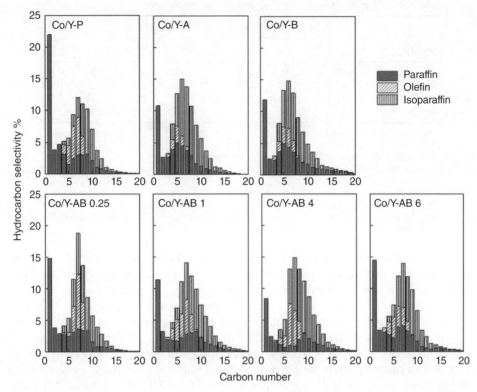

Figure 7.5 *Product distribution for the prepared catalysts.* (See color plate section for the color representation of this figure.)

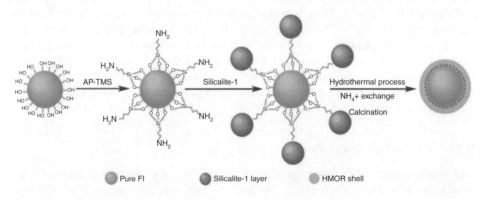

Scheme 7.10   *The synthesis procedure of the HMOR/FI capsule catalyst without a template.*

Table 7.5 *FT synthesis performances of various catalysts.*

Catalyst[a]	CO conv. [%]	Sel. [%]		Hydrocarbon sel. [%]				Iso/n (n ≥ 4)[c]
		CH_4[b]	CO_2	C_n	$C_=$	C_{iso}	C12+	
FI[d]	19	19.4	36.7	51.1	45.9	3	1.8	0.11
Silicalite-1/FI[e]	21.4	19.9	37	52.5	45	2.5	2.1	0.09
HMOR+FI[f]	26.5	14.6	37.2	47.6	42.7	9.7	0.6	0.37
HMOR/FI[g]	47.5	16	30.5	45	40.8	14.2	0.5	0.87

[a]FT synthesis reaction conditions: 0.5 g; $H_2/CO = 1/1$; 1.0 MPa; 300 °C; $W_{FI}/F_{(CO+H2)} = 10$ g_{FI} h mol^{-1}; TOS = 20 h.
[b]The calculation of CH_4 selectivity was based on all the hydrocarbons.
[c]iso/n was the ratio of iso-paraffin to n-paraffin with carbon number more than 3.
[d]Fused iron was denoted as FI.
[e]Silicalite-1 was attached on the surface of FI via AP-TMS.
[f]Physical mixture of FI and HMOR zeolite.
[g]Capsule catalyst with HMOR zeolite as shell and FI as core; this result was the mean value of repeated experiments.

FI surface to adsorb silicalite-1 zeolite. Subsequently, a HMOR shell-coated Silicalite-1/FI core was synthesized via a conventional hydrothermal synthetic method. The FT synthesis performances of various catalysts are listed in Table 7.5. The pure FI catalyst exhibited a CO conversion of 19.0%, which was slightly lower than that of Silicalite-1/FI. This difference was possibly attributed to the higher adsorbing availability of hydrogen on the surface decoration with silicate group and followed silicalite-1, further promoting the CO conversion [125]. The HMOR/FI capsule catalyst showed the highest CO conversion of 47.5% of all catalysts, while the physical mixture catalyst reduced the CO conversion to 26.5%. It was interesting that the CO conversion of the HMOR/FI capsule catalyst was higher than that of the physical mixture counterpart. This may have occurred because the intimate contact of the acidic zeolite and core catalyst allowed the decomposition of any waxy hydrocarbons that had accumulated in a layer at the surface much more efficiently than had the physical mixture catalyst, thereby accelerating reactant diffusion and adsorption. The acidic HMOR zeolite, as an important cracking and shape-selectivity catalyst, affects not only the catalytic activity but also FT synthesis product selectivity, which in turn provides acidic sites for the hydrocracking and isomerization of the primary hydrocarbon products. The linear hydrocarbons formed in the FI core (FT catalyst) must diffuse through the acidic zeolite shell. This may occur more efficiently due to a quicker access to the active sites in micropores over the capsule structure, resulting in a higher isoparaffin selectivity.

7.3.5 Conclusions and Outlook

The transformation of syngas to gasoline is one of the most practicable routes for the sustainable production of especially bio-gasoline synthesis from bio-syngas. Although details of the bio-syngas to bio-gasoline route have been reports only rarely, much interest will surely be aroused during at the next decade with the development of bifunctional FT catalysts and major advances in FT technology. Bio-syngas can be produced abundantly by biomass gasification technologies. Yet, a major problem or challenge of the FT reaction remains selectivity control, and the development of efficient bifunctional catalysts with controlled selectivities or tuned product distribution is a highly desirable goal. A combination of FT catalysts and acidic zeolite catalysts – as bifunctional catalysts – will provide both hydrocracking and isomerization reactions for tandem reactions in the synthesis of target products.

References

1. Wang, Y.L., Zhou, J.X., Guo, X.W. (2015) Catalytic hydrogenolysis of glycerol to propanediols: a review. *RSC Adv.*, **5**, 74611–74628.
2. Zhou, C.H.C., Beltramini, J.N., Fan, Y.X., Lu, G.Q.M. (2008) Chemoselective catalytic conversion of glycerol as a biorenewable source to valuable commodity chemicals. *Chem. Soc. Rev.*, **37**, 527–549.
3. Lee, C.S., Aroua, M.K., Daud, W.M.A.W., Cognet, P., Peres-Lucchese, Y., Fabre, P.L., Reynes, O., Latapie, L. (2015) A review: Conversion of bioglycerol into 1,3-propanediol via biological and chemical method. *Renew. Sust. Energ. Rev.*, **42**, 963–972.
4. Zope, B.N., Hibbitts, D.D., Neurock, M., Davis, R.J. (2010) Reactivity of the gold/water interface during selective oxidation catalysis. *Science*, **330**, 74–78.
5. Villa, A., Veith, G.M., Prati, L. (2010) Selective oxidation of glycerol under acidic conditions using gold catalysts. *Angew. Chem. Int. Ed.*, **49**, 4499–4502.
6. Villa, A., Gaiassi, A., Rossetti, I., Bianchi, C.L., van Benthem, K., Veith, G.M., Prati, L. (2010), Au on MgAl$_2$O$_4$ spinels: The effect of support surface properties in glycerol oxidation. *J. Catal.*, **275**, 108–116.
7. Crotti, C., Kaspar, J., Farnetti, E. (2010) Dehydrogenation of glycerol to dihydroxyacetone catalyzed by iridium complexes with P-N ligands. *Green Chem.*, **12**, 1295–1300.
8. Zhang, Y.H., Zhang, N., Tang, Z.R., Xu, Y.J. (2013) Identification of Bi$_2$WO$_6$ as a highly selective visible-light photocatalyst toward oxidation of glycerol to dihydroxyacetone in water. *Chem. Sci.*, **4**, 1820–1824.
9. Painter, R.M., Pearson, D.M., Waymouth, R.M. (2010) Selective catalytic oxidation of glycerol to dihydroxyacetone. *Angew. Chem. Int. Ed.*, **49**, 9456–9459.
10. Lari, G.M., Mondelli, C., Perez-Ramirez, J. (2015) Gas-phase oxidation of glycerol to dihydroxyacetone over tailored iron zeolites. *ACS Catal.*, **5**, 1453–1461.
11. Kwon, Y., Birdja, Y., Spanos, I., Rodriguez, P., Koper, M.T.M. (2012) Highly selective electro-oxidation of glycerol to dihydroxyacetone on platinum in the presence of bismuth. *ACS Catal.*, **2**, 759–764.
12. Hirasawa, S., Watanabe, H., Kizuka, T., Nakagawa, Y., Tomishige, K. (2013) Performance, structure and mechanism of Pd-Ag alloy catalyst for selective oxidation of glycerol to dihydroxyacetone. *J. Catal.*, **300**, 205–216.
13. Hirasawa, S., Nakagawa, Y., Tomishige, K. (2012) Selective oxidation of glycerol to dihydroxyacetone over a Pd-Ag catalyst. *Catal. Sci. Technol.*, **2**, 1150–1152.
14. Katryniok, B., Paul, S., Dumeignil, F. (2013) Recent developments in the field of catalytic dehydration of glycerol to acrolein. *ACS Catal.*, **3**, 1819–1834.
15. Alhanash, A., Kozhevnikova, E.F., Kozhevnikov, I.V. (2010) Gas-phase dehydration of glycerol to acrolein catalysed by caesium heteropoly salt. *Appl. Catal. A. Gen.*, **378**, 11–18.
16. Jia, C.J., Liu, Y., Schmidt, W., Lu, A.H., Schuth, F. (2010) Small-sized HZSM-5 zeolite as highly active catalyst for gas phase dehydration of glycerol to acrolein. *J. Catal.*, **269**, 71–79.
17. Ulgen, A., Hoelderich, W.G. (2009) Conversion of glycerol to acrolein in the presence of WO$_3$/ZrO$_2$ Catalysts. *Catal. Lett.*, **131**, 122–128.
18. Tao, L.Z., Yan, B., Liang, Y., Xu, B.Q. (2013) Sustainable production of acrolein: catalytic performance of hydrated tantalum oxides for gas-phase dehydration of glycerol. *Green Chem.*, **15**, 696–705.

19. Vaidya, P.D., Rodrigues, A.E. (2009) Glycerol reforming for hydrogen production: a review. *Chem. Eng. Technol.*, **32**, 1463–1469.

20. Zhou, J.X., Guo, L.Y., Guo, X.W., Mao, J.B., Zhang, S.G. (2010) Selective hydrogenolysis of glycerol to propanediols on supported Cu-containing bimetallic catalysts. *Green Chem.*, **12**, 1835–1843.

21. Ma, L., He, D.H. (2010) Influence of catalyst pretreatment on catalytic properties and performances of Ru-Re/SiO$_2$ in glycerol hydrogenolysis to propanediols. *Catal. Today*, **149**, 148–156.

22. Li, Y.M., Liu, H.M., Ma, L., He, D.H. (2014) Glycerol hydrogenolysis to propanediols over supported Pd-Re catalysts. *RSC Adv.*, **4**, 5503–5512.

23. Guo, L.Y., Zhou, J.X., Mao, J.B., Guo, X.W., Zhang, S.G. (2009) Supported Cu catalysts for the selective hydrogenolysis of glycerol to propanediols. *Appl. Catal. A. Gen.*, **367**, 93–98.

24. Gandarias, I., Arias, P.L., Requies, J., Guemez, M.B., Fierro, J.L.G. (2010) Hydrogenolysis of glycerol to propanediols over a Pt/ASA catalyst: The role of acid and metal sites on product selectivity and the reaction mechanism. *Appl. Catal. B. Environ.*, **97**, 248–256.

25. Daniel, O.M., DeLaRiva, A., Kunkes, E.L., Datye, A.K., Dumesic, J.A., Davis, R.J. (2010) X-ray absorption spectroscopy of bimetallic Pt-Re catalysts for hydrogenolysis of glycerol to propanediols. *ChemCatChem*, **2**, 1107–1114.

26. Xia, S.X., Zheng, L.P., Ning, W.S., Wang, L.N., Chen, P., Hou, Z.Y. (2013) Multiwall carbon nanotube-pillared layered Cu-0.4/Mg5.6Al$_2$O8.6: an efficient catalyst for hydrogenolysis of glycerol. *J. Mater. Chem. A*, **1**, 11548–11552.

27. Miyazawa, T., Kusunoki, Y., Kunimori, K., Tomishige, K. (2006) Glycerol conversion in the aqueous solution under hydrogen over Ru/C plus an ion-exchange resin and its reaction mechanism. *J. Catal.*, **240**, 213–221.

28. Nakagawa, Y., Tomishige, K. (2011) Heterogeneous catalysis of the glycerol hydrogenolysis. *Catal. Sci. Technol.*, **1**, 179–190.

29. Akiyama, M., Sato, S., Takahashi, R., Inui, K., Yokota, M. (2009) Dehydration-hydrogenation of glycerol into 1,2-propanediol at ambient hydrogen pressure. *Appl. Catal. A. Gen.*, **371**, 60–66.

30. Nakagawa, Y., Shinmi, Y., Koso, S., Tomishige, K. (2010) Direct hydrogenolysis of glycerol into 1,3-propanediol over rhenium-modified iridium catalyst. *J. Catal.*, **272**, 191–194.

31. Amada, Y., Shinmi, Y., Koso, S., Kubota, T., Nakagawa, Y., Tomishige, K. (2011) Reaction mechanism of the glycerol hydrogenolysis to 1,3-propanediol over Ir-ReO$_x$/SiO$_2$ catalyst. *Appl. Catal. B. Environ.*, **105**, 117–127.

32. Chaminand, J., Djakovitch, L., Gallezot, P., Marion, P., Pinel, C., Rosier, C. (2004) Glycerol hydrogenolysis on heterogeneous catalysts. *Green Chem.*, **6**, 359–361.

33. Huang, L., Zhu, Y.L., Zheng, H.Y., Ding, G.Q., Li, Y.W. (2009) Direct conversion of glycerol into 1,3-propanediol over Cu-H$_4$SiW$_{12}$O$_{40}$/SiO$_2$ in vapor phase. *Catal. Lett.*, **131**, 312–320.

34. Xia, S.X., Nie, R.F., Lu, X.Y., Wang, L.N., Chen, P., Hou, Z.Y. (2012) Hydrogenolysis of glycerol over Cu$_{0.4}$/Zn$_{5.6-x}$Mg$_x$Al$_2$O$_{8.6}$ catalysts: The role of basicity and hydrogen spillover. *J. Catal.*, **296**, 1–11.

35. Feng, J., Fu, H., Wang, J., Li, R., Chen, H., Li, X. (2008) Hydrogenolysis of glycerol to glycols over ruthenium catalysts: Effect of support and catalyst reduction temperature. *Catal. Commun.*, **9**, 1458–1464.

36. Chai, S.H., Wang, H.P., Liang, Y., Xu, B.Q. (2008) Sustainable production of acrolein: gas-phase dehydration of glycerol over 12-tungstophosphoric acid supported on ZrO$_2$ and SiO$_2$. *Green Chem.*, **10**, 1087–1093.

37. Chai, S.H., Tao, L.Z., Yan, B., Vedrine, J.C., Xu, B.Q. (2014) Sustainable production of acrolein: effects of reaction variables, modifiers doping and ZrO$_2$ origin on the performance of WO$_3$/ZrO$_2$ catalyst for the gas-phase dehydration of glycerol. *RSC Adv.*, **4**, 4619–4630.

38. Chai, S.H., Wang, H.P., Liang, Y., Xu, B.Q. (2007) Sustainable production of acrolein: investigation of solid acid-base catalysts for gas-phase dehydration of glycerol. *Green Chem.*, **9**, 1130–1136.

39. Atia, H., Armbruster, U., Martin, A. (2008) Dehydration of glycerol in gas phase using heteropolyacid catalysts as active compounds. *J. Catal.*, **258**, 71–82.
40. Tsukuda, E., Sato, S., Takahashi, R., Sodesawa, T. (2007) Production of acrolein from glycerol over silica-supported heteropoly acids. *Catal. Commun.*, **8**, 1349–1353.
41. Tao, L.Z., Chai, S.H., Zuo, Y., Zheng, W.T., Liang, Y., Xu, B.Q. (2010) Sustainable production of acrolein: Acidic binary metal oxide catalysts for gas-phase dehydration of glycerol. *Catal. Today*, **158**, 310–316.
42. Ma, L., He, D.H. (2009) Hydrogenolysis of glycerol to propanediols over highly active Ru-Re bimetallic catalysts. *Top. Catal.*, **52**, 834–844.
43. Kusunoki, Y., Miyazawa, T., Kunimori, K., Tomishige, K. (2005) Highly active metal-acid bifunctional catalyst system for hydrogenolysis of glycerol under mild reaction conditions. *Catal. Commun.*, **6**, 645–649.
44. D'Hondt, E., de Vyver, S.V., Sels, B.F., Jacobs, P.A. (2008) Catalytic glycerol conversion into 1,2-propanediol in absence of added hydrogen. *Chem. Commun.*, 6011–6012.
45. Jiang, N., Yang, G., Zhang, X., Wang, L., Shi, C.Y., Tsubaki, N. (2011) A novel silicalite-1 zeolite shell encapsulated iron-based catalyst for controlling synthesis of light alkenes from syngas. *Catal. Commun.*, **12**, 951–954.
46. Yang, G., He, J., Zhang, Y., Yoneyama, Y., Tan, Y., Han, Y., Vitidsant, T., Tsubaki, N. (2008) Design and modification of zeolite capsule catalyst, a confined reaction field, and its application in one-step isoparaffin synthesis from syngas. *Energy Fuel*, **22**, 1463–1468.
47. He, J., Liu, Z., Yoneyama, Y., Nishiyama, N., Tsubaki, N. (2006) Multiple-functional capsule catalysts: A tailor-made confined reaction environment for the direct synthesis of middle isoparaffins from syngas. *Chem. Eur. J.*, **12**, 8296–8304.
48. Bao, J., He, J., Zhang, Y., Yoneyama, Y., Tsubaki, N. (2008) A core/shell catalyst produces a spatially confined effect and shape selectivity in a consecutive reaction. *Angew. Chem. Int. Ed.*, **47**, 353–356.
49. Li, X., He, J., Meng, M., Yoneyama, Y., Tsubaki, N. (2009) One-step synthesis of H-Beta zeolite-enwrapped Co/Al$_2$O$_3$ Fischer–Tropsch catalyst with high spatial selectivity. *J. Catal.*, **265**, 26–34.
50. Bao, J., Yang, G., Okada, C., Yoneyama, Y., Tsubaki, N. (2011) H-type zeolite coated iron-based multiple-functional catalyst for direct synthesis of middle isoparaffins from syngas. *Appl. Catal. A. Gen.*, **394**, 195–200.
51. Lin, Q., Yang, G., Li, X., Yoneyama, Y., Wan, H., Tsubaki, N. (2013) A catalyst for one-step isoparaffin production via Fischer–Tropsch synthesis: Growth of a H-Mordenite shell encapsulating a fused iron core. *ChemCatChem*, **5**, 3101–3106.
52. Yang, G., Tsubaki, N., Shamoto, J., Yoneyama, Y., Zhang, Y. (2010) Confinement effect and synergistic function of H-ZSM-5/Cu-ZnO-Al$_2$O$_3$ capsule catalyst for one-step controlled synthesis. *J. Am. Chem. Soc.*, **132**, 8129–8136.
53. Yang, G., Thongkam, M., Vitidsant, T., Yoneyama, Y., Tan, Y., Tsubaki, N. (2011) A double-shell capsule catalyst with core–shell-like structure for one-step exactly controlled synthesis of dimethyl ether from CO$_2$-containing syngas. *Catal. Today*, **171**, 229–235.
54. Yang, G., Wang, D., Yoneyama, Y., Tan, Y., Tsubaki, N. (2012) Facile synthesis of H-type zeolite shell on a silica substrate for tandem catalysis. *Chem. Commun.*, **48**, 1263–1265.
55. Pinkaew K., Yang, G., Vitidsant, T., Jin, Y., Zeng, C., Yoneyama, Y., Tsubaki, N. (2013) A new core–shell-like capsule catalyst with SAPO-46 zeolite shell encapsulated Cr/ZnO for the controlled tandem synthesis of dimethyl ether from syngas. *Fuel*, **111**, 727–732.
56. Sun, J., Yang, G.H., Yoneyama, Y., Tsubaki, N. (2014) Catalysis chemistry of dimethyl ether synthesis. *ACS Catal.*, **4**, 3346–3356.
57. Yang, G., Kawata, Lin, Q., Wang, J., Jin, Y., Zeng, C., Yoneyama, Y., Tsubaki, N. (2013) Oriented synthesis of target products in liquid-phase tandem reaction over a tripartite zeolite capsule catalyst. *Chem. Sci.*, **4**, 3958–3964.

58. Jin, Y., Yang, R., Mori, Y., Sun, J., Taguchi, A., Yoneyama, Y., Abe, T., Tsubaki, N. (2013) Preparation and performance of Co-based capsule catalyst with the zeolite shell sputtered by Pd for direct isoparaffin synthesis from syngas. *Appl. Catal. A. Gen.*, **456**, 75–81.

59. ten Dam, J., Hanefeld, U. (2011) Renewable chemicals: dehydroxylation of glycerol and polyols. *ChemSusChem*, **4**, 1017–1034.

60. Christensen, C.H., Rass-Hansen, J., Marsden, C.C., Taarning, E., Egeblad, K. (2008) The renewable chemicals industry. *ChemSusChem*, **1**, 283–289.

61. Vennestrom, P.N.R., Osmundsen, C.M., Christensen, C.H., Taarning, E. (2011) Beyond petrochemicals: the renewable chemicals industry. *Angew. Chem. Int. Ed.*, **50**, 10502–10509.

62. Chum, H.L., Overend, R.P. (2001) Biomass and renewable fuels. *Fuel Process. Technol.*, **71**, 187–195.

63. Bhaumik, P., Dhepe, P.L. (2014) Exceptionally high yields of furfural from assorted raw biomass over solid acids. *RSC Adv.*, **4**, 26215–26221.

64. Kato, Y., Sekine, Y. (2013) One pot direct catalytic conversion of cellulose to hydrocarbon by decarbonation using Pt/H-beta zeolite catalyst at low temperature. *Catal. Lett.*, **143**, 418–423.

65. Taarning, E., Osmundsen, C.M., Yang, X.B., Voss, B., Andersen, S.I., Christensen, C.H. (2011) Zeolite-catalyzed biomass conversion to fuels and chemicals. *Energy Environ. Sci.*, **4**, 793–804.

66. Carlson, T.R., Jae, J., Lin, Y.C., Tompsett, G.A., Huber, G.W. (2010) Catalytic fast pyrolysis of glucose with HZSM-5: The combined homogeneous and heterogeneous reactions. *J. Catal.*, **270**, 110–124.

67. Lappas, A.A., Bezergianni, S., Vasalos, I.A. (2009) Production of biofuels via co-processing in conventional refining processes. *Catal. Today*, **145**, 55–62.

68. Carlson, T.R., Vispute, T.R., Huber, G.W. (2008) Green gasoline by catalytic fast pyrolysis of solid biomass-derived compounds. *ChemSusChem*, **1**, 397–400.

69. Carlson, T.R., Jae, J., Huber, G.W. (2009) Mechanistic insights from isotopic studies of glucose conversion to aromatics over ZSM-5. *ChemCatChem*, **1**, 107–110.

70. Zhang, Q., Qiu, K., Li, B., Jiang, T., Zhang, X.H., Ma, L.L., Wang, T.J. (2011) Isoparaffin production by aqueous phase processing of sorbitol over the Ni/HZSM-5 catalysts: Effect of the calcination temperature of the catalyst. *Fuel*, **90**, 3468–3472.

71. Huber, G.W., Cortright, R.D., Dumesic, J.A. (2004) Renewable alkanes by aqueous-phase reforming of biomass-derived oxygenates. *Angew. Chem. Int. Ed.*, **43**, 1549–1551.

72. Binod, P., Sindhu, R., Singhania, R.R., Vikram, S., Devi, L., Nagalakshmi, S., Kurien, N., Sukumaran, R.K., Pandey, A. (2010) Bioethanol production from rice straw: An overview. *Bioresource Technol.*, **101**, 4767–4774.

73. Garrote, G., Dominguez, H., Parajo, J.C. (2002) Autohydrolysis of corncob: study of non-isothermal operation for xylooligosaccharide production. *J. Food Eng.*, **52**, 211–218.

74. Saha, B.C. (2003) Hemicellulose bioconversion. *J. Ind. Microbiol. Biotechnol.*, **30**, 279–291.

75. Huang, Y.F., Chiueh, P.T., Shih, C.H., Lo, S.L., Sun, L.P., Zhong, Y., Qiu, C.S. (2015) Microwave pyrolysis of rice straw to produce biochar as an adsorbent for CO_2 capture. *Energy*, **84**, 75–82.

76. Zhang, H.Y., Xiao, R., Jin, B.S., Shen, D.K., Chen, R., Xiao, G.M. (2013) Catalytic fast pyrolysis of straw biomass in an internally interconnected fluidized bed to produce aromatics and olefins: Effect of different catalysts. *Bioresource Technol.*, **137**, 82–87.

77. Zhang, X.J., Zhao, T.S., Hara, N., Jin, Y.Z., Zeng, C.Y., Yoneyama, Y. (2014) Direct conversion of rice straw catalyzed by solid acid supported-Pt catalyst using in situ H_2 by ethanol steam reforming. *Fuel*, **116**, 34–38.

78. Zhang, Q., Cheng, K., Kang, J., Deng, W., Wang, Y. (2014) Fischer–Tropsch catalysts for the production of hydrocarbon fuels with high selectivity. *ChemSusChem*, **7**, 1251–1264.

79. Lv, P.M., Yuan, Z.H., Wu, C.Z., Ma, L.L., Chen, Y., Tsubaki, N. (2007) Bio-syngas production from biomass catalytic gasification. *Energy Conserv. Manage.*, **48**, 1132–1139.

80. Xing, C., Yang, G., Wu, M., Yang, R., Tan, L., Zhu, P., Wei, Q., Li, J., Mao, J., Yoneyama, Y., Tsubaki, N. (2015) Hierarchical zeolite Y-supported cobalt bifunctional catalyst for facilely tuning the product distribution of Fischer–Tropsch synthesis. *Fuel*, **148**, 48–57.

81. Zhang, Q.H., Kang, J.C., Wang, Y. (2010) Development of novel catalysts for Fischer–Tropsch synthesis: Tuning the product selectivity. *ChemCatChem*, **2**, 1030–1058.

82. Cheng, K., Kang, J., Huang, S., You, Z., Zhang, Q., Ding, J., Hua, W., Lou, Y., Deng, W., Wang, Y. (2012) Mesoporous beta zeolite-supported ruthenium nanoparticles for selective conversion of synthesis gas to C_5–C_{11} isoparaffins. *ACS Catal.*, **2**, 441–449.

83. Hamelinck, C.N., Faaij, A.P.C. (2002) Future prospects for production of methanol and hydrogen from biomass. *J. Power Sources*, **111**, 1–22.

84. Chmielniak, T., Sciazko, M. (2003) Co-gasification of biomass and coal for methanol synthesis. *Appl. Energ.*, **74**, 393–403.

85. Tijmensen, M.J.A., Faaij, A.P.C., Hamelinck, C.N., van Hardeveld, M.R.M. (2002) Exploration of the possibilities for production of Fischer–Tropsch liquids and power via biomass gasification. *Biomass Bioenerg.*, **23**, 129–152.

86. Dong, Y.J., Steinberg, M. (1997) Hynol – An economical process for methanol production from biomass and natural gas with reduced CO_2 emission. *Int. J. Hydrogen Energ.*, **22**, 971–977.

87. Magrini-Bair, K.A., Czernik, S., French, R., Parent, Y.O., Chornet, E., Dayton, D.C., Feik, C., Bain, R. (2007) Fluidizable reforming catalyst development for conditioning biomass-derived syngas. *Appl. Catal. A. Gen.*, **318**, 199–206.

88. Spivey, J.J., Egbebi, A. (2007) Heterogeneous catalytic synthesis of ethanol from biomass-derived syngas. *Chem. Soc. Rev.*, **36**, 1514–1528.

89. Kumar, A., Jones, D.D., Hanna, M.A. (2009) Thermochemical biomass gasification: A review of the current status of the technology. *Energies*, **2**, 556–581.

90. Xing, C., Sun, J., Chen, Q., Yang, G., Muranaka, N., Lu, P., Shen, W., Zhu, P., Wei, Q., Li, J., Mao, J., Yang, R., Tsubaki, N. (2015) Tunable isoparaffin and olefin yields in Fischer–Tropsch synthesis achieved by a novel iron-based micro–capsule catalyst. *Catal. Today*, **251**, 41–46.

91. Zhao, T.S., Chang, J., Yoneyama, Y., Tsubaki, N. (2005) Selective synthesis of middle isoparaffins via a two-stage Fischer–Tropsch reaction: Activity investigation for a hybrid catalyst. *Ind. Eng. Chem. Res.*, **44**, 769–775.

92. Li, X.H., Liu, X.H., Liu, Z.W., Asami, K., Fujimoto, K. (2005) Supercritical phase process for direct synthesis of middle iso-paraffins from modified Fischer–Tropsch reaction. *Catal. Today*, **106**, 154–160.

93. Sun, B., Qiao, M., Fan, K., Ulrich, J., Tao, F. (2011) Fischer–Tropsch synthesis over molecular sieve supported catalysts. *ChemCatChem*, **3**, 542–550.

94. Yao, M., Yao, N., Liu, B., Li, S., Xu, L.J., Li, X.N. (2015) Effect of SiO_2/Al_2O_3 ratio on the activities of CoRu/ZSM-5 Fischer–Tropsch synthesis catalysts. *Catal. Sci. Technol.*, **5**, 2821–2828.

95. Sartipi, S., Alberts, M., Santos, V.P., Nasalevich, M., Gascon, J., Kapteijn, F. (2014) Insights into the catalytic performance of mesoporous H-ZSM-5-supported cobalt in Fischer–Tropsch synthesis. *ChemCatChem*, **6**, 142–151.

96. Li, J., Tan, Y., Zhang, Q., Han, Y. (2010) Characterization of an HZSM-5/MnAPO-11 composite and its catalytic properties in the synthesis of high-octane hydrocarbons from syngas. *Fuel*, **89**, 3510–3516.

97. Yoneyama, Y., He, J., Morii, Y., Azuma, S., Tsubaki, N. (2005) Direct synthesis of isoparaffin by modified Fischer–Tropsch synthesis using hybrid catalyst of iron catalyst and zeolite. *Catal. Today*, **104**, 37–40.

98. Li, X., Asami, K., Luo, M., Michiki, K., Tsubaki, N., Fujimoto, K. (2003) Direct synthesis of middle iso-paraffins from synthesis gas. *Catal. Today*, **84**, 59–65.

99. Li, Y., Wang, T., Wu, C., Lv, Y., Tsubaki, N. (2008) Gasoline-range hydrocarbon synthesis over cobalt-based Fischer–Tropsch catalysts supported on SiO_2/HZSM-5. *Energy Fuel*, **22**, 1897–1901.

100. Yang G., Xing, C., Hirohama, W., Jin, Y., Zeng, C., Suehiro, Y., Wang, T., Yoneyama, Y., Tsubaki, N. (2013) Tandem catalytic synthesis of light isoparaffin from syngas via Fischer–Tropsch synthesis by newly developed core–shell-like zeolite capsule catalysts. *Catal. Today*, **215**, 29–35.

101. Jin, Y., Yang, G., Chen, Q., Niu, W., Lu, P., Yoneyama, Y., Tsubaki, N. (2015) Development of dual-membrane coated Fe/SiO$_2$ catalyst for efficient synthesis of isoparaffins directly from syngas. *J. Membr. Sci.*, **475**, 22–29.

102. Sun, J., Li, X., Taguchi, A., Abe, T., Niu, W., Lu, P., Yoneyama, Y., Tsubaki, N. (2014) Highly-dispersed metallic Ru nanoparticles sputtered on H-Beta zeolite for directly converting syngas to middle isoparaffins. *ACS Catal.*, **4**, 1–8.

103. Simonetti, D.A., Rass-Hansen, J., Kunkes, E.L., Soares, R.R., Dumesic, J.A. (2007) Coupling of glycerol processing with Fischer–Tropsch synthesis for production of liquid fuels. *Green Chem.*, **9**, 1073–1083.

104. Kang, J., Zhang, S., Zhang, Q., Wang, Y. (2009) Ruthenium nanoparticles supported on carbon nan-otubes as efficient catalysts for selective conversion of synthesis gas to diesel fuel. *Angew. Chem. Int. Ed.*, **48**, 2565–2568.

105. Davis, B.H. (2007) Fischer–Tropsch synthesis: Comparison of performances of iron and cobalt cata-lysts. *Ind. Eng. Chem. Res.*, **46**, 8938–8945.

106. Christensen, C.H. (2008) Templating mesoporous zeolites. *Chem. Mater.*, **20**, 946–960.

107. Meng, X., Xiao, F.S. (2014) Green routes for synthesis of zeolites. *Chem. Rev.*, **114**, 1522–1544.

108. Tao, Y.S., Kanoh, H., Abrams, L., Kaneko, K. (2006) Mesopore-modified zeolites: Preparation, char-acterization, and applications. *Chem. Rev.*, **106**, 896–910.

109. Chal, R., Gerardin, C., Bulut, M., van Donk, S. (2011) Overview and industrial assessment of synthesis strategies towards zeolites with mesopores. *ChemCatChem*, **3**, 67–81.

110. Chen, L.H., Li, X.Y., Rooke, J.C., Zhang, Y.H., Yang, X.Y., Tang, Y., Xiao, F.S., Su, B.L. (2012) Hier-archically structured zeolites: synthesis, mass transport properties and applications. *J. Mater. Chem.*, **22**, 17381–17403.

111. Silaghi, M.-C., Chizallet, C., Raybaud, P. (2014) Challenges on molecular aspects of dealumination and desilication of zeolites. *Micropor. Mesopor. Mater.*, **191**, 82–96.

112. Meng, X., Nawaz, F., Xiao, F.S. (2009) Templating route for synthesizing mesoporous zeolites with improved catalytic properties. *Nano Today*, **4**, 292–301.

113. Sasaki, Y., Suzuki, T., Takamura, Y., Saji, A., Saka, H. (198) Structure analysis of the mesopore in dealuminated zeolite Y by high-resolution TEM observation with slow scan CCD camera. *J. Catal.*, **178**, 94–100.

114. Christensen, C.H., Schmidt, I., Carlsson, A., Johannsen, K., Herbst, K. (2005) Crystals in crystals-nanocrystals within mesoporous zeolite single crystals. *J. Am. Chem. Soc.*, **127**, 8098–8102.

115. Schmidt I., Boisen, A., Gustavsson, E., Stahl, K., Pehrson, S., Dahl, S., Carlsson, A., Jacobsen, C.J.H. (2001) Carbon nanotube templated growth of mesoporous zeolite single crystals. *Chem. Mater.*, **13**, 4416–4418.

116. Janssen A.H., Schmidt, I., Jacobsen, C.J.H., Koster, A.J., de Jong, K.P. (2003) Exploratory study of mesopore templating with carbon during zeolite synthesis. *Micropor. Mesopor. Mater.*, **65**, 59–75.

117. Song, Y., Hua, Z., Zhu, Y., Zhou, X., Wu, W., Zhang, L., Shi, J. (2012) An in situ carbonaceous mesoporous template for the synthesis of hierarchical ZSM-5 zeolites by one-pot steam-assisted crys-tallization. *Chem. Asian J.*, **7**, 2772–2776.

118. Tao, Y., Hattori, Y., Matsumoto, A., Kanoh, H., Kaneko, K. (2005) Comparative study on pore struc-tures of mesoporous ZSM-5 from resorcinol-formaldehyde aerogel and carbon aerogel templating. *J. Phys. Chem. B*, **109**, 194–199.

119. Sartipi, S., Parashar, K., Makkee, M., Gascon, J., Kapteijn, F. (2013) Breaking the Fischer–Tropsch synthesis selectivity: Direct conversion of syngas to gasoline over hierarchical Co/H-ZSM-5 catalysts. *Catal. Sci. Technol.*, **3**, 572–575.

120. Sartipi, S., van Dijk, J.E., Gascon, J., Kapteijn, F. (2013) Toward bifunctional catalysts for the direct conversion of syngas to gasoline range hydrocarbons: H-ZSM-5 coated Co versus H-ZSM-5 supported Co. *Appl. Catal. A. Gen.*, **456**, 11–22.

121. Sartipi, S., Parashar, K., Valero-Romero, M.J., Santos, V.P., van der Linden, B., Makkee, M., Kapteijn, F., Gascon, J. (2013) Hierarchical H-ZSM-5-supported cobalt for the direct synthesis of gasoline-range hydrocarbons from syngas: Advantages, limitations, and mechanistic insight. *J. Catal.*, **305**, 179–190.

122. Sun, B., Yu, G., Lin, J., Xu, K., Pei, Y., Yan, S., Qiao, M., Fan, K., Zhang, X., Zong, B. (2012) A highly selective Raney Fe@HZSM-5 Fischer–Tropsch synthesis catalyst for gasoline production: one-pot synthesis and unexpected effect of zeolites. *Catal. Sci. Technol.*, **2**, 1625–1629.

123. Pour A.N., Zamani, Y., Tavasoli, A., Shahri, S.M.K., Taheri, S.A. (2008) Study on products distribution of iron and iron–zeolite catalysts in Fischer–Tropsch synthesis. *Fuel*, **87**, 2004–2012.

124. Fujimoto, K., Adachi, M., Tominaga, H. (1985) Direct synthesis of isoparaffins from synthesis gas. *Chem. Lett.*, 783–786.

125. Mogorosi, R.P., Fischer, N., Claeys, M., van Steen, E. (2012) Strong-metal-support interaction by molecular design: Fe-silicate interactions in Fischer–Tropsch catalysts. *J. Catal.*, **289**, 140–150.

8

Depolymerization of Lignin with Nanoporous Catalysts

Zhicheng Luo, Jiechen Kong, Liubi Wu, and Chen Zhao

Shanghai Key Laboratory of Green Chemistry and Chemical Process, School of Chemistry and Molecular Engineering, East China Normal University, China

8.1 Introduction

As the second most abundant polymer next to cellulose in Nature, lignin has a three-dimensional structure with the basic unit of C_9 propyl-phenol that is randomly linked by C–O and C–C bonds [1]. It is also the solo non-petroleum natural polymer resource, which contains abundant valuable renewable aromatics in its inherent structure. Because of the complicated molecular structure, lignin is considered as the most difficult polymer to be utilized by research groups, and in 2004 for example, only 2% of the 50 million tons of lignin available (extracted from the pulping and paper industry) was used to produce fuels with a low heating value [2]. The majority of lignin is discharged as waste.

Lignin is not well dissolved in common solvents due to its rigid structure, which consequently prevents contact between the active sites of catalysts and the inner structures of lignin [3]. Furthermore, the structures of lignin are prone to vigorous chemical transformations by alkaline or sulfonation treatment in the paper industry [4], and hence the modified lignin is relatively inert for conversion. The complexity of the lignin structure and further modifications by alkaline or sulfonation lead to a low level of utilization for lignin conversion. However, it should be noted that the potential high- energy density of benzene rings in lignin provides an available bioprecursor for the manufacture of aromatics and derivatives.

Nanoporous Catalysts for Biomass Conversion, First Edition. Edited by Feng-Shou Xiao and Liang Wang.
© 2018 John Wiley & Sons Ltd. Published 2018 by John Wiley & Sons Ltd.

On the other hand, lignin is attracting increasing attention due to its unique properties of abundance and sustainability, and an awareness of resource crises and environmental protection.

The main methods of lignin depolymerization include reductive and oxidative deconstruction, hydrolysis, catalytic cracking, and photonic, electronic or enzyme-induced depolymerization [5,6]. In general, the latter three approaches can be conducted under mild conditions, which are more energy-efficient and sustainable. The activities from these three techniques are relatively low, however, due to the low solubility of photocatalysts in electrolytes, the limited types of suitable enzymes with low activities, and so forth. By comparison, metal-catalyzed reduction and oxidation have shown to be more promising for lignin depolymerization, with the products formed typically being syringyl, guaiacyl, and phenyl derivatives from which phenolic aldehydes and acids are usually obtained when the oxidative technique is used.

Based on different resources and extraction techniques, lignin can be mainly classified as four types: sulfonate; alkali; enzymatic-hydrolysis; and organosolv. Sulfonate and alkali lignins are byproducts from the paper industry, and are generated by two different approaches. During the lignosulfonation process, the side chains of lignin are sulfonated and condensed (the condensation reaction also occurs during the alkali treatment). By comparison, organosolv and enzymatic lignins are obtained by extraction from organic solvents and enzyme-catalyzed hydrolysis, respectively. Unlike organosolv and enzymatic lignins, the structures of sulfonate and alkali lignins are dramatically changed. Consequently, the different properties of lignin structures require the development of diverse methods and catalysts for their depolymerization. For example, sulfur-tolerant catalysts should be used in the catalytic depolymerization of sulfonate lignin.

Previously, Huber *et al.* [5], Zakzeski *et al.* [6] and Li *et al.* [7] have reviewed several lignin valorization processes. The aim of this chapter is to provide a review of lignin conversion methods, especially using the approaches of reductive deconstruction in the presence and absence of hydrogen, as well as oxidation and hydrolysis in the liquid phase. Attention will be focused on aspects of the catalyst system, the reaction conditions, and mechanisms, and their advantages and problems in lignin depolymerization will be discussed. Moreover, some suggestions will be provided as to current techniques and the proposed uses for lignin in the future.

8.2 Developed Techniques for Lignin Depolymerization

8.2.1 Heterogeneous Noble Metal Catalyst System in the Presence of Hydrogen

The heterogeneous noble metals such as Ru, Pd, Pt are widely used in the direct hydrogenolysis of raw and pretreated lignin, due to their excellent capabilities in breaking down the C–O linkages in the presence of hydrogen (Table 8.1). Yan *et al.* [8] investigated the selective degradation of birth wood lignin over a number of noble catalysts including Ru/C, Rh/C, Pd/C, and Pt/C, via a two-step process in a solvent mixture of dioxane and water with an initial 4 MPa H_2, yielding around 42 wt% C_8–C_9 alkanes, 10 wt% C_{14}–C_{18} alkanes, and 11 wt% methanol. First, catalytic cleavage of the C–O–C bonds in lignin without disrupting the C–C linkages and the methoxy groups, led to the formation of phenolic dimers (see Figure 8.1). In a second step, these aromatic species were further hydrogenated

Table 8.1 Data of the depolymerization of lignin with noble metal catalysts.

Catalyst	S/C^a ratio (g g^{-1})	Reaction conditions		Solvent	Substrate	Major products	Conv. (%)	Reference
		T (°C)	P (MPa)					
Pt/C, H$_3$PO$_4$	40:3	200	4	dioxane/H$_2$O	Birch wood sawdust	C$_8$–C$_9$ cycloalkanes	42	[8]
Pt/C	20:1	350	3	ethanol	Asian lignin	Lignin-oil	77	[12]
Pt/Al$_2$O$_3$, NaOH	2:1	225	–	ethanol/H$_2$O	Lignin	Alkyl-phenolics	9	[15]
Pd/C, Hf(OTf)$_4$	4:1	150	4	Hf(OTf)$_4$	Oxydicyclo-hexane	Cyclohexane	93	[17]
Ru/TiO$_2$	20:1	400	10	-	Organosolv lignin	Lignin-oil	78	[19]
Ru/C, ZnCl$_2$	4:1	200	4	H$_2$O	Lignin	Phenol derivatives	62	[20]
Pd/C, ZnCl$_2$	100:1	225	3.5	methanol	Milled poplar wood	Propylguaiacol, propylsy-ringol	57	[22]

^{a}S/C : Substrate/catalyst ratio.

and concomitantly underwent hydrogenolysis of the C–O bonds, thereby transforming the lignin to C$_8$–C$_9$ and C$_{14}$–C$_{18}$ alkanes. Furthermore, synergistic results were also reported on bimetallic NiM (M = Au and Ru) catalysts for organosolv lignin hydrogenolysis in water by Zhang *et al.* [9]. This synergistic effect can be attributed to the increased surface atoms, enhanced H$_2$ dissociation, substrate activation (compared to Ni), and inhibited benzene ring hydrogenation. Similar to this catalytic system, 3.1% 4-ethylphenol and 1.3% 4-ethyguaiacol could be obtained from hydrogenolysis of corn stalk lignin at 275 °C and 2 MPa H$_2$ with ethanol-water as the solvent and Ru/C as the catalyst [10].

In an alternative approach, Laskar *et al.* [11] reported an aqueous phase one-pot process to depolymerize lignin, attaining a 35–60 wt% conversion and 65–70 wt% product selectivity for aromatics in the presence of noble metals (Ru, Rh, Pd, and Pt) over Al$_2$O$_3$ (or C) supports and solid zeolites (Y) catalyst under conditions of 250 °C and 5 MPa H$_2$. Using the supercritical alcohols (methanol, ethanol, and 2-propanol) as solvents, Kim *et al.* [12] showed that lignin was directly depolymerized with noble catalysts (Ru/C, Ni/C, Pd/C, and Pt/C) to produce lignin oil, in which the combination of ethanol and Pt/C proved to be an excellent catalyst for producing the highest yield of lignin oil (77.4 wt%). The produced bio-oil was mainly composed of a phenolic mixture, with a small amount of char.

Later, Van den Bosch *et al.* [13] reported a liquid reductive processing of birch wood with commercial Ru/C or Pd/C catalysts in a solvent of methanol, yielding about 50% conversion with a select set of phenolic monomers and a variety of phenolic dimers and oligomers. Changing the catalyst from Ru/C to Pd/C led to a drastic increase in the OH-content of the lignin-derived products, in particular for the phenolic monomers. In addition, a proposed catalytic lignocellulose bio-refinery process was presented [14] (see Figure 8.2) in which both polysaccharide and lignin components were valorized into a handful of chemicals. The selective delignification of lignocellulose in methanol through simultaneous solvolysis and catalytic hydrolysis resulted in a lignin oil that was rich in phenolic monomers. Meanwhile,

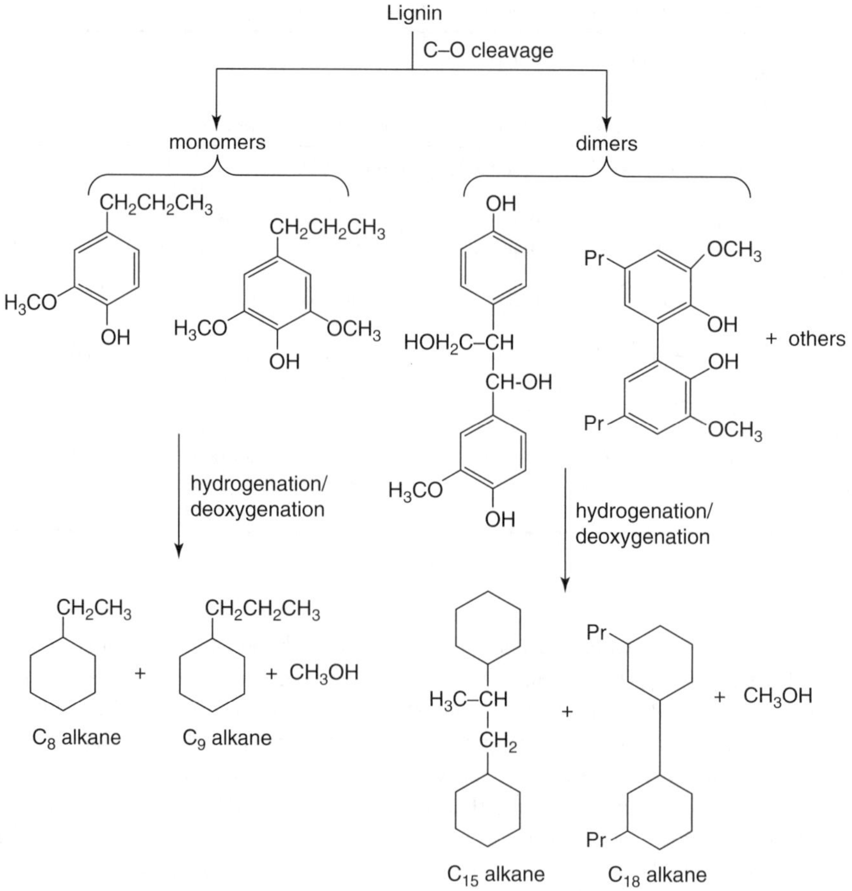

Figure 8.1 *A two-step route for the depolymerization of lignin and further hydrodeoxygenation to alkanes and methanol in the liquid phase.*

an almost quantitative retention of original cellulose and a large fraction of hemicellulose was obtained. The lignin oil consisted of a select set of methoxy-phenol monomers and dimers, which could be further upgraded into downstream products, including the possibility of producing methanol. The recovered solvent fraction also contained methyl acetate, formed from the transesterification of hemicellulose acetyl groups. Catalytic reductive splitting converts the pulp into bioderived sugar polyols, also demonstrating multiple use of the Ru/C catalyst.

A two-step approach was also reported by Weckhuysen's group [15] (see Figure 8.3). Lignin was first depolymerized in an alkaline ethanol-water (1:1) solution over Pt/Al$_2$O$_3$ at 225 °C, and subsequent hydrodeoxygenation of the produced lignin-oil via liquid-phase reforming was performed in dodecane with catalysts of CoMo/Al$_2$O$_3$ and Mo$_2$C/CNF at 300 °C. The products after hydrodeoxygenation were oxygen-free aromatics such as benzene, toluene and xylenes, and mono-, bis-, and tris-oxygenated products. Furthermore, Jongerius *et al.* [16] investigated the stability of Pt/Al$_2$O$_3$ catalysts in solution under

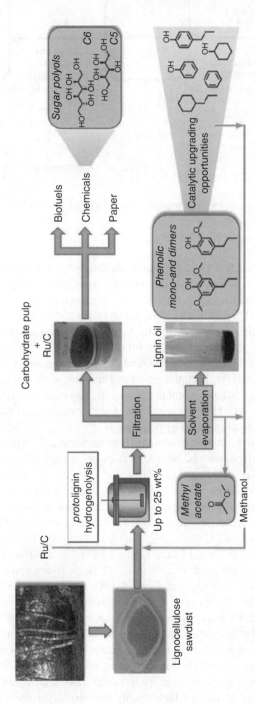

Figure 8.2 *Schematic representation of an integrated bio-refinery process for the full utilization of lignocellulose.* (See color plate section for the color representation of this figure.)

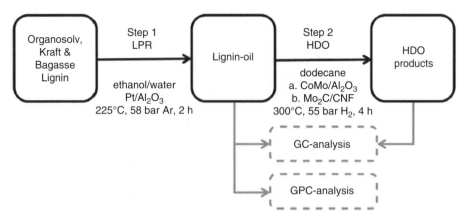

Figure 8.3   *Two-step approach for the hydrodeoxygenation of organosolv lignin, including depolymeriza-tion and the liquid-phase reforming (LPR) reaction.*

liquid-phase reforming reaction conditions and showed that, in the absence of reactants, Al_2O_3 was converted into böhmite that caused the formation of a crystalline phase and a loss of surface area, while catalysts treated in the presence of lignin showed almost no formation of böhmite. This result was interpreted as an interaction between the larger number of oxygen functionalities obtained by lignin depolymerization with the alumina that led to a slower transformation of the support oxide.

For investigation of the lignin model compound, using heterogeneous Pd/C and homogeneous $M(OTf)_n$ tandem catalysts [17], the ethers and alcohols were efficiently converted to saturated hydrocarbons at 4 MPa H_2 without the solvent. The ytterbium(III) triflate complex broke the O–alkyl bond via C–H to O–H hydrogen transfer to form an alkene intermediate, which was subsequently hydrogenated by palladium.

Most studies on the depolymerization of lignin have been carried out in a solvent. In many cases, the presence of solvent may be beneficial to disperse the lignin in the liquid phase, to promote contacts between lignin and the active sites of catalysts, and to serve as an additional hydrogen source, while the absence of a solvent would reduce the cost. Meier *et al.* [18] screened various catalysts (Pd/C, Raney Ni, and supported NiMo catalysts) for the catalytic hydro-treatment of organosolv lignin in the absence of a solvent. Pd/C obtained the highest yield, which mainly consisted of methyl, ethyl, and propyl-cyclohexanols at conditions of 380 °C, 10 MPa H_2, and 15 min. Later, Kloekhorst and Heeres [19] conducted a catalyst screening study in a batch reactor in which Acell lignin underwent hydrotreatment using supported Ru (C, Al_2O_3, and TiO_2), Pd (C, Al_2O_3), and Cu/ZrO_2 catalysts at 400 °C and 10 MPa for 4 h. In this case, Ru/TiO_2 showed the best performance with 78% conversion, together with 9.1% alkyl-phenols, 2.5% aromatics, and 3.5% catechols.

Highly concentrated $ZnCl_2$ solutions were noted to have been widely used in coal liquefaction for the disruption of ether bonds and the removal of heteroatoms. Based on lignin's similar chemical structure to coal, its degradation in aqueous $ZnCl_2$ solution under mild conditions was investigated by Wang *et al.* [20]. In this case, almost half of the softwood lignin was catalyzed to oil products which mainly included alkyl-phenols. Moreover, by adding Ru/C and hydrogen as a co-catalyst, the phenolic products were further converted into more stable cyclic hydrocarbons via hydrodeoxygenation.

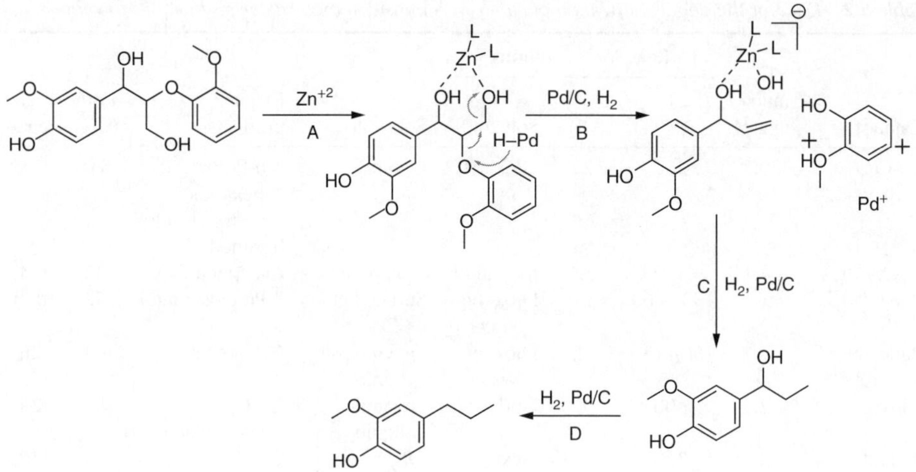

Scheme 8.1 *Proposed mechanism for the cleavage and HDO of β-O-4 ether linkage using Pd/C and Zn²⁺ catalysts, and the role of Zn²⁺ and synergy effect with a palladium hydride catalyst.*

Parsell *et al.* [21] and Klein *et al.* [22] also reported a similar bimetallic Pd/C and Zn²⁺ catalytic system that can selectively cleave recalcitrant ether bonds of lignin model dimers and polymers under relatively mild conditions at 150 °C and 2 MPa H₂, using methanol as a solvent (Table 8.1). The proposed mechanism for cleavage and HDO of the β-O-4 ether linkage by Pd/C and Zn²⁺ catalysis is displayed in Scheme 8.1, in which Zn²⁺ played the role of activating and facilitating removal of the hydroxyl group at the C$_\gamma$ position of β-O-4 linkages. The bifunctional Pd/C and Zn²⁺ catalyst system was also applied to the selective cleavage and HDO of milled poplar wood, attaining 15% propylguaiacol and 29% propylsyringl yields. Notably, a high substrate/catalyst ratio (100:1) was used. These results all suggested that the introduction of Zn²⁺ into noble-based catalysts is far more effective than using the noble metal alone for the hydrotreatment of lignin.

8.2.2 Heterogeneous Transition Metal Catalyst System in the Presence of Hydrogen

Compared to noble metal catalysts, the transition metals possess relatively low hydrogenolysis capabilities (Table 8.2). Early studies were mainly focused on the structure elucidation of lignin, using heterogeneous metal catalysts. In 1938, Harris *et al.* [23] reported that a copper–chromium oxide catalyst catalyzed the hydrogenolysis of hardwood lignin in the presence of hydrogen under harsh conditions (260 °C, 22 MPa H₂), yielding fully hydrogenated products with some monomeric propylcyclohexanols and methanol.

The metal sulfide catalysts were widely applied in hydrosulfurizaton (HDS) and hydrodenitrogeneration (HDN) processes in the petroleum industry, and also used for the hydrodeoxygenation of biomass. Narani *et al.* [24] described a sulfided S-NiW/AC catalyst for the depolymerization of lignin at 320 °C in the presence of 3.5 MPa H₂, with 35 wt% monomer yields and a product mixture that contained up to 26 wt% alkyl-phenolics. Notably, competing ring hydrogenation or severe repolymerization to generate insoluble char was not detected. However, the loss of sulfur can lead to deactivation of the catalyst.

Table 8.2   *Data for the depolymerization of lignin with transition metal catalysts in liquid solvents.*

Catalyst	S/C ratio (g g^{-1})	Reaction conditions				Major products	Conv. (%)	Reference
		T (°C)	P (MPa)	Solvent	Substrate			
Cu-CrO$_x$	80:7	260	22	Dioxane	Lignin	4-n-Propylcyclo-hexanol, 4-n-Propylcyclo-hexanediol	70	[23]
S-NiW/AC	4:1	320	3.5	Methanol	Kraft lignin	Alkylphenolics	35	[24]
Raney Ni	1:1	150–200	3.4	Dioxane–water	Spruce lignin	4-Propylsyringol	52	[25]
Raney Ni, NaOH	6:1	160–180	2	Dioxane–water	Enzymolysis lignin	Oligomers	83	[26]
Ni/ASA	2:1	300	6	Dodecane	Enzamatic lignin	C$_3$–C$_{17}$ cycloalkanes	80	[29]
Ni/HBEA	2:1	320	2	Hexa-decane	Organosov lignin	C$_6$–C$_9$ hydrocarbons	99	[30]
S-NiW/MgO-La$_2$O$_3$	20:1	350	12	None	Kraft lignin	Alkyl-phenolics	87	[33]
TiN-Ni	–	150	2.5	CH$_3$OH	Kraft lignin	Lignin-oil	3.2	[34]
Ni-W$_2$C/AC	5:2	235	1	Water	Raw woody biomass	Monophenols	47	[35]
Co nano-particles	–	140	–	Toluene	4-O-5	Cleaved products	100	[36]
Fe nano-particles	–	140	5	Toluene	β-O-4	Cleaved products	69	[37]

Metal sulfide catalysts can maintain high activities during long-term catalytic testing with continuous sulfur supplements from petroleum ingredients, but suffer severe catalyst deactivation due to sulfur–oxygen exchange and few sulfur supplements from the biomass. Although the hydrotreatment activity of the catalysts could be held stable by adding H$_2$S to the system, the products generated would be contaminated by sulfur compounds. Hence, more environment-friendly and cost-effective, non-sulfide catalysts should be urgently developed for the hydrogenolysis of lignin.

There is a long history of nickel-based catalysts used for lignin hydrogenation and hydrogenolysis. During the 1960s, Raney Ni was used for the hydrogenolysis of lignin, as reported by Pepper and Steck [25], the main products being 4-propyl-syringol, dihydroxysinapyl alcohol, and dihydroconiferyl alcohol. The reaction achieved a 52 wt% conversion at 150–200 °C in the presence of 3.4 MPa H$_2$ in dioxane:water (1:1). Subsequently, Xin *et al.* [26] also showed that Raney Ni in 3% NaOH solution could convert enzymolysis lignin to low-molecular-weight oligomers at 160–180 °C and 2 MPa H$_2$ in a mixed dioxane:water solvent. In a later report, Wang and Rinaldi [27] addressed such solvent effects on the hydrogenolysis of diphenyl ether and lignin with Raney Ni, and showed Raney Ni in non-basic solvents with Lewis basicity to be an active catalyst for hydrogenolysis and hydrogenation.

Supported Ni catalysts have been shown to be highly active for selective hydrogenolysis of the aryl ether C–O bonds, even when treating real lignin as a feedstock. He *et al.*

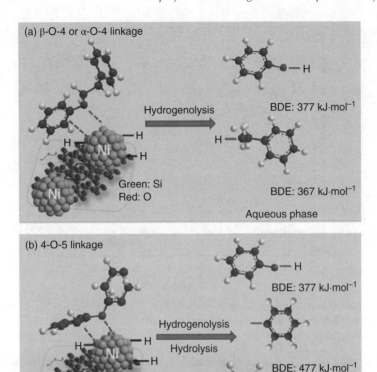

Figure 8.4 *The proposed different mechanisms of cleavage of aryl ether C—O bonds in the α-O-4, β-O-4, and 4-O-5 model compounds of lignin over Ni/SiO₂ in the aqueous phase. (See color plate section for the color representation of this figure.)*

[28] described the Ni/SiO$_2$-catalyzed selective cleavage of the ether bonds of lignin-derived aromatic ethers in the aqueous phase, as shown in Figure 8.4. The C–O aryl ether bonds of α-O-4 and β-O-4 linkages were cleaved by Ni, while the C–O bond of 4-O-5 was cleaved by parallel hydrogenolysis and hydrolysis. Later, Kong *et al.* [29] and Kasakov *et al.* [30] developed a new and simple route for the direct production of hydrocarbons from lignin with Ni-supported amorphous Si-Al and HBEA composites in an one-pot process in an apolar solvent. This catalytic route produced a 46% yield of liquid products, with 100% selectivity of cyclic alkanes and alkanes with Ni/ASA in dodecane. The naphthene and paraffin hydrocarbons formed were considered to be promising precursors in the further manufacture of clean fuels such as gasoline, kerosene, and diesel, as well as aromatic chemicals.

Song *et al.* [31] reported a series of heterogeneous nickel catalysts for the hydrogenolysis of lignosulfonate lignin to phenols and alkanes. The screened catalysts could be separated into three categories that included zeolite-supported Ni catalysts, Raney catalysts

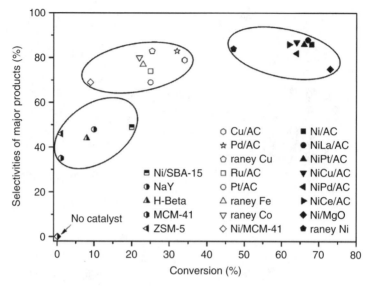

Figure 8.5   *Conversion of lignosulfonate lignin over various catalysts, using methanol as solvent. The major products included 4-propyl-guaiacol (PA) and 4-ethyl-guaiacol (EA).*

(Ni, Cu, Fe, and Co), and precious metal catalysts (Pd/AC, Ru/AC, and Pt/AC), as well as transition metal catalysts (Cu/AC, and Ni/MCM-41) (Figure 8.5). Among these catalysts, the Ni-based ones supported on either activated carbon (Ni/AC, NiLa/AC, NiPt/AC, NiCu/AC, NiPd/AC and NiCe/AC) or on magnesium oxide (Ni/MgO) or Raney Ni offered high conversions (>76%) and selectivities of 75–95% for substituted guaiacols, 4-propylguaiacol (PA), and 4-ethyl-guaiacol (EA).

Following the Ni-derived catalysts, NiMoP/γ-Al$_2$O$_3$ was used to break down lignin into monomeric units in a semi-continuous flow reactor at high temperatures of 320–380 °C and 4–7 MPa hydrogen pressure [32]. The lignin was fully converted to gaseous, liquid, and solid products, with the liquid products mainly consisting of aromatics and naphthenes in the organic phase, and phenols in the aqueous phase. Kumar *et al.* [33] used the conventional sulfide NiMo and CoMo catalysts on different acidic and basic supports (Al$_2$O$_3$, ZSM-5, activated carbon, and MgO-La$_2$O$_3$) in a fixed-bed reactor for the catalytic hydrotreatment of Kraft lignin in the absence of solvent. In this case, the best results were obtained with the sulfide NoMo/MgO-La$_2$O$_3$ catalyst giving 87 wt% lignin conversion with 26.4 wt% monomers, 15.7 wt% phenolics, and 5.9 wt% aromatics.

Notably, it was suggested that the introduction of an additional element to the Ni main center might enhance the catalytic reactions. When Molinari *et al.* [34] used a continuous-flow reactor for the hydrogenolysis of Kraft lignin over a TiN-Ni heterogeneous catalyst, the products formed included substituted phenols (3.2%), together with many aromatic fragments of smaller molecular weight (up to 60%) in comparison to pristine lignin. A similar synergistic effect was also reported for Ni-W$_2$C/AC by Li and coworkers [35].

In addition to Ni, Ren *et al.* [36] described for the first time a cobalt nanoparticle catalyst (prepared by LiAlH$_4$, Co(acac)$_2$, *t*BuONa, and toluene under a nitrogen atmosphere) that was effective in the reductive cleavage of inert aromatic C–O bonds, with high selectivity (Table 8.2). These authors suggested that the heterogeneous Co species from the reduction of Co(acac)$_2$ might be the true catalytically active species, similar to previous reports using Ni(COD)$_2$. Later, Ren *et al.* [37] described another highly selective iron-based catalyst for the selective cleavage of lignin-derived β-O-4 to phenolics, which employs [Fe(acac)$_3$] as the precursor under a hydrogen atmosphere. The catalytically active species may consist of heterogeneous iron clusters or particles resulting from reduction of the [Fe(acac)$_3$].

8.2.3 Homogeneous Catalyst System for Lignin Depolymerization in the Presence of H$_2$

As lignin has a very complex three-dimensional structure, it is very difficult to achieve a good contact between the polymer and the accessible active metal center. Compared to heterogeneous catalysts, homogeneous catalysts offer the advantage of less steric limitation when contacting with the C–O bond linkages of lignin. Among the reported homogeneous catalytic systems, the effective metal centers mainly include Ni, Ru, Ir, and Mo.

Sergeev and Hartwig [38] designed a nickel complex catalyst that is stabilized by an N-heterocyclic carbene (NHC) for the hydrogenolysis of diaryl ethers. Although the reactions were conducted at only 0.1 MPa H$_2$ at about 100 °C, this afforded high yields (54–99%) of the corresponding phenols and arenes, without arene hydrogenation. Among possible mechanisms for the nickel-catalyzed hydrogenolysis of aryl ethers, it can be envisioned that the favorable effect of a strong base in the hydrogenolysis may lead to the formation of anionic nickel complexes that are more active for the cleavage of C–O bonds, or for the activation of coordinated dihydrogen. Furthermore, a broad lignin-derived model substrate scope could be suitable for this catalyst system. Following these studies, the same group developed an in-situ-formed heterogeneous nickel catalyst from the well-defined soluble precursor Ni(COD)$_2$ or Ni-(CH$_2$TMS)$_2$(TMEDA), at loadings of 0.25 mol.%, without any additional dative ligand in the presence of a base additive such as ^{t}BuONa, for the selective cleavage of C$_{Ar}$–O bonds in aryl ether models of lignin, without the hydrogenation of aromatic rings at 0.1 MPa H$_2$ pressure [39]. Samanat and Kabalka [40] also reported the details of homogeneous nickel catalysts combined with NaO^{t}Bu, and LiAl(O^{t}Bu)$_3$H as a hydrogen source, in which aryl ether bond cleavage was also realized, but this led to a full reduction of the aromatic systems. However, it is worth noting that these reported catalyst systems were usually found to be water-insensitive, despite water being ubiquitously present in the biomass and biomass processing course.

Apart from nickel, other transition metal complexes were also investigated for C$_{Alkyl}$–O bond cleavage. Kusumoto and Nozaki [41] reported a direct and selective hydrogenolysis of sp^2 C–OH bonds in substituted phenols and naphthols, and sp^3 C–O bonds in aryl methyl ethers by hydroxycyclopentadienyl iridium complexes. Nichols *et al.* [42] used Ru(CO)H$_2$(PPh$_3$)$_3$ in combination with xantphos to selectively cleave the C–O bond of a β-O-4 model compound by a tandem α-alcohol dehydrogenation and reductive ether cleavage. A subsequent density functional theory (DFT) and kinetics study confirmed that the sequential dehydrogenation was followed by hydrogenolysis [43]. Inspired by the findings of Nichols *et al.*, Wu *et al.* [44] applied the ruthenium–xantphos complex in the presence of hydrogen to a direct hydrogenolysis of the β-O-4 model compound, which contained a ketone at the benzylic carbon instead of an alcohol.

Few examples of these homogeneous catalysts have been applied directly to the conversion of true lignin. Dating back to 1933, Oasmaa and Johansson [45] used a water-soluble ammonium heptamolybdate catalyst (0.8% Mo compared to the weight of lignin) in the hydrotreatment of Kraft lignin (at 430 °C for 60 min), and obtained a high a yield (61 wt% of the original lignin) of low-molecular-weight oil that was mainly composed of phenols (8.7% of the original lignin), cyclohexanes (5.0%), benzenes (3.8%), naphthalenes (4.0%), and phenanthrenes (1.2%). The catalyst was equally effective as a solid sulfided NiMo and Cr_2O_3 catalyst mixture when lignin was hydrotreated to form an oil-like product.

8.2.4 Cleavage of C–O Bonds in Lignin with Metals and Hydrogen-Donor Solvents in the Absence of Hydrogen

In view of the very high costs associated with the introduction of high-pressure H_2 into reactions, the use of a hydrogen-donating solvent can offer a 'green' and sustainable way to achieve lignin depolymerization over metal catalysts. Hence, attention will be focused here on five common hydrogen-donating solvent systems, including the use utilization of tetralin, formic acid, methanol, ethanol, and isopropanol solvents for lignin conversion (Table 8.3).

Table 8.3 Data for the depolymerization of lignin with metal catalysts in the absence of H_2.

Catalyst	S/C ratio (g g^{-1})	T (°C)	Solvent	Substrate	Major products	Conv. (%)	Reference
Ni/Al-SBA-15	1:1	140	Tetralin	Lignin	Desitol syringaldehyde	17	[46]
Pd/C, Nafion SAC-13	3:20	300	Formic acid	Spruce lignin	Guaiacol, pyrocatechol, resorcinol	11	[47]
Pt/C	1:10	350	Formic acid	Switchgrass	Gaiacol, p-methylguaiacol, p-ethylguaiacol	21	[48]
Ni/AC	1:20	200	Methanol	Birch sawdust	Propylguaiacol, propylsyringol	50	[49]
Cu-PMO	1:1	320	Methanol	Pine sawdust	Aliphatic alcohols and methylated derivatives	99	[50]
Cu-PMO	1:1	300	Methanol	DHBF	2-Ethylphenol, 2-Ethylcyclohexanol	68	[51]
Cu-La-PMO	1:1	310	Methanol	Lignin	Methanol-soluble oligomers	100	[52]
CuMgAlO$_x$	1:2	300	Ethanol	Soda lignin	Monomers	23	[53]
CuMgAlO$_x$	1:2	340	Ethanol	Soda lignin	Monomers	36	[54]
a-MoC/AC	1:2	280	Ethanol	Kraft lignin	C_6–C_{10} alcohols	N. R.	[55]
Raney Ni, HBEA	1:2	160	2-PrOH	Phenol	Benzene	100	[56]
Raney Ni	1:1.6	160	2-PrOH	Organosolv lignin	Phenols, cyclohexanols	55	[57]

DHBF: Dihydro-benzofuran.

The advantages of tetralin as a hydrogen-donor solvent include its high boiling point, as well as its ready release of hydrogen atoms under hydrocracking conditions, leading to the formation of naphthalene, a relatively stable compound. Toledano *et al.* [46] disclosed a novel microwave-assisted mild hydrogenolytic methodology to depolymerize organo-solv lignin into phenolic compounds with noble metals (Ru, Rh, Pd, and Pt) supported on Al-SBA-15 at 150 °C in tetralin. A 24% conversion was achieved from lignin, with major products of demethoxylated phenolic compounds.

Formic acid, which can be manufactured from the decomposition of monosaccharides obtained from biomass, is a freely available 'green' reagent in a bio-refinery. Liguori and Barth [47] reported a novel procedure utilizing formic acid as the hydrogen source in combination with Pd/C and Nafion-SAC-13 as solid catalysts in water, where guaiacol, pyrocatechol, and resorcinol were the major products from the catalytic pyrolysis of various types of lignin. Xu *et al.* [48] reported the combination of formic acid and Pt/C to produce low-molecular-weight compounds in liquid products, with a 76% yield from lignin.

Methanol is also considered a promising hydrogen-releasing solvent, and this has been confirmed using isotopic tracing experiments. When Song *et al.* [49] depolymerized birch wood lignin using Ni/AC in methanol in the absence of hydrogen, the results showed that lignin could be selectively cleaved into propyguaiacol and propylsyringol (total selectivity >90%), with a conversion of 50%. Using methanol as a hydrogen resource, Ford and col-leagues [50,51a,b] conducted a series of studies on the catalytic depolymerization of lignin to bio-oil over Cu-doped metal oxide as catalyst. The Cu catalyst first catalyzed the decom-position of methanol to produce CO and H_2, such that methanol would serve as a liquid syngas to catalyze any diverse hydrogenolysis and hydrogenation processes. This first step was successfully applied to conversion of the selective cleavage of the aromatic ether bond of dihydrobenzofuran (DHBF) as lignin model compounds, renewable organosolv lignin, and even lignocellulose composites such as sawdust (see Figure 8.6). Subsequently, the same group developed another Cu- and La-doped metal oxide system to hydrogenolyze lignin in methanol, in the absence of hydrogen [52].

Following on from Ford's studies, Huang *et al.* [53] reported the details of a one-pot valorization of soda lignin in ethanol using a $CuMgAlO_x$ catalyst, which resulted in a high monomer yield (23%) without char formation. Two-dimensional heteronuclear single quantum correlation spectroscopy (HSQC) nuclear magnetic resonance (NMR) analysis showed that ethanol reacted with lignin fragments by alkylation and esterification, and

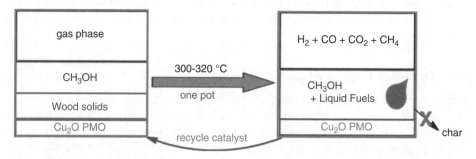

Figure 8.6 *Conversion of wood over a Cu catalyst in methanol in the absence of hydrogen. PMO: porous metal oxides.*

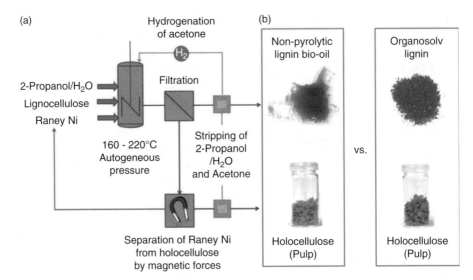

Figure 8.7   *(a) Schematic representation of the catalytic bio-refining method over Raney Ni in the mixed solvent of 2-propanol and H$_2$O. (b) Comparison of lignin-derived bio-oil and the traditional organosolv process. (See color plate section for the color representation of this figure.)*

that this was found to play an important role in suppressing repolymerization. Later, the same authors investigated the roles of Cu-Mg-Al mixed oxides in the depolymerization of soda lignin in supercritical ethanol [54]. Ma *et al.* [55] reported the ethanolysis of Kraft lignin over an α-MoC$_{1-x}$ catalyst to produce low-molecular-weight compounds composed of C$_6$–C$_{10}$ esters, alcohols, arenes, and phenols in an inert atmosphere. These authors noted that ethanol performed best among the selected alcohols with the α-MoC$_{1-x}$ catalyst in terms of yields of liquid products, and the reaction without hydrogen proceeded much more efficiently than that with hydrogen.

Isopropanol was also used as a hydrogen donor solvent for the dehydroxylation of phenols to arenes by Wang and Rinaldi [56] and Ferrini and Rinaldi [57], using Raney Ni and H$^+$-exchanged beta polymorph zeolite (HBEA) catalysts. Such catalytic systems were introduced to deoxygenate phenol to benzene, but were then successfully extended to realistic substrates such as bio-oil and organosolv lignin (see Figure 8.7). The catalytic evaluation results showed that a lignin bio-oil could be effectively used as a feedstock to produce alkanes or arenes, and these formed products were confirmed by 2D-GC and NMR analysis methods.

8.3   Oxidative Depolymerization of Lignin

Compared to the hydrogenolysis technique for the cleavage of C–O bonds of lignin, which produces diverse phenols, the oxidative depolymerization of lignin offers some attractive advantages, such as the production of more valuable functional aromatic compounds and no consumption of expensive extra hydrogen. The aromatic compounds produced, such as vanillin, syringaldehyde, and 4-hydroxybenzaldehyde, can be used as precursors for organic synthesis or employed as chemicals after some purification steps. Catalyst systems for the oxidation of lignin can be classified into four categories: metal-supported oxide catalysts; polyoxometalate catalysts; organometallic catalysts; and ionic liquid catalysts. A summary of lignin oxidation catalysts, including the selected reaction conditions and the catalytic results obtained, are listed in Table 8.4.

8.3.1   Metal-Supported Oxide Catalysts

The oxidative deconstruction of lignin has been performed with a variety of oxide-based catalysts in aqueous or alcoholic media, in the presence of O_2 or H_2O_2. The selected catalysts were precious metals (Pd, Ru) supported on oxides with oxygen vacancies, such as CeO_2 and TiO_2 [58,59]. Such approaches were reported to efficiently transform lignin and lignin model compounds into monomeric aromatic compounds. The Pd catalyst supported on CeO_2 [59] resulted in an aromatic aldehyde yield of 5.2% in the oxidative conversion of lignin. The metal center played a crucial role in the oxidation of C–OH, C=O, and the linkage of lignin (e.g., β-O-4 bond). In addition, no significant char deposition was observed on the catalyst surface. Pd/γ-Al_2O_3 [60] was applied and tested for the oxidative conversion of alkaline lignin extracted from sugarcane bagasse in a batch slurry reactor and a continuous, three-phase fluidized-bed reactor under reaction conditions of 140 °C and 1 MPa O_2 pressure. After a series/parallel reaction network, 0.65 g vanillin and 11.48 g syringaldehyde were obtained from 300 g lignin in a continuous operation at 140 °C. Although noble metal catalysts were widely applied during the wet aerobic oxidation processes, the high price of these catalysts limited their further commercial application. Consequently, the highly active and stable perovskite-type oxides (ABO_3) have attracted much attention.

The perovskite-type catalysts, including $LaMnO_3$ [61], $LaFeO_3$ [62] and $LaCoO_3$ [63] have each shown good activity in the lignin oxidation process. Furthermore, the catalytic activity can be enhanced by the partial substitution of B-site cations. Deng and coworkers [64] prepared different contents of Cu-doped $LaCo_{1-x}Cu_xO_3$ catalysts using the sol–gel method, and maximum yields of vanillin and syringaldehyde were attained at 5.3% and 12.8% with $LaCo_{0.8}Cu_{0.2}O_3$, respectively. The improved activity of the catalyst was more pronounced with the increased Cu content. This improvement of catalytic activity was not considered to have occurred because of the leached Cu^{2+} or CuO after the Cu substitutes, but rather was attributed to the fact that the Cu incorporation promoted the formation of more anion vacancies in the $LaCoO_3$ catalyst, thereby increasing the amount of adsorbed oxygen surface active site $[Co_{(surf)}^{3+}O^{2-}]$ species and producing activated species $[Cu_{(surf)}^{2+}O^{2-}]$. This phenomenon was confirmed by Zhang *et al.* [65]. In another case, the perovskite-type oxide catalyst such as $LaFe_{1-x}Cu_xO_3$ ($x=0$, 0.1, 0.2) was tested in the wet aerobic oxidation of lignin (obtained from the enzymatic hydrolysis of steam-exploded cornstalks) into aromatic aldehydes. With the partial replacement of Fe^{3+} by Cu^{2+}, the amount of activated species $Fe_{(surf)}^{3+}O_2^-$ and $Cu_{(surf)}^{2+}O_2^-$ were both increased (see Figure 8.8). The high stability of the perovskite catalyst structures caused the lignin conversion and the yield of aromatic aldehydes to remain at the same level for five successive cycles of the reaction.

Table 8.4 Data for the oxidative depolymerization of lignin.

Entry	Catalyst	Solvent	Oxidant	Experimental conditions			Substrate	Products	Conv. (%)	Yield (%)	Reference(s)
				T (°C)	P (MPa)	t (h)					
1	Ru/TiO$_2$	H$_2$O	O$_2$	190	5.5		Plant effluents	Acetic acid Glutaric acid	Not provided	Not provided	[58]
2	Pd/CeO$_2$	MeOH	O$_2$	185	0.1	24	Organosolv lignin	Vanillin Guaiacol 4-hydroxybenzaldehyde	Not provided	5.2 0.87 2.4	[59]
3	Pd/Al$_2$O$_3$	H$_2$O	O$_2$	140	1	8	Alkaline lignin	Vanillin Syringaldehyde p-hydroxybenzaldehyde	80	4.4 6.7 7.3	[60]
4	LaMnO$_3$	NaOH/H$_2$O	O$_2$	120	2	3	Enzymatic hydrolysis lignin	Vanillin Syringaldehyde p-hydroxybenzaldehyde	57	4.3 9.3 1.3	[61]
5	LaFeO$_3$	NaOH/H$_2$O	O$_2$	120	2	3	Enzymatic hydrolysis lignin	Vanillin Syringaldehyde p-hydroxybenzaldehyde	19	4.1 8.7 1.1	[62]
6	LaCoO$_3$	NaOH/H$_2$O	O$_2$	120	2	3	Steam-exploded lignin	Vanillin Syringaldehyde p-hydroxybenzaldehyde	57	4.55 9.99 2.23	[63]
7	LaCo$_{1-x}$Cu$_x$O$_3$ (x = 0, 0.1, 0.2)	NaOH/H$_2$O	O$_2$	120	2	3	Lignin	Vanillin; syringaldehyde; p-hydroxybenzaldehyde	65	5.3 12.8 2.88	[64]
8	LaFe$_{1-x}$Cu$_x$O$_3$	NaOH/H$_2$O	O$_2$	120	2	3	Enzymatic hydrolysis lignin	Vanillin; syringaldehyde; p-hydroxybenzaldehyde	68	4.6 12 2.4	[65]
10	H$_5$PV$_2$M$_{10}$O$_{40}$	PhMe	H$_2$O$_2$	R.T.		3.35	Kraft lignin	MMBQ DMBQ	8	2 3	[66]

#	Catalyst	Solvent	Oxidant	Temp		Time	Lignin/Substrate	Products	Yield	Yield	Ref.
11	$H_5PMo_{12}V_2O_{40}$	MeOH	H_2O_2	R.T.	3.4	12h	Kraft lignin (softwoods)	Dimethoxybenzoquinone	Not provided	3	[67,68]
12	$H_3PMo_{12}O_{40}$	MeOH/H_2O	O_2	170	0.5	0.3	Spruce lignin	Vanillin, acetovanillone, methyl vanillate, syringaldehyde, acetosyringone	3.25	Not provided	[69]
13	Vanadium-oxo complex	ethyl acetate	air	80		24	(1*R**,2*S**)-1-(4-ethoxy-3-methoxyphenyl)-2-(2-methoxyphenoxy)propane-1,3-diol	1-(4-ethoxy-3-methoxyphenyl)prop-2-en-1-one	>95	71	[70]
14	Vanadium-oxo complex	Triethylamine	air	80		48	1-(4-hydroxy-3,5-dimethoxyphenyl)-2-(2-methoxyphenoxy)1,3-propanediol-[2,3-$^{13}C_2$]	Ketone	80	57	[71]
15	Vanadium-oxo complex	Triethylamine	air	100		4	Ethanosolv- lignin Dioxasolv lignin	Vanillin syringic acid syringaldehyde 4-Hydroxybenzaldehyde Vanillic aicd	Not provided	0.78 0.67 0.59 0.33 0.31	[72]
16	Co-Schiff base catalyst	MeOH/DMSO	O_2	R.T.	0.34		Poplar lignin	2,6-Dimethoxy-benzoquinone benzaldehydes	Not provided	1.6 1.2	[73]
17	Co(DPCys)	H_2O	H_2O_2	180		10	Enzymolysis lignin	H-type compounds G-type compounds S-type compounds Aliphatics	39	3.4 14.7 9.5 8.7	[74]
19	Methyl-rhenium trioxide	acetic acid	H_2O_2	R.T.		24	Sugarcane lignin, Kraft lignin	Aliphatic OH Condensed OH Guaiacyl OH *p*-Hydroxy phenyl COOH units	Not provided	Not provided	[75,76]

(continued overleaf)

Table 8.4 (continued)

Entry	Catalyst	Solvent	Oxidant	Experimental conditions T (°C)	P (MPa)	t (h)	Substrate	Products	Conv. (%)	Yield (%)	Reference(s)
20	$Mn(NO_3)_2$	[EMIM][CF$_3$SO$_3$]	air	100		24	Organosolv lignin	Syringol Vanillin DMBQ syringaldehyde	27.1–54.6	Not provided	[77]
21	$CuSO_4$	[mmim][Me$_2$PO$_4$]	O_2	195	3	2	Organosolv lignin	Vanillin syringaldehyde p-hydroxybenzaldehyde	86–100	14.7 8.8 6.2	[78]
22	CuO, Co_2O_3, Fe_2O_3, MnO_2	[HMIM][Cl]	O_2	150	8	4	Oignin of eucalyptus woods	Vanillic acid Vanilline Syringic acid Syringaldehyde Acetosyringone	Not provided	8 8 16 20 8	[79]
23	H_3PO_4	[BnMIm][NTf$_2$]	O_2	130	1	3	Organosolv lignin	Benzoic acid Benzaldehyde Some other compounds	75	43	[80]
24	None	[Et$_3$NH][HSO$_4$] [HC$_4$im][HSO$_4$]	H_2O_2	120		22	Lignin from black liquors	Guaiacol Catechol Phenol Vanillic acid	19	Not provided	[81]

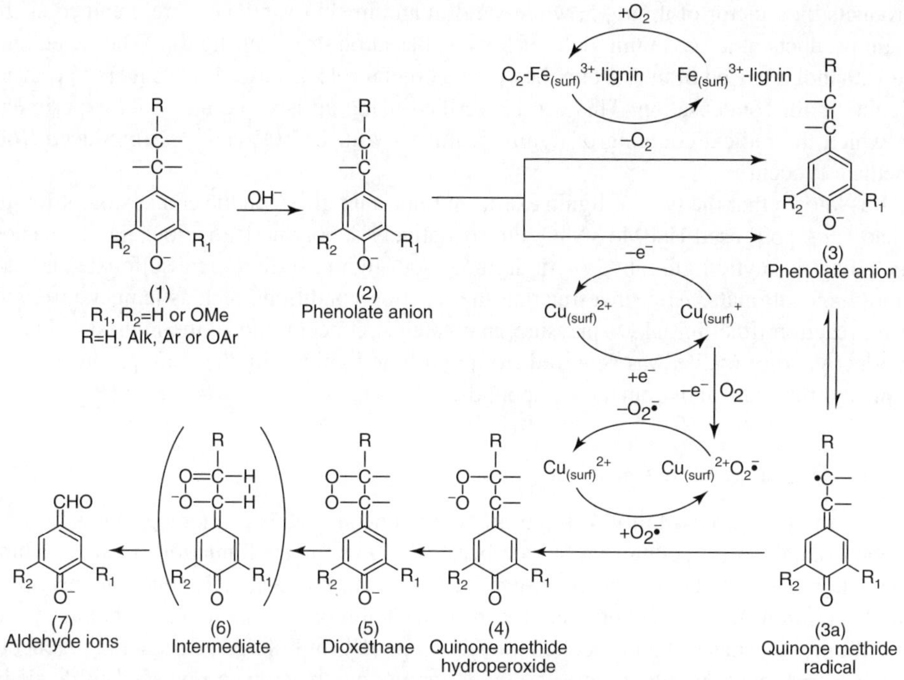

Figure 8.8 *Proposed mechanism of oxidation of lignin to aromatic aldehydes for the $LaFe_{1-x}Cu_xO_3$ (x = 0, 0.1, 0.2) catalyst. Reprinted with permission from Ref. [65]. Copyright 2009, Molecular Diversity Preservati.*

8.3.2 Polyoxometalate Catalysts

The three-dimensional structure of lignin leads to less accessibility to the designed heterogeneous catalyst. Therefore, polyoxometalates (POMs) with properties of reversible oxidation and easy dissolution in aqueous and organic media, provide a promising alternative application in the oxidative depolymerization of lignin. The oxidative degradation of lignin with oxygen and POMs can be summarized by Eqns (8.1) and (8.2).

$$Lignin + POM^{n-} \rightarrow Lignin_{(ox)} + POM^{(n+a)-} \tag{8.1}$$

$$POM^{(n+a)-} + (a/4)O_2 + aH^+ \rightarrow POM^{n-} + (a/2)H_2O \tag{8.2}$$

Bozell and coworkers [66] carried out extensive investigations into the oxidation of various lignin model compounds and lignin to benzoquinones with various POM catalysts. The yields of quinones (8%) were observed when POMs were used as catalysts in the presence of hydrogen peroxide; however, a low yield was obtained when such a catalyst was used for the oxidation of real lignin samples, with yields of only 2–3 wt% of 2,6-dimethoxy-benzoquinone being achieved.

It has been observed that repolymerization reactions become one of the main problems in the production of monomers. Voitl and Rohr [67,68] used aqueous POMs ($H_3PMo_{12}O_{40}$) as catalysts to oxidatively degrade lignin to aromatic chemicals, attaining a low yield of 0.2%. The addition of methanol to the reaction resulted in an increase in yield of the monomeric products by a factor of up to 15, while vanillin and methyl vanillate were obtained as the main products at a maximum yield of 5 wt%, based on dry Kraft lignin. When methanol and ethanol are used in the system they play a crucial role as radical scavengers to prevent lignin–lignin condensation. This can be verified using an isotopic methanol experiment, in which the radical coupling of lignin fragments with $CH_3O^\bullet$ and $^\bullet H_3$ (produced from methanol) occurs.

It is known that the type of lignin exerts an important effect on the conversion of lignin. Zhao *et al*. p69] used $H_5PMo_{10}V_2O_{40}$ to oxidatively depolymerize various lignin species, including pyrolytic lignin, hydrolytic lignin, alkali lignin, sodium lignosulfonate and calcium lignosulfonate. After investigating the reaction conditions such as temperature, solvent, reaction time, initial O_2 pressure, and catalyst concentrations, the highest oxidative product yield of 65.2% was achieved from pyrolytic lignin, with the main products being dimethyl fumarate and aromatic compounds.

8.3.3   Organometallic Catalysts

Due to the poor accessibility of lignin to heterogeneous catalysts, homogeneous organometallic catalysts attract much attention because of their space limitations in approaching the bulky substrate. Unfortunately, organometallic catalysts generally function best under mild conditions that are not efficient for the conversion of polymeric lignin, but are effective for the cleavage of lignin-derived model compounds or linkages. Recent homogeneous catalyst treatments of lignin and its model compounds have been reviewed in detail by Zakzeski *et al*. [6] and Li *et al*. [7]. Among these reactions, the Re, V, and Co metals were most widely used.

Son and Toste [70] applied a vanadium-oxo complex catalyst to achieve a non-oxidative C–O cleavage of a dimeric lignin model compounds at 80 °C in ethyl acetate as solvent. A high conversion (>95%) was achieved, accompanied with good selectivity of a highly functionalized aryl enone. When Hanson and coworkers [71] investigated vanadium-oxo complexes as catalysts for oxidative C–C bond cleavage of lignin models under air in [D_5]-pyridine, a conversion of 87% was observed for the β-O-4 lignin model compound after 72 h at 100 °C. The 8-quinolinate-based vanadium (V) complex was used by Chan and coworkers [72] to catalyze the non-oxidative degradation of lignin model compounds and organosolv-lignin in triethylamine under air conditions. Aromatic degradation products of vanillin, syringic acid, and syringaldehyde were found to be the three major products, with 0.78%, 0.67% and 5.9% yields, respectively.

A new Co-Schiff base catalyst bearing a bulky heterocyclic nitrogen base as a substituent was recently developed by Biannic and Bozell [73], and used to successfully oxidize lignin phenolic model monomers and dimers to benzoquinones in high yields (60–80%) with molecular oxygen in methanol at room temperature.

Zhu and coworkers [74] designed and synthesized a series of metallo (Fe, Co, Mn, Ru)-deuteroporphyrins derived from hemin. These authors noted that the modified cobalt deuteroporphyrin that has no substituents at the meso sites, but does have the disulfide linkage in the propionate side chains at the β sites, exhibited much higher activity and stability than the synthetic tetraphenylporphyrin. These novel oxidative catalysts can further convert real lignin into depolymerized products, including a significant portion of well-defined aromatic monomers in a solvent of H_2O and methanol using oxone as oxidant.

Crestini and coworkers [75,76] showed that methyltrioxorhenium (MTO) was a powerful and promising catalyst for the oxidation of lignin model compounds and lignin by use of hydrogen peroxide as primary oxidant. The MTO catalysts oxidized lignin to more soluble lignin fragments with, lower amounts of aliphatic and condensed OH groups, and higher amounts of carboxylic acid moieties of more soluble lignin fragments. In addition, the lower Lewis acidity of immobilized MTO catalysts with respect to homogeneous MTO would direct their reactivity towards aliphatic C–H insertion and Dakin-like reactions rather than aromatic ring oxidation, thus producing products with higher levels of free phenolic guaiacyl groups.

8.3.4 Ionic Liquid Catalysts

It has been reported that many ionic liquid solvents can readily dissolve lignin and so favor carbocation-forming reactions, and consequently much interest has arisen in the use of ionic liquids in the oxidative conversion of lignin. The developed catalytic system includes metal-combined ionic liquids and metal-free ionic liquids for the oxidative degradation of lignin and its model compounds.

A multiparallel batch reactor system was applied to screen different ionic liquids and metal catalysts for the oxidative depolymerization of lignin by Stark and coworkers [77]. The most effective catalytic system, which consisted of $Mn(NO_3)_2$ and 1-ethyl-3-methylimidazolium trifluoromethane-sulfonate [EMIM][CF_3SO_3] with synthetic air, resulted in a maximum conversion of 66.3% at 100 °C for 24 h. By adjusting the reaction conditions and catalyst loading, 2,6-dimethoxy-1,4-benzoquinone (DMBQ) could be formed as the main product with 21.0% selectivity, and can be isolated as a pure substance in 11.5 wt% yield, using a simple extraction/crystallization process.

Liu and coworkers [78] investigated the $CuSO_4$-catalyzed oxidation of organosolv lignin in an ionic liquid [Mmim][Me_2PO_4], where the major products were vanillin, syringaldehyde, and p-hydroxybenzaldehyde. The authors noted that the addition of methylisobutylketone agent into the reaction system could avoid the deep oxidation of the extracted products by the ionic liquid layer, and significantly improved the lignin conversion and the yield of aromatic aldehydes.

As well as a combination of metal catalyst and ionic liquids, metal-free ionic liquids can be also used for the oxidative deconstruction of lignin. Jia and coworkers [79] used an acidic ionic liquid ([HMIM]Cl) to realize the oxidative depolymerization of lignin model compounds (guaiacylglycerol-b-guaiacyl ether, and veratrylglycerol-b-guaiacyl ether) in water at 150 °C, and more than 70% guaiacol was obtained by cleaving β-O-4 bonds. An increase in available water can lead to additional β-O-4 bond-cleavage products as a result of the GG (guaiacylglycerol-b-guaiacyl ether) and VG (veratrylglycerol-b-guaiacyl ether) reacted with water to produce guaiacol. The [HMIM]$^+$ cation was recognized as a Brønsted acid site to promote the hydrolysis step. Another example [80] of metal-free ionic liquid utilization was the use of a catalytic amount of H_3PO_4 in 1-benzyl-3-methylimidazolium bis(trifluoromethylsulfonyl)-imide ([BnMIm][NTf$_2$]) to produce benzoic acid and phenol from 2-phenoxyacetophenone (lignin β-O-4 model compound), using O_2 as the oxidant. Based on UV-visible spectroscopy analysis and electron paramagnetic resonance, it can be shown that ionic liquids can promote the generation of ·OOH free radicals, which may accelerate the reactions. An oil yield of 75% was obtained when treating with an organosolv lignin.

Prado and coworkers [81] also successfully depolymerized the lignin directly in the black liquor by oxidation using the ionic liquids (butylimidazolium hydrogensulfate $[HC_4im][HSO_4]$ and triethylammonium hydrogensulfate $[Et_3NH][HSO_4]$) at 120 °C with H_2O_2 as oxidation agent, under metal-free conditions. The former ionic liquid led to a high activity but was itself oxidized, whereas the latter ionic liquid was less reactive but was not affected by H_2O_2. The main degradation products found in the extracted oils were vanillic acid, benzoic acid, and 1,2-benzenedicarboxylic acid.

8.4   Hydrolysis of Lignin with Base and Acid Catalysts

With the attached phenolic –OH functional groups, lignin is very soluble in alkali solutions, and this may accelerate its further depolymerization over heterogeneous or homogeneous catalysts (Table 8.5). During the late 1990s, Miller and coworkers [82] conducted extensive studies on the depolymerization of Kraft- and organosolv lignin by KOH, NaOH, CsOH, LiOH, Ca(OH)$_2$, and Na$_2$CO$_3$ in supercritical methanol or ethanol. Their results indicated that strong bases such as NaOH and Ca(OH)$_2$ gave a superior conversion, while other bases such as LiOH or CsOH showed negative synergistic effects. When Watanabe and coworkers [83] used NaOH and ZrO$_2$ for the partial oxidative gasification of organosolv lignin in supercritical water, the synergistic effects of NaOH and ZrO$_2$ inhibited char formation and promoted CO and H$_2$ formation. Karagöz and coworkers [84] also pointed out that the use of base catalysts hindered the formation of char after exploring the effect of rubidium (Rb) and cesium (Cs) carbonate solutions on the hydroliquefaction of wood biomass in water. Gas chromatography-mass spectrometry (GC-MS) analyses showed that 4-methyl-phenol, 2-furan carboxaldehyde and 2-methoxy-phenol were the major compounds.

When Ehara and coworkers [85] treated softwood and hardwood in supercritical water (380 °C, 100 MPa) for 8 s, a water-soluble portion, a methanol-soluble portion, and methanol-insoluble residues were obtained after the liquefaction of lignin. The products mainly included guaiacyl and syringyl monomeric compounds. Wahyudiono and coworkers [86] converted lignin in near and supercritical water at temperatures of 350 °C and 400 °C using a batch-type reactor. In this case, the conversion of lignin increased with increasing temperature, and catechol (28.4 wt%), phenol (7.5 wt%), *m,p*-cresol (7.9 wt%) and *o*-cresol (3.8 wt%) were identified as the major products. When Nenkova and colleagues [87] depolymerized lignin with a 5% NaOH solution at a temperature of 180 °C for 6 h, the species of the selected lignin substrate was seen to have a major influence on the liquefaction products. For poplar wood bark, 29.0% insoluble residue remained and 53.2% low-molecular-weight products was obtained. In the case of poplar wood sawdust, the proportions of insoluble residues and low-molecular-weight products were 44.2% and 32.2%, respectively.

Onwudili and Williams [88] used combined catalysts of formic acid (FA) and Pd/C to catalyze the depolymerization of alkali lignin in subcritical water. Compared to lignin in subcritical water alone, the presence of FA increased the yield from 22.3 wt% to 33.1 wt%,

Table 8.5 Data for the hydrolysis of lignin with base and acid catalysts.

Entry	Catalyst	Experimental conditions				Substrate	Products	Conv. (%)	Yield (%)	Reference
		Solvent	T (°C)	P (MPa)	t (min)					
1	CsOH, KOH, NaOH, LiOH, Na$_2$CO$_3$	Ethanol	290		60	Alkali lignin	Not provided	93	Not provided	[82]
2	NaOH, ZrO$_2$	H$_2$O	300		60	Organic lignin	Not provided	40	Not provided	[83]
3	CsCO$_3$, RbCO$_3$	H$_2$O	280		15	Sawdust from pine	2-methoxy-phenol 4-methyl-phenol 4-Oxo-pentanoic acid 2-Furan carboxaldehyde	88	Not provided	[84]
4	None	H$_2$O	380	100	0.13	Japanese cedar, Japanese beech	Propylguaiacol Syringol Propenylsyringol Syringaldehyde	35	Not provided	[85]
5	None	H$_2$O	400		24	Alkaline lignin	Catechol Phenol o-Cresol m,p-Cresol	95	35 4.6 3.8 8.0	[86]
6	NaOH	H$_2$O	180		120	Poplar wood sawdust	2-Methoxyphenol 2,6-Dimethoxyphenol 1-[4-hydroxy-3-methoxyphenyl]ethanone	32	Not provided	[87]
7	HCOOH Pd/C	H$_2$O	265		360	Alkali lignin	Guaiacol 1,2-Benzenediol Methylguaiacol Propylguaiacol	62	13 18 3 4	[88]
8	HCOOH HCOONa	H$_2$O	110		1440	Aspen enzyme lignin	Syringyl compounds Guaiacyl compounds p-Hydroxyphenyl compounds	61	30 18 4.4	[89]

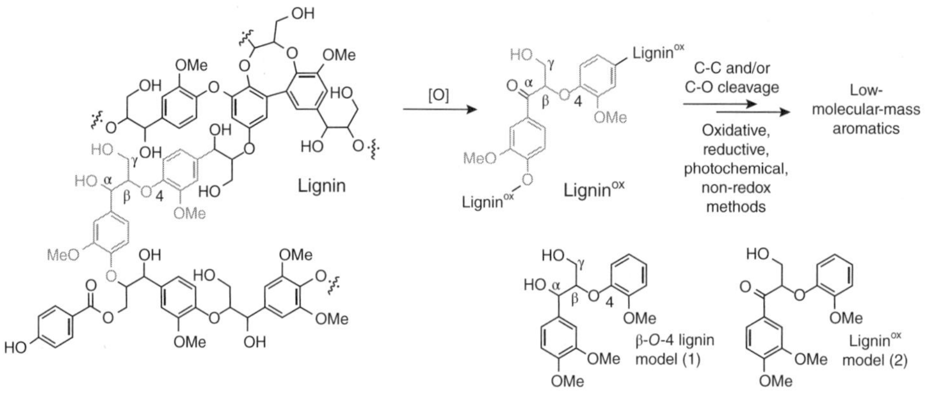

Figure 8.9   *The proposed strategies for depolymerization of oxidized lignin via two steps.*

and changed the predominant product from guaiacol to catechol. The yield of oil products was slightly decreased after the addition of Pd/C (25.7%), but the conversion of catechol was increased because Pd/C benefited from the hydrogenolysis of aryl–O ether bonds as well as the hydrogenation of C=C bonds. In order to obtain high yields of phenolic products, Rahimi and coworkers [89] designed a two-step strategy for the depolymerization of oxidative lignin under mild conditions in aqueous FA solution. The first step involved the chemoselective aerobic oxidation of C_a alcohols to ketones, followed by the cleavage of C−O bonds of lignin with FA (see Figure 8.9). In this way, a yield of more than 60 wt% of low-molecular-mass aromatics was obtained. It should be noted that, without the chemoselective aerobic oxidation step, no product was observed with FA.

8.5   Other Depolymerization Techniques (Cracking, Photocatalysis, Electrocatalysis, and Biocatalysis)

Apart from reduction and oxidation-induced lignin depolymerization, some other depolymerization techniques also exist. Pyrolysis is a commonly used simple approach for the direct production of bio-oil from lignin. The process was conducted with a rapid heating rate up to 450–600 °C, achieving C–C bond cleavage without a catalyst. Catalysts used in the reaction include silica-alumina catalysts (zeolite) [90], alkaline materials (NaOH or KOH) [91,92], Ni/W-supported on zeolites [93], and metal catalysts supported on TiO_2/CeO_2 [94,95]. Similar to a technique used in the petroleum industry, a large amount of coke was produced at high working temperatures, which seriously suppressed any further reaction. In investigating the influencing factors for coke formation, Chantal and coworkers [96] showed that the amount of coke was related to the flow velocity of feedstock, as well as the temperature and gas flow rate. Sharma and Bakhshi [93] suggested that the structure of the catalyst may impede H transfer during pyrolysis, and hence reduce the coke formation. A rapid pyrolysis (heating rate >100 °C s^{-1}) attracts great attention due to the positive effect for preventing coke formation, and the reaction was often complete within a few minutes or seconds such that the pyrolysis temperatures had a major influence on product distribution. Cleavage of the C–O aryl ether bonds was the major reaction at low temperatures, while the C–C bond could be cleaved when high temperatures are attained.

Photocatalysis can be achieved under mild conditions with lower catalyst costs, and this can be considered an alternative means for the depolymerization of lignin using sustainable and abundant solar energy sources. The reactions were mostly actualized using artificial or solar radiation, with TiO_2 as the catalyst. Thus, the recovery of a fine TiO_2 powder is a key issue in industrial implementation. In order to enhance the photocatalytic ability of TiO_2 many additives, such as noble metals, transition metals, and nanostructured oxides (CeO_2, La_2O_3), were added. Portjanskaja and Preis [97] studied the influence of Fe^{2+} ions on the photocatalytic oxidation (PCO) of UV-irradiated aqueous solutions containing lignin over TiO_2. Low concentrations of iron would accelerate the PCO limitation stage by decreasing the rate of recombination between positively charged holes and conductivity band electrons, whereas high concentrations of Fe^{2+} ions had a negative effect that was caused by the adsorption of Fe^{2+} onto the TiO_2 surface so as to block the active sites. Nguyen and coworkers [98] proposed a two-step strategy for lignin degradation using Bobbitt's salt ([4-AcNH-TEMPO]BF_4) mediating as a benzylic oxidation, and a chemoselective visible-light mediated reductive C–O bond cleavage promoted by the photocatalyst [Ir(ppy)$_2$(dtbbpy)]PF_6. Although the lignin was successful depolymerized by the photocatalysis, aspects of scalability, stoichiometric waste and cost remained to be emphasized.

Electrochemical depolymerization may be used as an alternative method, because of its controllability and environmental friendliness. Parpot and coworkers [99] investigated the ability of anodes of Ni, Au, Pt, Cu, DSA-O_2, and PbO_2 to electrochemically oxidize Kraft lignin in batch reactor and flow cells in NaOH solution at room temperature. A better yield of vanillin (9.4%) was obtained using nickel and DSA-O_2 anodes. The partial pressure of O_2 caused by the applied current density had a major influence on the conversion and chemical yields. Currently, Pt and Ni are frequently used as electrodes, with Pt commonly used in electrochemical reactions due to its high capability for the oxidation of organic compounds. Meanwhile, the relatively cheap, abundant-resource Ni was also advantageous as an electrode. Zhu and coworkers [100] designed an electrolytic cell with a cylindrical graphite felt cathode inside and a RuO_2-IrO_2/Ti mesh anode outside, for lignin depolymerization in alkaline aqueous solution. The current density and temperature were the two major factors for lignin depolymerization.

Biocatalysis for the cleavage of lignin in an environment-friendly way has aroused wide interest. Crestini and coworkers [101] investigated layer-by-layer immobilized laccases and laccase microcapsules to oxidize the lignin in an acetate buffer solution (pH = 6) at 40 °C using air as oxidant. The addition of 1-hydroxybenzotriazole or violuric acid as the oxidation mediator enhanced the oxidation efficiency of laccases, with an increased yield up to 94%. Three enzymes – LigD (C_α-dehydrogenase), LigF (β-etherase), and LigG (glutathione lyase) – were applied by Reiter and coworkers [102] to release lignin monomers from a lignin model substrate [1-(4-hydroxy-3-methoxyphenyl)-2-(2-methoxyphenoxy)-1,3-propanediol] in optimum buffer solution. Combined with LigDFG and *Allochromatium vinosum* (AVR), an efficient system was obtained for the degradation of lignin, which gave 89.8 mg l^{-1} vanillin after 50 h. Picart and coworkers [103] also used LigG, LigG-NS and LigG-TD catalysts for the depolymerization of lignin at 147 °C, and indicated that the stereospecificity of these enzymes can be modulated by active-site mutations, thereby demonstrating the possibility of altering the selectivity of the enzymes (as well as their activity) by protein engineering.

8.6 Conclusions

Currently, the main methods for lignin depolymerization are metal-catalyzed reductive and oxidative deconstruction, hydrolysis, and other techniques. During the early stages, thermal catalytic cracking was widely used for lignin depolymerization, but was found to be unsuitable for the depolymerization of macromolecular solid lignin, due to poor selectivity and low liquid yields, as well as rapid catalyst deactivation. In addition, lignin-derived $C_{aliphatic}$–OR and $C_{aromatic}$–OR bonds can be broken down in the process of hydrolysis with concentrated NaOH or KOH solution, although the hydrolysis reaction rates are very low when treating real lignin systems. The yields of phenols following simple alkali hydrolysis were less than 20–23 wt% under diverse reaction conditions. The main advantage of processing lignin by alkali solution is that the lignin can be easily dissolved in such a system, and the aryl ether bonds of lignin are prone to be cleaved under basic conditions with NaOH or KOH. While hydrolysis with a strong base can be considered an efficient method for deconstruction, it still faces serious problems in product separation from the liquid phase, and the disposal of highly concentrated alkali solutions and environmental pollution. More seriously, phenol monomers generated from the hydrolysis of alkali solution were easy to polymerize again in strong basic solutions through nucleophilic reactions.

By comparison, the techniques of hydrogenation and oxidation to convert lignin by metal catalysts are much 'greener' and more efficient. The secondary structure units of lignin include aryl ethers of β-O-4, α-O-4, 4-O-5 linkages, and the C–O bonds of linkages can be effectively cleaved by noble metals such as Pd, Ru, Pt, Rh, Ir, as well as transition metal catalysts such as MoC_2, Ni_2P, MoO_3, FeMoP, Mo, Ni, Fe, Co, Cu, and W via the route of hydrogenolysis in gas and liquid phases. However, when true lignin is treated as the substrate, the activities of these catalysts remain very low, mainly because the active sites (metal center) cannot access the functional sites of lignin, as the latter is not soluble in common solvents. While methanol and dioxane solvents can easily dissolve lignin, the selected metal centers were less active in these solvents. Currently, various research groups have failed to identify a better balance or combination of good lignin dissolution solvent systems and catalytic circumstances for C–O bond cleavage. Therefore, the keys to building a highly effective catalytic system for the hydrogenolysis of lignin are to select the correct solvent and catalyst centers. The development of this crucial point should be addressed urgently. In addition, as hydrogen is a very expensive resource, hydrogen-donor solvents become a highly attractive option. Indeed, by using a common solvent to generate hydrogen in situ, the approach to the hydrogenolysis of lignin becomes more attractive.

The oxidation method can also provide opportunities for producing valuable chemicals such as vanillin and vanillic acid from lignin. The catalyst system includes metal-supported oxide catalysts (Pd and Ru supported on CeO_2, TiO_2, and Al_2O_3), as well as perovskite-type catalysts, polyoxometalate catalysts, organometallic catalysts (vanadium, cobalt, iron, manganese and ruthenium-oxo complex), and ionic liquid catalysts. However, as the yield from lignin oxidation is currently only about 20%, it is not only the efficiency of such techniques that needs to be improved, but simple and efficient separation methods should also be developed towards products and catalytic systems. Attention should also be paid to other techniques such as photocatalysis, electrocatalysis and biocatalysis for lignin depolymerization. Today, research groups must focus on these key points, starting with a fundamental understanding to achieve major innovations and breakthroughs in these significant scientific areas.

Acknowledgments

This research was supported by the Recruitment Program of Global Young Experts in China, National Natural Science Foundation of China (Grant No. 21573075) and Shanghai Pujiang Program (PJ1403500).

References

1. Achyuthan, K.E., Achyuthan, A.M., Adams, P.D., Dirk, S.M., Harper, J.C., Simmons, B.A., Singh, A.K. (2010) Supramolecular self-assembled chaos: polyphenolic lignin's barrier to cost-effective lignocellulosic biofuels. *Molecules*, **15** (12), 8641–8888.
2. Hu, L.H., Pan, H., Zhou, Y.H., Zhang, M. (2011) Methods to improve lignin's reactivity as a phenol substitute and as replacement for other phenolic compounds: a brief review. *Bioresource Technol.*, **6** (3), 3515–3525.
3. Rinaldi, R. (2015) *Catalytic Hydrogenation for Biomass Valorization*. The Royal Society of Chemistry, pp. 74–98.
4. Biermann, C.J. (1993) *Essentials of Pulping and Papermaking*. Academic Press, San Diego.
5. Huber, G.W., Iborra, S., Corma, A. (2006) Synthesis of transportation fuels from biomass: chemistry, catalysts, and engineering. *Chem. Rev.*, **106** (9), 4044–4098.
6. Zakzeski, J., Bruijnincx, P.C.A., Jongerius, A.L., Weckhuysen, B.M. (2010) The catalytic valorization of lignin for the production of renewable chemicals. *Chem. Rev.*, **110** (6), 3552–3599.
7. Li, C., Zhao, X., Wang, A., Huber, G.W., Zhang, T. (2015) Catalytic Transformation of Lignin for the Production of Chemicals and Fuels. *Chem. Rev.*, **115** (21), 11559–11624.
8. Yan, N., Zhao, C., Dyson, P.J., Wang, C., Liu, L.T., Kou, Y. (2008) Selective degradation of wood lignin over noble-metal catalysts in a two-step process. *ChemSusChem.*, **1** (7), 626–629.
9. Zhang, J., Teo, J., Chen, X., Asakura, H., Tanaka, T., Teramura, K., Yan, N.A (2014) Series of NiM (M = Ru, Rh, and Pd) bimetallic catalysts for effective lignin hydrogenolysis in water. *ACS Catal.*, **4** (5), 1574–1583.
10. Ye, Y., Zhang, Y., Fan, J., Chang, J. (2012) Selective production of 4-ethylphenolics from lignin via mild hydrogenolysis. *Bioresource Technol.*, **118**, 648–651.
11. Laskar, D.D., Tucker, M.P., Chen, X., Helms, G.L., Yang, B. (2014) Noble-metal catalyzed hydrodeoxygenation of biomass-derived lignin to aromatic hydrocarbons. *Green Chem.*, **16** (2), 897.
12. Kim, J.-Y., Park, J., Kim, U.-J., Choi, J.W. (2015) Conversion of lignin to phenol-rich oil fraction under supercritical alcohols in the presence of metal catalysts. *Energ. Fuels*, **29** (8), 5154–5163.
13. Van den Bosch, S., Schutyser, W., Koelewijn, S.F., Renders, T., Courtin, C.M., Sels, B.F. (2015) Tuning the lignin oil OH-content with Ru and Pd catalysts during lignin hydrogenolysis on birch wood. *Chem. Commun.*, **51** (67), 13158–13161.
14. Van den Bosch, S., Schutyser, W., Vanholme, R., Driessen, T., Koelewijn, S.F., Renders, T., De Meester, B., Huijgen, W.J.J., Dehaen, W., Courtin, C.M., Lagrain, B., Boerjan, W., Sels, B.F. (2015) Reductive lignocellulose fractionation into soluble lignin-derived phenolic monomers and dimers and processable carbohydrate pulps. *Energ. Environ. Sci.*, **8** (6), 1748–1763.
15. Jongerius, A.L., Bruijnincx, P.C.A., Weckhuysen, B.M. (2013) Liquid-phase reforming and hydrodeoxygenation as a two-step route to aromatics from lignin. *Green Chem.*, **15** (11), 3049.

16. Jongerius, A.L., Copeland, J.R., Foo, G.S., Hofmann, J.P., Bruijnincx, P.C.A., Sievers, C., Weckhuysen, B.M. (2013) Stability of Pt/γ-Al₂O₃ catalysts in lignin and lignin model compound solutions under liquid phase reforming reaction conditions. *ACS Catal.*, **3** (3), 464–473.

17. Li, Z., Assary, R.S., Atesin, A.C., Curtiss, L.A., Marks, T.J. (2014) Rapid ether and alcohol C-O bond hydrogenolysis catalyzed by tandem high-valent metal triflate + supported Pd catalysts. *J. Am. Chem. Soc.*, **136** (1), 104–107.

18. Meier, D., Ante, R., Faix, O. (1992) Catalytic hydropyrolysis of lignin – Influence of reaction conditions on the formation and composition of liquid products. *Bioresource Technol.*, **40**, 171–177.

19. Kloekhorst, A., Heeres, H.J. (2015) Catalytic hydrotreatment of Alcell lignin using supported Ru, Pd, and Cu catalysts. *ACS Sustain. Chem. Eng.*, **3** (9), 1905–1914.

20. Wang, H., Zhang, L., Deng, T., Ruan, H., Hou, X., Cort, J.R., Yang, B. (2016) ZnCl₂ induced catalytic conversion of softwood lignin to aromatics and hydrocarbons. *Green Chem.*, **18** (9), 2802–2810.

21. Parsell, T.H., Owen, B.C., Klein, I., Jarrell, T.M., Marcum, C.L., Haupert, L.J., Amundson, L.M., Kenttämaa, H.I., Ribeiro, F., Miller, J.T., Abu-Omar, M.M. (2013) Cleavage and hydrodeoxygenation (HDO) of C–O bonds relevant to lignin conversion using Pd/Zn synergistic catalysis. *Chem. Sci.*, **4** (2), 806–813.

22. Klein, I., Marcum, C., Kenttämaa, H., Abu-Omar, M.M. (2016) Mechanistic investigation of the Zn/Pd/C catalyzed cleavage and hydrodeoxygenation of lignin. *Green Chem.* **18** (8), 2399-2405.

23. Harris, E.E., D'Ianni, J., Adkins, H. (1938) Reaction of hardwood lignin with hydrogen. *J. Am. Chem. Soc.*, **60** (6), 1467–1470.

24. Narani, A., Chowdari, R.K., Cannilla, C., Bonura, G., Frusteri, F., Heeres, H.J., Barta, K. (2015) Efficient catalytic hydrotreatment of Kraft lignin to alkylphenolics using supported NiW and NiMo catalysts in supercritical methanol. *Green Chem.*, **17** (11), 5046–5057.

25. Pepper, J.M., Steck, W. (1963) The effect of time and temperature on the hydrogenation of aspen lignin. *Can. J. Chem.*, **41** (11), 2867–2875.

26. Xin, J., Zhang, P., Wolcott, M.P., Zhang, X., Zhang, J. (2014) Partial depolymerization of enzymolysis lignin via mild hydrogenolysis over Raney nickel. *Bioresource Technol.*, **155**, 422–426.

27. Wang, X., Rinaldi, R. (2012) Solvent effects on the hydrogenolysis of diphenyl ether with Raney nickel and their implications for the conversion of lignin. *ChemSusChem.*, **5** (8), 1455–1466.

28. He, J., Zhao, C., Lercher, J.A. (2012) Ni-catalyzed cleavage of aryl ethers in the aqueous phase. *J. Am. Chem. Soc.*, **134** (51), 20768–20775.

29. Kong, J., He, M., Lercher, J.A., Zhao, C. (2015) Direct production of naphthenes and paraffins from lignin. *Chem. Commun.*, **51** (99), 17580–17583.

30. Kasakov, S., Shi, H., Camaioni, D.M., Zhao, C., Baráth, E., Jentys, A., Lercher, J.A. (2015) Reductive deconstruction of organosolv lignin catalyzed by zeolite supported nickel nanoparticles. *Green Chem.*, **17** (11), 5079–5090.

31. Song, Q., Wang, F., Xu, J. (2012) Hydrogenolysis of lignosulfonate into phenols over heterogeneous nickel catalysts. *Chem. Commun.*, **48** (56), 7019–7021.

32. Horáček, J., Homola, F., Kubičková, I., Kubička, D. (2012) Lignin to liquids over sulfided catalysts. *Catal. Today*, **179** (1), 191–198.

33. Kumar, C.R., Anand, N., Kloekhorst, A., Cannilla, C., Bonura, G., Frusteri, F., Barta, K., Heeres, H.J. (2015) Solvent free depolymerization of Kraft lignin to alkyl-phenolics using supported NiMo and CoMo catalysts. *Green Chem.*, **17** (11), 4921–4930.

34. Molinari, V., Clavel, G., Graglia, M., Antonietti, M., Esposito, D. (2016) Mild continuous hydrogenolysis of Kraft lignin over titanium nitride–nickel catalyst. *ACS Catal.*, **6** (3), 1663–1670.

35. Li, C., Zheng, M., Wang, A., Zhang, T. (2012) One-pot catalytic hydrocracking of raw woody biomass into chemicals over supported carbide catalysts: simultaneous conversion of cellulose, hemicellulose and lignin. *Energ. Environ. Sci.*, **5** (4), 6383–6390.

36. Ren, Y.-L., Tian, M., Tian, X.-Z., Wang, Q., Shang, H., Wang, J., Zhang, Z.C. (2014) Highly selective reductive cleavage of aromatic carbon–oxygen bonds catalyzed by a cobalt compound. *Catal. Commun.*, **52**, 36–39.

37. Ren, Y., Yuan, M., Wang, J., Zhang, Z.C., Yao, K. (2013) Selective reductive cleavage of inert aryl C–O bonds by an iron catalyst. *Angew. Chem. Int. Ed.*, **125** (48), 12906–12910.

38. Sergeev, A.G., Hartwig, J.F. (2011) Selective, nickel-catalyzed hydrogenolysis of aryl ethers. *Science*, **332** (6028), 439–443.

39. Sergeev, A.G., Webb, J.D., Hartwig, J.F. (2012) A heterogeneous nickel catalyst for the hydrogenolysis of aryl ethers without arene hydrogenation. *J. Am. Chem. Soc.*, **134** (50), 20226–20229.

40. Samant, B.S., Kabalka, G.W. (2012) Hydrogenolysis-hydrogenation of aryl ethers: selectivity pattern. *Chem. Commun.*, **48** (69), 8658–8660.

41. Kusumoto, S., Nozaki, K. (2015) Direct and selective hydrogenolysis of arenols and aryl methyl ethers. *Nature Commun.*, **6**, 6296.

42. Nichols, J.M., Bishop, L.M., Bergman, R.G., Ellman, J.A. (2010) Catalytic C–O bond cleavage of 2-aryloxy-1-arylethanols and its application to the depolymerization of lignin-related polymers. *J. Am. Chem. Soc.*, **132** (36), 12554–12555.

43. Chmely, S.C., Kim, S., Ciesielski, P.N., Jiménez-Osés, G., Paton, R.S., Beckham, G.T. (2013) Mechanistic study of a Ru-Xantphos catalyst for tandem alcohol dehydrogenation and reductive aryl-ether cleavage. *ACS Catal.*, **3** (5), 963–974.

44. Wu, A., Patrick, B.O., Chung, E., James, B.R. (2012) Hydrogenolysis of beta-O-4 lignin model dimers by a ruthenium-xantphos catalyst. *Dalton Trans.*, **41** (36), 11093–11106.

45. Oasmaa, A., Johansson, A. (1993) Catalytic hydrotreating of lignin with water-soluble molybdenum catalyst. *Energ. Fuels*, **7** (3), 426–429.

46. Toledano, A., Serrano, L., Pineda, A., Romero, A.A., Luque, R., Labidi, J. (2014) Microwave-assisted depolymerisation of organosolv lignin via mild hydrogen-free hydrogenolysis: Catalyst screening. *Appl. Catal. B: Environ.*, **145**, 43–55.

47. Liguori, L., Barth, T. (2011) Palladium-Nafion SAC-13 catalysed depolymerisation of lignin to phenols in formic acid and water. *J. Anal. Appl. Pyrol.*, **92** (2), 477–484.

48. Xu, W., Miller, S.J., Agrawal, P.K., Jones, C.W. (2012) Depolymerization and hydrodeoxygenation of switchgrass lignin with formic acid. *ChemSusChem*, **5** (4), 667–675.

49. Song, Q., Wang, F., Cai, J., Wang, Y., Zhang, J., Yu, W., Xu, J. (2013) Lignin depolymerization (LDP) in alcohol over nickel-based catalysts via a fragmentation–hydrogenolysis process. *Energ. Environ. Sci.*, **6** (3), 994–1007.

50. Matson, T.D., Barta, K., Iretskii, A.V., Ford, P.C. (2011) One-pot catalytic conversion of cellulose and of woody biomass solids to liquid fuels. *J. Am. Chem. Soc.*, **133** (35), 14090–14097.

51. (a) Macala, G.S., Matson, T.D., Johnson, C.L., Lewis, R.S., Iretskii, A.V., Ford, P.C. (2009) Hydrogen transfer from supercritical methanol over a solid base catalyst: a model for lignin depolymerization. *ChemSusChem.*, **2** (3), 215–217; (b) Barta, K., Ford, P.C. Catalytic conversion of nonfood woody biomass solids to organic liquids. *Acc. Chem. Res.*, **47**, 1503–1512.

52. Warner, G., Hansen, T.S., Riisager, A., Beach, E.S., Barta, K., Anastas, P.T. (2014) Depolymerization of organosolv lignin using doped porous metal oxides in supercritical methanol. *Bioresource Technol.*, **161**, 78–83.

53. Huang, X., Koranyi, T.I., Boot, M.D., Hensen, E.J. (2014) Catalytic depolymerization of lignin in supercritical ethanol. *ChemSusChem*, **7** (8), 2276–2288.

54. Huang, X., Atay, C., Korányi, T.I., Boot, M.D., Hensen, E.J.M. (2015) Role of Cu–Mg–Al mixed oxide catalysts in lignin depolymerization in supercritical ethanol. *ACS Catal.*, **5** (12), 7359–7370.

55. Ma, R., Hao, W., Ma, X., Tian, Y., Li, Y. (2014) Catalytic ethanolysis of Kraft lignin into high-value small-molecular chemicals over a nanostructured alpha-molybdenum carbide catalyst. *Angew. Chem. Int. Ed.*, **53** (28), 7310–7315.

56. Wang, X., Rinaldi, R. (2013) A route for lignin and bio-oil conversion: dehydroxylation of phenols into arenes by catalytic tandem reactions. *Angew. Chem. Int. Ed.*, **52** (44), 11499–11503.

57. Ferrini, P., Rinaldi, R. (2014) Catalytic biorefining of plant biomass to non-pyrolytic lignin bio-oil and carbohydrates through hydrogen transfer reactions. *Angew. Chem. Int. Ed.*, **53** (33), 8634–8639.

58. Pintar, A., Besson, M., Gallezot, P. (2001) Catalytic wet air oxidation of Kraft bleach plant effluents in a trickle-bed reactor over a Ru/TiO$_2$ catalyst. *Appl. Catal. B: Environ.*, **31** (4), 275–290.

59. Deng, W., Zhang, H., Wu, X., Li, R., Zhang, Q., Wang, Y. (2015) Oxidative conversion of lignin and lignin model compounds catalyzed by CeO$_2$-supported Pd nanoparticles. *Green Chem.*, **17** (11), 5009–5018.

60. Sales, F.G., Maranhão, L.C.A., Filho, N.M.L., Abreu, C.A.M. (2007) Experimental evaluation and continuous catalytic process for fine aldehyde production from lignin. *Chem. Eng. Sci.*, **62** (18-20), 5386–5391.

61. Deng, H., Lin, L., Sun, Y., Pang, C., Zhuang, J., Ouyang, P., Li, Z., Liu, S. Perovskite-type oxide LaMnO$_3$: An efficient and recyclable heterogeneous catalyst for the wet aerobic oxidation of lignin to aromatic aldehydes. *Catal. Lett.*, **126** (1-2), 106–111.

62. Deng, H.B., Lin, L., Sun, Y., Pang, C.S., Zhuang, J.P., Ouyang, P.K., Li, J.J. (2008) Activity and stability of LaFeO$_3$ catalyst in lignin catalytic wet oxidation to aromatic aldehydes. *Chin. J. Catal.*, **29**, 753–757.

63. Deng, H., Lin, L., Sun, Y., Pang, C., Zhuang, J., Ouyang, P., Li, J., Liu, S. (2009) Activity and stability of perovskite-type oxide LaCoO$_3$ catalyst in lignin catalytic wet oxidation to aromatic aldehydes process. *Energ. Fuels*, **23** (1), 19–24.

64. Deng, H., Lin, L., Liu, S. (2010) Catalysis of Cu-doped Co-based perovskite-type oxide in wet oxidation of lignin to produce aromatic aldehydes. *Energ. Fuels*, **24** (9), 4797–4802.

65. Zhang, J., Deng, H., Lin, L. (2009) Wet aerobic oxidation of lignin into aromatic aldehydes catalysed by a perovskite-type oxide: LaFe$_{(1-x)}$Cu$_{(x)}$O$_3$ (x = 0, 0.1, 0.2). *Molecules*, **14** (8), 2747–2457.

66. Bozell, J.J., Hoberg, J.O., Dimmel, D.R. (2000) Heteropolyacid catalyzed oxidation of lignin and lignin models to benzoquinones. *J. Wood Chem. Technol.*, **20** (1), 19–41.

67. Voitl, T., Rohr, P.R.v. (2008) Oxidation of lignin using aqueous polyoxometalates in the presence of alcohols. *ChemSusChem.*, **1** (8-9), 763–769.

68. Voitl, T., Rohr, P.R.v. (2010) Demonstration of a process for the conversion of Kraft lignin into vanillin and methyl vanillate by acidic oxidation in aqueous methanol. *Ind. Eng. Chem. Res.*, **49** (2), 520–525.

69. Zhao, Y., Xu, Q., Pan, T., Zuo, Y., Fu, Y., Guo, Q.-X. (2013) Depolymerization of lignin by catalytic oxidation with aqueous polyoxometalates. *Appl. Catal. A: Gen.*, **467**, 504–508.

70. Son, S., Toste, F.D. (2010) Non-oxidative vanadium-catalyzed C–O bond cleavage: application to degradation of lignin model compounds. *Angew. Chem. Int. Ed.*, **49** (22), 3791–3794.

71. Hanson, S.K., Wu, R., Silks, L.A.P. (2012) C–C or C–O bond cleavage in a phenolic lignin model compound: selectivity depends on vanadium catalyst. *Angew. Chem. Int. Ed.*, **51** (14), 3410–3413.

72. Chan, J.M.W., Bauer, S., Sorek, H., Sreekumar, S., Wang, K., Toste, F.D. (2013) Studies on the vanadium-catalyzed nonoxidative depolymerization of *Miscanthus giganteus*-derived lignin. *ACS Catal.*, **3** (6), 1369–1377.

73. Biannic, B., Bozell, J.J. (2013) Efficient cobalt-catalyzed oxidative conversion of lignin models to benzoquinones. *Org. Lett.*, **15** (11), 2730–2733.

74. Zhu, C., Ding, W., Shen, T., Tang, C., Sun, C., Xu, S., Chen, Y., Wu, J., Ying, H. (2015) Metallo-deuteroporphyrin as a biomimetic catalyst for the catalytic oxidation of lignin to aromatics. *ChemSusChem*, **8** (10), 1768–1778.

75. Crestini, C., Pro, P., Neri, V., Saladino, R. (2005) Methyltrioxorhenium: a new catalyst for the activation of hydrogen peroxide to the oxidation of lignin and lignin model compounds. *Bioorg. Med. Chem.*, **13** (7), 2569–2578.

76. Crestini, C., Caponi, M.C., Argyropoulos, D.S., Saladino, R. (2006) Immobilized methyltrioxo rhenium (MTO)/H$_2$O$_2$ systems for the oxidation of lignin and lignin model compounds. *Bioorg. Med. Chem.*, **14** (15), 5292–5302.

77. Stark, K., Taccardi, N., Bosmann, A., Wasserscheid, P. (2010) Oxidative depolymerization of lignin in ionic liquids. *ChemSusChem*, **3** (6), 719–723.

78. Liu, S., Shi, Z., Li, L., Yu, S., Xie, C., Song, Z. (2013) Process of lignin oxidation in an ionic liquid coupled with separation. *RSC Adv.* **3** (17), 5789–5793.

79. Jia, S., Cox, B.J., Guo, X., Zhang, Z.C., Ekerdt, J.G. (2010) Cleaving the beta-O-4 bonds of lignin model compounds in an acidic ionic liquid, 1-H-3-methylimidazolium chloride: an optional strategy for the degradation of lignin. *ChemSusChem*, **3** (9), 1078–1084.

80. Yang, Y., Fan, H., Song, J., Meng, Q., Zhou, H., Wu, L., Yang, G., Han, B. (2015) Free radical reaction promoted by ionic liquid: a route for metal-free oxidation depolymerization of lignin model compound and lignin. *Chem. Commun.*, **51** (19), 4028–4031.

81. Prado, R., Brandt, A., Erdocia, X., Hallet, J., Welton, T., Labidi, J. (2016) Lignin oxidation and depolymerisation in ionic liquids. *Green Chem.*, **18** (3), 834–841.

82. Miller, J.E., Evans, L., Littlewolf, A., Trudell, D.E. (1999) Batch microreactor studies of lignin and lignin model compound depolymerization by bases in alcohol solvents. *Fuel*, **78** (11), 1363–1366.

83. Watanabe, M., Inomata, H., Osada, M., Sato, T., Adschiri, T., Arai, K. (2003) Catalytic effects of NaOH and ZrO$_2$ for partial oxidative gasification of n-hexadecane and lignin in supercritical water. *Fuel*, **82** (5), 545–552.

84. Karagöz, S., Bhaskar, T., Muto, A., Sakata, Y. (2004) Effect of Rb and Cs carbonates for production of phenols from liquefaction of wood biomass. *Fuel*, **83** (17-18), 2293–2299.

85. Ehara, K., Takada, D., Saka, S. (2005) GC-MS and IR spectroscopic analyses of the lignin-derived products from softwood and hardwood treated in supercritical water. *J. Wood Sci.*, **51** (3), 256–261.

86. Wahyudiono, Sasaki, M., Goto, M. (2008) Recovery of phenolic compounds through the decomposition of lignin in near and supercritical water. *Chem. Eng. Process.*, **47** (9-10), 1609–1619.

87. Nenkova, S., Vasileva, T., Stanulov, K. (2008) Production of phenol compounds by alkaline treatment of technical hydrolysis lignin and wood biomass. *Chem. Nat. Comp.*, **44** (1573-8388), 182–185.

88. Onwudili, J.A., Williams, P.T. (2014) Catalytic depolymerization of alkali lignin in subcritical water: influence of formic acid and Pd/C catalyst on the yields of liquid monomeric aromatic products. *Green Chem.*, **16** (11), 4740–4748.

89. Rahimi, A., Ulbrich, A., Coon, J.J., Stahl, S.S. (2014) Formic-acid-induced depolymerization of oxidized lignin to aromatics. *Nature*, **515** (7526), 249–252.

90. Adjaye, J.D., Bakhshi, N.N. (1995) Production of hydrocarbons by catalytic upgrading of a fast pyrolysis bio-oil. Part I: Conversion over various catalysts. *Fuel Process. Technol.*, **45** (3), 161–183.

91. Adjaye, J.D.; Bakhshi, N.N. (1995) Production of hydrocarbons by catalytic upgrading of a fast pyrolysis bio-oil. Part II: Comparative catalyst performance and reaction pathways. *Fuel Process. Technol.*, **45** (3), 185–202.

92. Toledano, A., Serrano, L., Labidi, J. (2012) Organosolv lignin depolymerization with different base catalysts. *J. Chem. Technol. Biotechnol.*, **87** (11), 1593–1599.

93. Sharma, R.K., Bakhshi, N.N. (1991) Upgrading of wood-derived bio-oil over HZSM-5. *Bioresource Technol.*, **35** (1), 57–66.

94. Jackson, M.A., Compton, D.L., Boateng, A.A. (2009) Screening heterogeneous catalysts for the pyrolysis of lignin. *J Anal. Appl. Pyrol.*, **85** (1-2), 226–230.

95. Lu, Q., Zhang, Y., Tang, Z., Li, W.-z., Zhu, X.-f. (2010) Catalytic upgrading of biomass fast pyrolysis vapors with titania and zirconia/titania based catalysts. *Fuel*, **89** (8), 2096–2103.

96. Chantal, P., Kaliaguine, S., Grandmaison, J.L., Mahay, A. (1984) Production of hydrocarbons from aspen poplar pyrolytic oils over H-ZSM5. *Appl. Catal.*, **10** (3), 317–332.

97. Portjanskaja, E., Preis, S. (2007) Aqueous photocatalytic oxidation of lignin: the influence of mineral admixtures. *Int. J. Photoenergy*, **2007**, 1–7.

98. Nguyen, J.D., Matsuura, B.S., Stephenson, C.R. (2014) A photochemical strategy for lignin degradation at room temperature. *J. Am. Chem. Soc.*, **136** (4), 1218–1221.

99. Parpot, P., Bettencourt, A.P., Carvalho, A.M., Belgsir, E.M. (2000) Biomass conversion: attempted electrooxidation of lignin for vanillin production. *J. Appl. Electrochem.*, **30**, 727–731.

100. Zhu, H., Wang, L., Chen, Y., Li, G., Li, H., Tang, Y., Wan, P. (2014) Electrochemical depolymerization of lignin into renewable aromatic compounds in a non-diaphragm electrolytic cell. *RSC Adv.*, **4** (56), 29917–29924.
101. Crestini, C., Perazzini, R., Saladino, R. (2010) Oxidative functionalisation of lignin by layer-by-layer immobilised laccases and laccase microcapsules. *Appl. Catal. A: Gen.*, **372** (2), 115–123.
102. Reiter, J., Strittmatter, H., Wiemann, L.O., Schieder, D., Sieber, V. (2013) Enzymatic cleavage of lignin β-O-4 aryl ether bonds via net internal hydrogen transfer. *Green Chem.*, **15** (5), 1373–1381.
103. Picart, P., Sevenich, M., Domínguez de María, P., Schallmey, A. (2015) Exploring glutathione lyases as biocatalysts: paving the way for enzymatic lignin depolymerization and future stereoselective applications. *Green Chem.*, **17** (11), 4931–4940.

9

Mesoporous Zeolite for Biomass Conversion

Liang Wang, Shaodan Xu, Xiangju Meng, and Feng-Shou Xiao

Key Laboratory of Applied Chemistry of Zhejiang Province, Department of Chemistry, Zhejiang University, China

9.1 Introduction

The discovery and application of crude oil provided an inexpensive energy source that has endured since the nineteenth century, and has assisted hugely in the development of industries worldwide, as well as improving the quality of human life [1–6]. Today, however, with decreasing amounts of crude oil available and increased demands for petroleum resources, it is vital that alternative sources are identified for the production of fuels and fine chemicals. Plant biomass, produced by plants via the conversion of CO_2 and water, catalyzed by sunlight, provides huge amounts of carbohydrates each year; current annual availability is almost 130 billion metric tons [7–12]. Consequently, biomass is regarded as a highly promising feedstock for producing fuels and fine chemicals, on the basis of its excellent features of renewability, low N and S concentrations, and sustainability with low CO_2 emission. Clearly, biomass potentially has a highly important role as an energy source [13–17]. Since, at present, only about 4% of the organic carbon in plant biomass is used, the development of efficient strategies for further applications of biomass is a very important topic, both for laboratory research and industrial applications.

The biomass mainly includes cellulose formed from glucose, hemicellulose with monomers of glucose and xylose, and lignin from monomers of various phenolic molecules, lipids, and terpenes [1]. These compounds can be directly converted into fuels and fine chemicals to partly replace crude oils. Until now, many reaction routes have been

Nanoporous Catalysts for Biomass Conversion, First Edition. Edited by Feng-Shou Xiao and Liang Wang.
© 2018 John Wiley & Sons Ltd. Published 2018 by John Wiley & Sons Ltd.

widely investigated, including the depolymerization of cellulose to glucose and cellobiose [18,19], and the rapid pyrolysis of cellulose and lignin topyrolysis oil [20–22], which can produce a variety of platform chemicals from sugars [23–25] and lipids [26]. In these processes, acid catalysts were generally necessary to catalyze the cracking, isomerization, alkylation, hydration, and dehydration steps. Conventionally, the homogeneous acids of phosphoric acid, sulphuric acid and acid ionic liquids were generally used. Notably, these homogeneous acids normally have difficulties in their separation and regeneration from the reaction system, which greatly hinders their wide application.

As one type of typical solid acids, aluminosilicate zeolites have been successfully applied in petrochemical industry [27–29]. Based on their obvious advantages such as high stability, ease of separation, strong acidity, and abundant acidic sites the aluminosilicate zeolites have been also used to catalyze the conversion of biomass [28,30–32]. Notably, most biomass molecules are larger than the diameters of the zeolite micropores, and therefore the challenge remains to realize the conversion of relatively bulky biomass molecules in zeolite micropores [33].

One efficient methodology for solving this problem is to introduce mesoporosity into the zeolite crystals, forming a hierarchical porous structure containing both micropores and mesopores; such materials are designated mesoporous zeolites [34–37]. These mesoporous zeolite-based catalysts have exhibited superior properties in hydrodesulphurization [38,39], Fischer–Tropsch syntheses [40,41], isomerization, and alkylation [42,43]. Recently, the use of mesoporous zeolites was extended to include the catalytic conversion of biomass [44,45]. In these cases, the catalyst mesoporosity offered clear advantages as follows: (i) a rapid mass transfer, which is favourable for the adsorption, desorption, and surface diffusion of reactants and products in catalysis; (ii) accessible acid sites to bulky molecules, which is helpful for the conversion of biomass molecules; and (iii) a synergistic effect between acids in mesopores and micropores. These features significantly enhance the catalytic activity, product selectivity, and catalyst life. The aim of this chapter is to discuss and evaluate mesoporous zeolite-based catalysts in the conversion of biomass molecules, mainly including the production of biofuels and transformation of glycerol to platform chemicals.

9.2 Production of Biofuels

9.2.1 Pyrolysis of Biomass

The conversion of biomass feedstocks to alkane fuels is one of the most important route in biomass conversion, as shown in Figure 9.1. Generally, the first step is a rapid pyrolysis of lignocellulose, producing a pyrolysis oil containing a variety of substituted phenolic molecules, acids, and sugars. The addition of a catalyst to the pyrolysis reactor has been regarded as an efficient method for controlling product selectivity. Many solid catalysts have been used in these pyrolysis reactions, of which the HZSM-5 catalysts are the most widely studied, because of their high selectivity towards gasoline-range hydrocarbons. However, heavy coke formation is typically observed in these cases. It is necessary, therefore, that catalyst anti-deactivation can be improved by the development of new catalysts [1].

The addition of Ce species into heterogeneous catalysts can significantly reduce the formation [46], and this approach has been widely used in a variety of reactions of gasification [47,48], hexane isomerization [49,50], and steam reforming [51,52]. The Ce-based catalysts

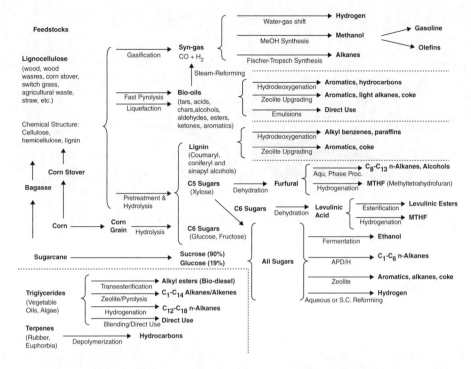

Figure 9.1   *Known routes for the production of fuels from biomass.*

have also been used in the pyrolysis of biomass where, in particular, the Ce-modified ZSM-5 has attracted much attention as it provides a significant reduction in coke formation compared with Ce-free catalysts. Mesoporosity is also introduced in the Ce-modified ZSM-5 zeolite, so as to obtain Ce-modified mesoporous zeolite catalysts [46]. The presence of Ce in catalysts can efficiently reduce the acidity (Figures 9.2 and 9.3), shifting selectivity from typical HZSM products (e.g., benzene, toluene and xylenes) to more valuable products (e.g., furans, ketone, and aldehydes). The introduction of mesoporosity in catalysts remarkably led to a simultaneous reduction in coke formation.

As biomass-feedstocks usually have bulky molecular diameters, the introduction of a mesostructure into the zeolite crystals could effectively enhance access of the acid sites to the substrates. Furthermore, the combination of other catalytically active sites with mesoporous zeolites to obtain multifunctional mesoporous zeolite-based catalysts would potentially be important in the future development of new catalysts with desirable catalytic properties.

9.2.2   Upgrading of Pyrolysis Oil

It is worth noting that the direct use of the pyrolysis oil is impossible because their high oxygen content leads to low energy density, high viscosity, and low stability. Right now, upgrading the pyrolysis oil by hydrodeoxygenation is an efficient strategy to obtain high-quality alkane oil (Figure 9.4), where the aluminosilicate zeolite (*e.g.* ZSM-5, Beta, Y) supported

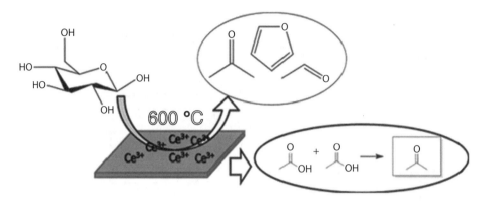

Figure 9.2   *The pyrolysis of sugars to various chemicals over the Ce-modified mesoporous ZSM-5 catalyst.*

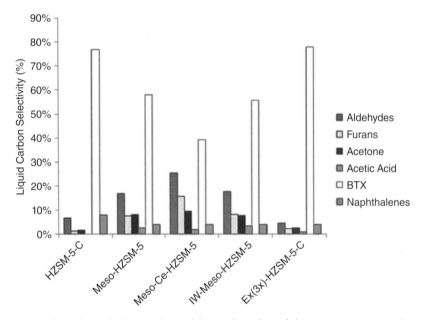

Figure 9.3   *Molar carbon selectivity in the catalytic rapid pyrolysis of glucose over various catalysts. BTX: benzene, toluene, xylene; Naphthalenes: naphthalene, methyl-naphthalene.*

metal nanoparticles (NPs) have been regarded as promising bifunctional catalysts. However, in the hydrodeoxygenation of relatively bulky molecules to obtain diesel-range alkanes, the catalytic activity and product selectivity are still unsatisfactory [53].

In order to solve this problem, Wang *et al.* employed a mesoporous ZSM-5 zeolite as support for loading Ru NPs (Ru/HZSM-5-M) [54]. In the hydrodeoxygenation of phenol with smaller molecule sizes than the micropores of the ZSM-5 zeolite, Ru/ZSM-5-M exhibited similar catalytic performances as the conventional ZSM-5 zeolite supported a Ru nanocatalyst (Ru/HZSM-5). However, in the hydrodeoxygenation of bulky 2,6-dimethoxyphenol,

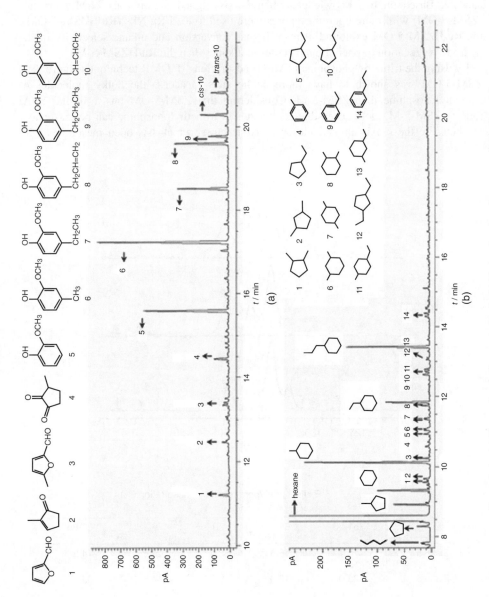

Figure 9.4 (a) The molecules in pyrolysis oil. (b) The molecules of corresponding alkane oil.

Ru/HZSM-5-M exhibited a much higher conversion and alkane product selectivity than Ru/HZSM-5, indicating the positive effect of mesoporous structure for the conversion of bulky biomass molecules in the catalysis. In this case, the accessible acidic sites to the bulky molecules are directly attributed to the open mesopores in the Ru/HZSM-5-M catalyst. Based on this knowledge, when *b*-axis-aligned mesoporous ZSM-5 crystals (ZSM-5-OM) with more open mesopores were used to load Ru NPs (Ru/HZSM-5-OM), the Ru/HZSM-5-OM exhibited a much higher conversion and alkane selectivity in the hydrodeoxygenation of phenolic molecules to alkanes than did Ru/HZSM-5-M.

By using the trimethylphosphine (TMPO)-adsorption ^{31}P NMR technique (Figure 9.5), ZSM-5-OM was shown to have more acidic sites to access the bulky molecules of trimethylphosphine than ZSM-5-M. Considering that ZSM-5-OM has a similar Si/Al ratio to ZSM-5-M, more accessible acidic sites to trimethylphosphine can be reasonably attributed to the *b*-axis-aligned mesoporous structure, which has open mesopores with

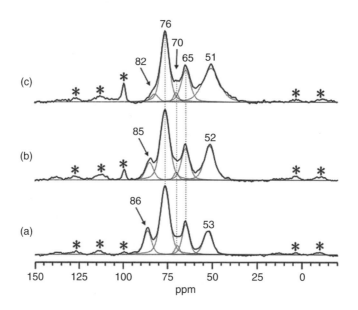

Samples	^{31}P NMR peak/ppm relative concentration (%)				
	51-53	65	70	76	82-86
H-ZSM-5	16.4	16.5	3.4	49.5	14.3
H-ZSM-5-M	25.0	18.0	2.7	44.5	9.9
H-ZSM-5-OM	36.5	19.3	3.1	37.1	4.1

Figure 9.5 ^{31}P NMR spectra of TMPO adsorbed on (a) HZSM-5, (b) HZSM-5-M, and (c) HZSM-5-OM. The peaks at 51–53 ppm correspond to the acid sites in the mesopores or on the external surface of zeolite crystals.

abundant exposed acidic sites. A greater number of open mesopores in ZSM-5-OM than in ZSM-5-M should give rise to a faster reaction rate in the hydrodeoxygenation of bulky phenolic molecules [54].

More recently, attention has been focused on the synthesis of zeolites with controllable mesoporous structures and their applications in the production of biofuels [55–58]. For example, Vu *et al.* synthesized mesoporous ZSM-5 zeolite via a post-synthesis involving the optimization of base treatment and subsequent strong acid washing of commercial Al-rich ZSM-5 [59]. The treatment conditions greatly influence the zeolite acidity; for example, a mild alkaline treatment could slightly increase the fraction of strong acid sites, which might be due to the removal of amorphous parts in the zeolite sample. In this case, a small amount of mesoporosity would be formed. Further increasing the severity of the alkaline treatment could lead to a loss of strong acid sites by dealumination, and rich mesopores ranging from 10 nm to 50 nm would be formed. These mesoporous ZSM-5 zeolites were used in the cracking of triglyceride-rich biomass under fluid catalytic cracking (FCC) conditions. By adjusting the alkaline treatment conditions to obtain the necessary zeolite acidity, gasoline-range hydrocarbons and light olefins could be efficiently obtained from triglyceride-rich biomass feedstocks (Figure 9.6). The superior catalytic properties of mesoporous ZSM-5 catalysts originate from the enhanced accessibility and mass transfer provided by substantial mesoporosity while retaining the intrinsic catalytic properties of ZSM-5. This enables hierarchical ZSM-5 to effectively convert triglycerides, independent of the degree of unsaturation, to gasoline hydrocarbons and light olefins.

Wang *et al.* reported the details of a simple route for synthesizing mesoporous ZSM-5 zeolite-supported Ni NPs, starting from conventional ZSM-5 zeolite crystals (Figure 9.7) [60]. Using this specially designed method, mesoporous ZSM-5 zeolite with

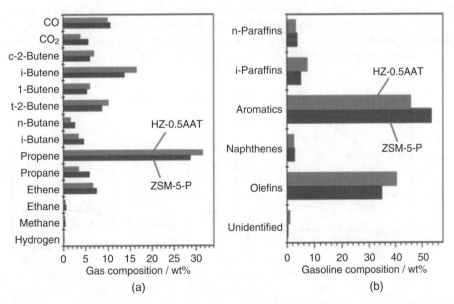

Figure 9.6 *Gas and gasoline composition from the cracking of cooking oil over conventional (ZSM-5-P) and mesoporous ZSM-5 (HZ-0.5AAT) catalysts.*

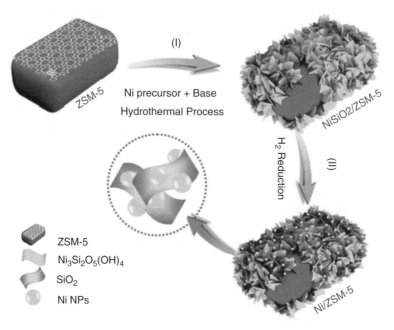

Figure 9.7   *Preparation procedures for Ni nanoparticles supported on mesoporous ZSM-5 zeolite crystals.* (See color plate section for the color representation of this figure.)

a flower-like morphology and well-dispersed ~5-nm Ni NPs could be obtained in one pot. When the hydrodeoxygenation of stearic acid was employed as a model reaction to obtain diesel-range alkanes, the conversion of stearic acid was up to 86.1%, with n-C_{18} alkane as a major product. By comparison, the conventional ZSM-5-supported Ni NPs exhibited a much lower conversion of stearic acid (<30%) and yield of n-C_{18} alkane (<20%) (Figure 9.8). In addition, this mesoporous ZSM-5 zeolite-supported Ni catalyst

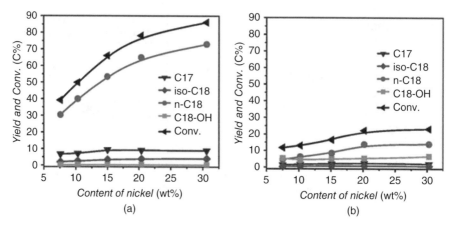

Figure 9.8   *Yield and conversion for stearic acid hydrodeoxygenation over (a) Ni/ZSM-5 and (b) Ni/ mesoporous ZSM-5 catalysts.*

maintained its catalytic activity and nanoparticle size. Under the same conditions, the Ni NPs supported on conventional ZSM-5 were clearly aggregated with each other, and lost 40% of their activity. This one-pot synthesis to simultaneous obtaining open mesopores and highly dispersed metal NPs is a universal route, which could be extended to synthesize other zeolite (MOR, Beta, Y, MCM-22, ZSM-5) -supported Cu, Co, and Ni catalysts.

9.2.3   Conversion of Lipids into Alkane Oil

In addition to pyrolysis oil from cellulose and lignin, lipids from animal oils (e.g., camelina oil, algae oil, waste restaurant oil) are also important sources of renewable fuels, which have rich triglyceride and free fatty acid contents [61–65]. These oils have been regarded as suitable sources for producing jet fuels, because of their low aromatic content. In order to obtain high-quality jet fuels with a high energy density, and a low freezing point and viscosity, a multi-step strategy has been claimed to convert lipids into jet-range hydrocarbons [61,62], which involves several reaction processes where the products are separated from each reaction. Compared with the multi-step strategy, a one-step route should be more efficient and favourable, where a combination of the metal with zeolites is suitable for the direct hydrodeoxygenation to produce alkanes.

Considering that the lipid molecules are relatively bulky, Verma *et al.* employed mesoporous SAPO-11 and ZSM-5 zeolites to support metal components (e.g., Pt, NiO, WO_3, MoO_3) [66] as composite catalysts. In these cases, the mesoporous zeolite properties, such as crystallinity, acidity and porosity, are tuneable. Yet, by adjusting the zeolite structure and metal composition, the yield of jet-fuel range alkanes (C_9–C_{15}) could reach as high as 78.5% from Jatropha oil as feedstock over a mesoporous ZSM-5-supported NiO-WO_3 composite catalyst (Figure 9.9). This result was much higher than that from the two-step processes (33.0%), indicating the significant advantage of mesoporous zeolite-based catalysts in one-step hydrodeoxygenation.

Later, Ma *et al.* systemically studied the mesoporous structure in the zeolites for the hydrodeoxygenation of palm oil [67]. Various post-treatments were carried out on the parent beta-zeolite to obtain mesoporous beta-zeolite with controllable mesopore structures. After loading the Ni species, the catalyst with Ni NPs in the inter-crystalline mesopores provided a greater accessibility towards bulky molecules, as well as restricting the aggregation of Ni NPs. This novel catalyst exhibited an initial reaction rate of 67 mmol g^{-1} h^{-1} in producing 85% n-C_{17}/C_{18} and 11% *iso*-C_{17}/C_{18} alkanes by stearic acid conversion. This was much higher than the conventional beta-supported Ni catalysts in terms of both activity and product selectivity (the mixed TPAOH/NaOH-treated case versus the untreated case in Figure 9.10). This hydrodeoxygenation route followed the major pathways of sequential hydrogenation and dehydration steps, affording a highly economical process for the production of high-grade diesel.

Notably, in the hydrodeoxygenation of pyrolysis oil or lipids as feedstocks, a combination of metallic sites and mesoporous zeolite is very important for the enhancement of catalytic properties in the hydrogenation/dehydration and hydration/dehydration/isomerization [68–70], where the synthesis of mesoporous structures and the preparation of metal NPs are particularly emphasized [71–74]. Thus, the construction of an efficient synergism between the metal sites and mesoporous zeolites should be important for the development of next-generation catalysts for the dehydrogenation of biomass.

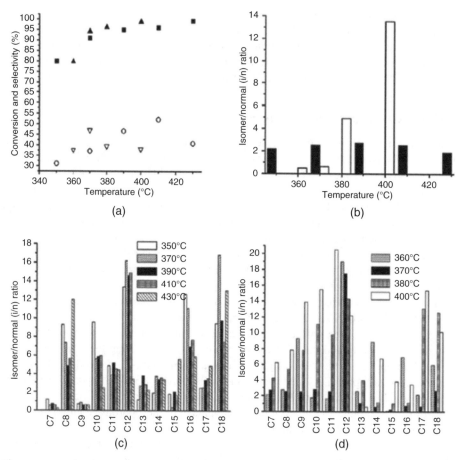

Figure 9.9 *Hydroconversion of Jatropha oil (triglycerides and free fatty acids) into hydrocarbons over sulphided Ni–Mo catalysts supported on high-surface-area semi-crystalline (HSASC) and low-surface-area crystalline (LSAC) hierarchical mesoporous H-ZSM-5. (a) Conversion (■, ▲) and C_9–C_{15} hydrocarbon yield (○,▽) over LSAC (▲,▽) and HSASC (■, ○) supports. (b) Distribution of isomer/normal alkane (C_9–C_{15}) ratio over LSAC (□) and HSASC (■) supports; and distribution of isomer/normal alkane (C_7–C_{18}) ratio at different reaction temperatures for HSASC (c) and LSAC (d) supports.*

9.2.4   Synthesis of Ethyl Levulinate Biofuel

Furfural, as an important biomass platform chemical, could be easily obtained via hydrolysis [75,76]. Hydrogenation of the furfural could produce furfuryl alcohol, which can be transformed into ethyl levulinate [77], an important renewable oxygenate fuel additive which was recognized as one of the top-10 biorefinery candidates in 2004 by the United States Department of Energy [78–82].

The conversion of furfuryl alcohol to ethyl levulinate usually proceeds via two steps which include the hydrolysis of furfuryl alcohol to levulinic acid and the esterification of

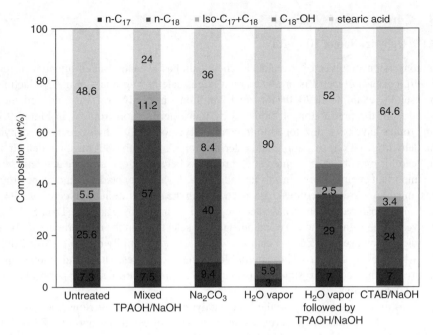

Figure 9.10 *Product composition in 1-h reaction for the hydrodeoxygenation of stearic acid over various beta-supported Ni catalysts. The treatment methods used to synthesize the mesoporous beta-catalyst are shown on the x-axis.*

levulinic acid with ethanol. The drawback of this route is the side reaction of furfuryl alcohol polymerization during the hydrolysis step, leading to a reduction in levulinic acid selectivity [77]. In order to overcome this problem a new one-step route has been developed which involves the ethanolysis of furfuryl alcohol to ethyl levulinate. This route is atom-economic and more beneficial than the conventional, two-step route.

Many homogeneous acids have been used in the ethanolysis of furfuryl alcohol, and have exhibited superior catalytic activity [83,84]. However, it is well known that these homogeneous acids are difficult to separate and regenerate from the reaction liquor. Although various solid acids such as resins and zeolites have been investigated in the reaction, the zeolites exhibited a higher product yield than other catalysts [84,85].

Nandiwale *et al.* compared the catalytic properties of USY, H-Beta, H-ZSM-5 and mesoporous ZSM-5 catalysts in the ethanolysis of furfuryl alcohol to ethyl levulinate [77]. Three intermediates, namely ethoxymethylfuran, 4,5,5-triethoxypentan-2-one and diethyl ether, have been identified during the reaction process. Among these catalysts, the mesoporous ZSM-5 exhibited the highest activity, where the catalysts followed an order of USY < H-Beta < H-ZSM-5 < mesoporous ZSM-5. This phenomenon is mainly related to the fact that mesoporous ZSM-5 is favourable for the access of substrates to the active sites, compared to the other catalysts. Clearly, the acid-catalyzed reactions and metal-catalyzed hydrogenations in the mesoporous zeolite-supported metal catalysts are important factors that operate synergistically to achieve excellent catalytic performances.

9.3 Conversion of Glycerol

9.3.1 Dehydration of Glycerol

The conversion of glycerol to valuable chemicals has attracted much attention recently [86–89], because glycerol is produced on a large scale as byproduct in the production of biodiesel. For example, in 2010 a total of 1.2 million tons of glycerol was produced, mostly derived from the production of biodiesel by transesterification (route 1 in Figure 9.11). Many routes have been developed to transform glycerol, such as hydrogenation, dehydration, dehydrogenation, oxidation, and esterification. Among these routes, the dehydration of glycerol is one of the most interesting processes because it can produce acrolein, which is an important chemical used in various industries. A variety of acids, including heteropoly acids, mixed oxides and zeolites, have been used in the dehydration of glycerol [90–94], of which aluminosilicate zeolites have been regarded as promising catalysts. The ZSM-5 zeolite gives a high catalytic activity and outstanding selectivity in the dehydration of glycerol to acrolein. However, the ZSM-5 catalyst was deactivated rapidly due to the formation of coke, which covers the active sites and/or blocks the micropores. The introduction of meso-porosity into the zeolite crystals was expected to improve their catalytic performances in glycerol dehydration.

Possato *et al.* systemically studied the catalytic properties of various catalysts in the gas-phase dehydration of glycerol, using alkaline-desilicated mesoporous ZSM-5 zeolite [95], where the samples exhibited similar catalytic performances to each other, and only the coke content was increased with the pore volume. This phenomenon might be attributed to the similar mesoporous structure from the same treatment method. Recently, Zhang *et al.* employed various methods to synthesize ZSM-5 zeolites with different mesoporous structures [90], including: (i) H-Z5-1, with larger crystals but less-closed intracrystal mesopores; (ii) H-Z5-2, with smaller crystals but more closed intracrystal mesopores;

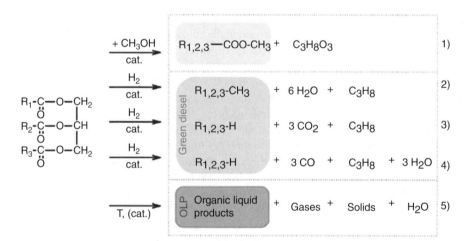

Figure 9.11 *A simplified scheme of the main reaction pathways in the transformation of triglycerides into biofuels via catalytic processes. (1) Transesterification; (2) Hydrodeoxygenation; (3) Hydrodecarboxylation; (4) Decarbonylation; and (5) Cracking/catalytic cracking.*

(iii) H-Z5-3, with abundant open intercrystal mesopores due to its nanocrystallite oriented self-assembled structure; and (iv) H-Z5-4, which combined the structures of intra- and intercrystal mesopores with special self-assembled structures of nanocrystallite containing intracrystal mesopores. As a result, the number of accessible acid sites was very low for samples with intracrystal mesopores because the small micropores limited the mass transfer of bulky *t*-butylamine molecule. In contrast, the number of accessible acid sites was very high for the H-Z5-3 and H-Z5-4 zeolites with open mesopores. In particular, H-Z5-3 had the highest percentage of accessible acid sites of the bulky molecules, owing to the abundant open mesopores structure.

In further investigations of glycerol dehydration, all samples exhibited a full conversion of glycerol at the start of the reaction, but the life-times of each catalyst were very different (Figure 9.12). The conventional ZSM-5 catalyst was shown to undergo the fastest deactivation, while the mesoporous zeolite catalysts exhibited significantly improved catalytic reaction life-times. H-Z5-3 exhibited the best life-time, because its open porous structure is favourable for mass transfer and hiding coke formation [96–98]. With kinetic studies, it was interesting to find that the mesoporous zeolites also exhibited higher product selectivities than the conventional ZSM-5 (Figure 9.13). These abnormal catalytic performances can be easily explained by proposing the pore condensation of glycerol or its heavier derivatives. The high catalytic activity, good selectivity and excellent catalyst life-time of the mesoporous zeolite supports its potential importance for widespread applications in the dehydration of glycerol in the future.

9.3.2 Etherification of Glycerol

Among various routes for the conversion of glycerol, etherification is important for the production of oxygenated blending stock for diesel [99–102]. The addition of glycerol

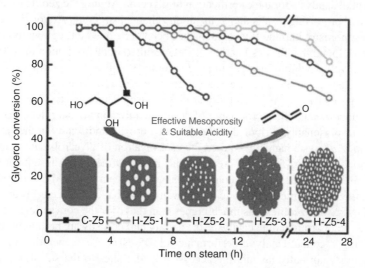

Figure 9.12 *Dependences of glycerol conversion on reaction time over different mesoporous zeolite catalysts.* (See color plate section for the color representation of this figure.)

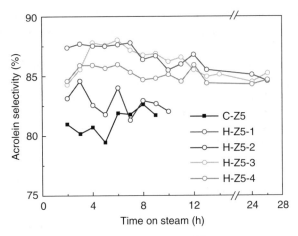

Figure 9.13 *Acrolein selectivity in glycerol conversion over different mesoporous zeolite catalysts.* (See color plate section for the color representation of this figure.)

ether enhances the cetane number and reduces the fumes and particulate mattered carbon monoxide emissions. Compared with the route of glycerol etherification with *t*-butyl alcohol, glycerol esterification with isobutylene could give better product yields. The esterification mainly involves the reaction of Brønsted acid sites with the hydroxyl groups of alcohols, followed by interaction with glycerol to form glycerol ethers. In this process, the acidic strength and catalyst porosity are sensitive to the yield of target products. Various acidic resins with strong acidity and rich porosity have been used, but the relatively low stability of the resins under high reaction temperature hinders their wide application. In contrast, zeolites with excellent thermal and hydrothermal stabilities have been considered to be potential catalysts for the esterification of glycerol. Among the zeolites, Beta catalyst has exhibited promising results due to its unique channel structure. Although the introduction of mesoporosity into the Beta crystals has been performed through various strategies, Saxena *et al.* found that the desilication of Beta resulted in the formation of ~20 nm range mesopores, with a twofold increase in pore volume as well as an increased acid intensity, compared to the parent Beta zeolite [103]. Very interestingly, the mesoporous Beta zeolite exhibited up to 98% glycerol selectivity to diesel-miscible oxygenates. These results indicated the superior catalytic performance of desilicated Beta in the esterification of glycerol, which can be reasonably attributed to the suitable porosity and acidity in the catalyst.

Arellano *et al.* systematically studied the esterification of glycerol with benzyl alcohol using mesoporous ZSM-5 zeolites prepared via alkali treatment and subsequent HCl washing [104]. In the esterification, the conventional and mesoporous ZSM-5 zeolites exhibited similar catalytic activities, but selectivity towards the bulky-molecule product over the mesoporous ZSM-5 zeolites was much higher than that over the conventional ZSM-5.

In these mesoporous zeolite catalysts, the acidity and porosity are both important factors for achieving excellent catalytic performances [105–107]. The generation of mesopores in the catalysts normally favours the access of acidic sites to the substrates, and therefore the acidity and mesoporosity should be considered simultaneously for the design of mesoporous zeolite catalysts.

9.3.3 Aromatization of Glycerol

The aromatic compounds benzene, toluene and xylene are well known as raw materials for the production of chemicals in daily life [108]. Conventionally, these aromatics are formed by the pyrolysis and reforming of crude oil, which has to proceed via complex reaction pathways [109]. With increasing requirements for these chemicals, the development of new routes without using crude oil has attracted much attention during recent years, with various reaction routes having been designed for the use of biomass and its derived chemicals. One promising route to obtain aromatics is the aromatization of glycerol (Figure 9.14) [110], which would not only solve the problems for the conversion of crude oil but also develop a value-adding avenue for the conversion of glycerol waste from the biodiesel industry.

Hoang *et al.* investigated the production of aromatics from the conversion of glycerol, where the zeolite catalysts of HZSM-5, HY, MOR, and ZSM-22 were employed. Notably, the HZSM-5 zeolite exhibited a higher yield of aromatics than the other catalysts, which should be due to the matching micropores and acidity. For the cases of liquid-phase aromatization in water solvent and gas-phase aromatization in a fixed-bed reactor, HZSM-5 exhibited high selectivity for aromatics. However, the catalysts would be deactivated within a short reaction time, due mainly to pore blocking by coke formation during the aromatization. As glycerol has a similar molecular size (0.5 nm) to the micropores of ZSM-5 zeolite (0.55 nm), it is possible that blocking of the micropores would occur very easily. Hence, the introduction of mesoporosity into the zeolite crystals with a fast mass transfer should be favourable in this reaction.

In the preparation of mesoporous zeolites, the acidity of the samples was also influenced by the synthetic conditions [110]. By increasing the concentration of NaOH solution from 0.1 to 0.7 M, the density of Brønsted acidity was significantly decreased with only a slight increase in Lewis acid density. These results were further confirmed by pyridine-adsorption

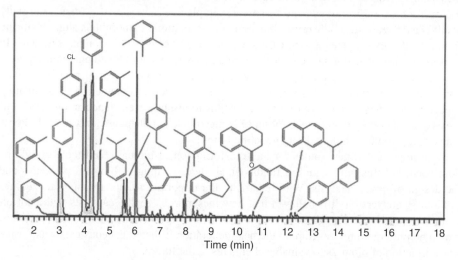

Figure 9.14 *Total chromatogram of liquid products collected between 1 and 2 h in the aromatization of 40% glycerol in methanol at 400 °C over the conventional HZSM-5 zeolite.*

infrared (IR) spectra, where the Brønsted sites were transformed into Lewis acid sites during the alkaline-treatment process. During aromatization, the Brønsted and Lewis acids functioned synergistically; therefore, an adjustment of the ratio of Brønsted/Lewis acid sites would be expected to achieve an excellent catalytic performance in the aromatization of glycerol.

In the aromatization of glycerol, the conversion of glycerol could reach 100% over all of the ZSM-5 catalysts [111]. Two types of product – liquid and gaseous – are obtained from the reaction, and an analysis of these indicated that the HZSM-5 zeolite, when treated in mild alkali solution, could improve the yields of xylene in the transformation of glycerol and methanol to aromatics. This would occur not only because the introduction of mesopores into the HZSM-5 crystals for molecule diffusion when the aromatization reaction occurred, but also the appropriate Brønsted/Lewis acid ratios benefiting the aromatization. These features make the conversion of reaction intermediates to aromatics very easy. In addition, the newly formed mesopores would reduce the transfer limitation of large molecules, shorten the diffusion pathway, and significantly improve the reaction life.

In these studies, the mesopores in the zeolite catalysts were shown to be important for catalytic activity, product selectivity, and catalyst life-time. Notably, the selective conversion of glycerol into valuable chemicals may consume the glycerol. Particularly, the selective formation of ethanol is excepted but still challenging. This exception requires the controllably selective cleavage of C–C bonds and C–O bonds in the glycerol, and the mesoporous zeolite may play an important role in this type of reaction in the future.

9.4 Overall Summary and Outlook

Compared to conventional zeolites, mesoporous zeolites with good mass transfer have been widely used in the conversion of various biomass molecules, and have provided excellent catalytic activities, selectivities, and catalyst life-times. The synthesis of mesoporous zeolites mainly involves post-treatments such as desilication and dealumination and templating routes, such as the use of organic mesoscale templates. These post-treatments normally result in a decrease in zeolite crystallinity and the formation of amorphous aluminosilicate fragments that exist in the mesopores, and which have a negative effect on mass transfer. In contrast, the use of an organic mesoscale template may lead to an adjustment of mesoporosity in zeolites, which would in turn greatly help to improve catalytic properties in biomass conversions. The relatively high costs of mesoporous zeolites obtained via a templating route remain a major challenge for practical applications, however. Hence, a 'one-pot' synthesis of mesoporous zeolites in the absence of mesoscale templates as catalysts for biomass conversion, would be highly desirable in the future.

More importantly, new routes for the transformation of biomass feedstocks should be developed, but this will require a fundamental understanding of biomass pathways and the design of novel catalysts. With regards to the important role of mesoporous zeolites in biomass conversion, it is believed that new mesoporous zeolite catalysts with controllable mesoporosity and acidic sites, as well as adjustable active metal sites/nanoparticles, would be preferable for developing the next generation of heterogeneous catalysts for the transformation of biomass to valuable chemicals in the future.

References

1. Huber, G., Iborra, S., Corma, A. (2006) Synthesis of transportation fuels from biomass: Chemistry, catalysts, and engineering. *Chem. Rev.*, **106** (9), 4044–4098.
2. Goossens, H., Deleeuw, J., Schenck, P., Brassell, S. (1984) Tocopherols as likely precursors of pristane in ancient sediments and crude oils. *Nature*, **312** (5993), 440–442.
3. Jobson, A., Westlake, D., Cook, F. (1972) Microbial utilization of crude-oil. *Appl. Microbiol.*, **23** (6), 1082.
4. Jones, D., Head, I., Gray, N., Adams, J., Rowan, A., Aitken, C., Bennett, B., Huang, H., Brown, A., Bowler, B., Oldenburg, T., Erdmann, M., Larter, S. (2008) Crude-oil biodegradation via methanogenesis in subsurface petroleum reservoirs. *Nature*, **451** (7175), 176–U6.
5. Qian, K., Rodgers, R., Hendrickson, C., Emmett, M., Marshall, A. (2001) Reading chemical fine print: Resolution and identification of 3000 nitrogen-containing aromatic compounds from a single electrospray ionization Fourier transform ion cyclotron resonance mass spectrum of heavy petroleum crude oil. *Energy Fuels*, **15** (2), 492–498.
6. Yakovlev, V., Khromova, S., Sherstyuk, O., Dundich, V., Ermakov, D., Novopashina, V., Lebedev, M., Bulavchenko, O., Parmon, V. (2009) Development of new catalytic systems for upgraded bio-fuels production from bio-crude-oil and biodiesel. *Catal. Today*, **144** (3-4), 362–366.
7. Calamari, D., Bacci, E., Focardi, S., Gaggi, C., Morosini, M., Vighi, M. (1991) Role of plant biomass in the global environmental partitioning of chlorinated hydrocarbons. *Environ. Sci. Technol.*, **25** (8), 1489–1495.
8. Flanagan, L., Johnson, B. (2005) Interacting effects of temperature, soil moisture and plant biomass production on ecosystem respiration in a northern temperate grassland. *Agr. Forest. Meteorol.*, **130** (3-4), 237–253.
9. Keiluweit, M., Nico, P., Johnson, M., Kleber, M. (2010) Dynamic molecular structure of plant biomass-derived black carbon (Biochar). *Environ. Sci. Technol.*, **44** (4), 1247–1253.
10. Moore, T., Bubier, J., Frolking, S., Lafleur, P., Roulet, N. (2002) Plant biomass and production and CO_2 exchange in an ombrotrophic bog. *J. Ecol.*, **90** (1), 25–36.
11. Smidansky, E., Martin, J., Hannah, L., Fischer, A., Giroux, M. (2003) Seed yield and plant biomass increases in rice are conferred by deregulation of endosperm ADP-glucose pyrophosphorylase. *Planta*, **216** (4), 656–664.
12. Steen, E., Kang, Y., Bokinsky, G., Hu, Z., Schirmer, A., McClure, A., del Cardayre, S., Keasling, J. (2010) Microbial production of fatty-acid-derived fuels and chemicals from plant biomass. *Nature*, **463** (7280), 559–U182.
13. Wang, L., Xiao, F.-S. (2015) Nanoporous catalysts for biomass conversion. *Green Chem.*, **17** (1), 24–39.
14. Vassilev, S., Baxter, D., Andersen, L., Vassileva, C. (2013) An overview of the composition and application of biomass ash. Part 2. Potential utilisation, technological and ecological advantages and challenges. *Fuel*, **105**, 19–39.
15. Chheda, J.N., Huber, G., Dumesic, J. (2007) Liquid-phase catalytic processing of biomass-derived oxygenated hydrocarbons to fuels and chemicals. *Angew. Chem. Int. Ed.*, **46** (38), 7164–7183.
16. Mosier, N., Wyman, C., Dale, B., Elander, R., Lee, Y., Holtzapple, M., Ladisch, M. (2005) Features of promising technologies for pretreatment of lignocellulosic biomass. *Biores. Technol.*, **96** (6), 673–686.
17. Tilman, D., Hill, J., Lehman, C. (2006) Carbon-negative biofuels from low-input high-diversity grassland biomass. *Science*, **314** (5805), 1598–1600.
18. Rinaldi, R., Meine, N., vom Stein, J., Palkovits, R., Schueth, F. (2010) Which controls the depolymerization of cellulose in ionic liquids: the solid acid catalyst or cellulose? *ChemSusChem*, **3** (2), 266–276.

19. de Oliveira, H., Fares, C., Rinaldi, R. (2015) Beyond a solvent: the roles of 1-butyl-3-methyl-imidazolium chloride in the acid-catalysis for cellulose depolymerisation. *Chem. Sci.*, **6** (9), 5215–5224.

20. Graham, R., Mok, L., Bergougnou, M., Delasa, H., Freel, B. (1984) Fast pyrolysis (ultrapyrolysis) of cellulose. *J. Anal. Appl. Pyrol.*, **6** (4), 363–374.

21. Lu, Q., Ye, X., Zhang, Z., Dong, C., Zhang, Y. (2014) Catalytic fast pyrolysis of cellulose and biomass to produce levoglucosenone using magnetic SO42-/TiO$_2$-Fe$_3$O$_4$. *Biores. Technol.*, **171**, 10–15.

22. Zhou, G., Li, J., Yu, Y., Li, X., Wang, Y., Wang, W., Komarneni, S. (2014) Optimizing the distribution of aromatic products from catalytic fast pyrolysis of cellulose by ZSM-5 modification with boron and co-feeding of low-density polyethylene. *Appl. Catal. A. Gen.*, **487**, 45–53.

23. Lee, S., Lee, Y., Wang, F. (1999) Chiral compounds from bacterial polyesters: Sugars to plastics to fine chemicals. *Biotechnol. Bioeng.*, **65** (3), 363–368.

24. Burk, M. (2010) Sustainable production of industrial chemicals from sugars. *Int. Sugar. J.*, **112** (1333), 30–35.

25. Ruppert, A., Weinberg, K., Palkovits, R. (2012) Hydrogenolysis goes bio: from carbohydrates and sugar alcohols to platform chemicals. *Angew. Chem. Int. Ed.*, **51** (11), 2564–2601.

26. Cheng, J., Feng, J., Ge, T., Yang, W., Zhou, J., Cen, K. (2015) Pyrolytic characteristics of biodiesel prepared from lipids accumulated in diatom cells with growth regulation. *J. Biosci. Bioeng.*, **120** (2), 161–166.

27. Xu, S., Sheng, H., Ye, T., Hu, D., Liao, S. (2016) Hydrophobic aluminosilicate zeolites as highly efficient catalysts for the dehydration of alcohols. *Catal. Commun.*, **78**, 75–79.

28. Yoshioka, M., Yokoi, T., Tatsumi, T. (2015) Development of the CON-type aluminosilicate zeolite and its catalytic application for the MTO reaction. *ACS Catal.*, **5** (7), 4268–4275.

29. Liu, M., Yokoi, T., Kondo, J., Tatsumi, T. (2014) Synthesis of SFH-type aluminosilicate zeolite with 14-membered ring and its applications as solid acidic catalyst. *Micropor. Mesopor. Mater.*, **193**, 166–172.

30. Triantafyllidis, K., Iliopoulou, E., Antonakou, E., Lappas, A., Wang, H., Pinnavaia, T. (2007) Hydrothermally stable mesoporous aluminosilicates (MSU-S) assembled from zeolite seeds as catalysts for biomass pyrolysis. *Micropor. Mesopor. Mater.*, **99** (1-2), 132–139.

31. McKendry, P. (2002) Energy production from biomass (part 1): overview of biomass. *Biores. Technol.*, **83** (1), 37–46.

32. Bridgwater, A. (1994) Catalysis in thermal biomass conversion. *Appl. Catal. A. Gen.*, **116** (1-2), 5–47.

33. Wang, L., Zhang, J., Yang, S., Sun, Q., Zhu, L., Wu, Q., Zhang, H., Meng, X., Xiao, F. (2013) Sulfonated hollow sphere carbon as an efficient catalyst for acetalisation of glycerol. *J Mater. Chem. A*, **1** (33), 9422–9426.

34. Liu, F., Willhammar, T., Wang, L., Zhu, L., Sun, Q., Meng, X., Carrillo-Cabrera, W., Zou, X., Xiao, F. (2012) ZSM-5 zeolite single crystals with b-axis-aligned mesoporous channels as an efficient catalyst for conversion of bulky organic molecules. *J. Am. Chem. Soc.*, **134** (10), 4557–4560.

35. Jacobsen, C., Madsen, C., Houzvicka, J., Schmidt, I., Carlsson, A. (2000) Mesoporous zeolite single crystals. *J. Am. Chem. Soc.*, **122** (29), 7116–7117.

36. Xiao, F., Wang, L., Yin, C., Lin, K., Di, Y., Li, J., Xu, R., Su, D., Schlogl, R., Yokoi, T., Tatsumi, T. (2006) Catalytic properties of hierarchical mesoporous zeolites templated with a mixture of small organic ammonium salts and mesoscale cationic polymers. *Angew. Chem. Int. Ed.*, **45** (19), 3090–3093.

37. Egeblad, K., Christensen, C., Kustova, M., Christensen, C. (2008) Templating mesoporous zeolites. *Chem. Mater.*, **20** (3), 946–960.

38. Sun, Y., Prins, R. (2008) Hydrodesulfurization of 4,6-dimethyldibenzothiophene over noble metals supported on mesoporous zeolites. *Angew. Chem. Int. Ed.*, **47** (44), 8478–8481.

39. Fu, W., Zhang, L., Tang, T., Ke, Q., Wang, S., Hu, J., Fang, G., Li, J., Xiao, F.-S. (2011) Extraordinarily high activity in the hydrodesulfurization of 4,6-dimethyldibenzothiophene over Pd supported on mesoporous zeolite Y. *J. Am. Chem. Soc.*, **133** (39), 15346–15349.

40. Kang, J., Cheng, K., Zhang, L., Zhang, Q., Ding, J., Hua, W., Lou, Y., Zhai, Q., Wang, Y. (2011) Mesoporous zeolite-supported ruthenium nanoparticles as highly selective Fischer–Tropsch catalysts for the production of C-5–C-11 isoparaffins. *Angew. Chem. Int. Ed.*, **50** (22), 5200–5203.

41. Kim, J.-C., Lee, S., Cho, K., Na, K., Lee, C., Ryoo, R. (2014) Mesoporous MFI zeolite nanosponge supporting cobalt nanoparticles as a Fischer–Tropsch catalyst with high yield of branched hydrocarbons in the gasoline range. *ACS Catal.*, **4** (11), 3919–3927.

42. Khitev, Y., Ivanova, I., Kolyagin, Y., Ponomareva, O. (2012) Skeletal isomerization of 1-butene over micro/mesoporous materials based on FER zeolite. *Appl. Catal. A. Gen.*, **441**, 124–135.

43. Christensen, C., Johannsen, K., Schmidt, I. (2003) Catalytic benzene alkylation over mesoporous zeolite single crystals: Improving activity and selectivity with a new family of porous materials. *J. Am. Chem. Soc.*, **125** (44), 13370–13371.

44. Perego, C., Bosetti, A. (2011) Biomass to fuels: The role of zeolite and mesoporous materials. *Micropor. Mesopor. Mater.*, **144** (1-3), 28–39.

45. Li, J., Li, X., Zhou, G., Wang, W., Wang, C., Komarneni, S., Wang, Y. (2014) Catalytic fast pyrolysis of biomass with mesoporous ZSM-5 zeolites prepared by desilication with NaOH solutions. *Appl. Catal. A. Gen.*, **470**, 115–122.

46. Neumann, G., Hicks, J. (2012) Novel hierarchical cerium-incorporated MFI zeolite catalysts for the catalytic fast pyrolysis of lignocellulosic biomass. *ACS Catal.*, **2** (4), 642–646.

47. Cheah, S., Gaston, K., Parent, Y., Jarvis, M., Vinzant, T., Smith, K., Thornburg, N., Nimlos, M., Magrini-Bair, K. (2013) Nickel cerium olivine catalyst for catalytic gasification of biomass. *Appl. Catal. B. Environ.*, **134**, 34–45.

48. Inaba, M., Murata, K., Saito, M., Takahara, I. (2006) Hydrogen production by gasification of cellulose over Ni catalysts supported on zeolites. *Energy Fuels*, **20** (2), 432–438.

49. Pinto, T., Dufaud, V., Lefebvre, F. (2014) Isomerization of n-hexane on heteropolyacids supported on SBA-15. 1. Monofunctional impregnated catalysts. *Appl. Catal. A. Gen.*, **483**, 103–109.

50. Musselwhite, N., Na, K., Sabyrov, K., Aayogu, S., Somorjai, G. (2015) Mesoporous aluminosilicate catalysts for the selective isomerization of n-hexane: the roles of surface acidity and platinum metal. *J. Am. Chem. Soc.*, **137** (32), 10231–10237.

51. Stagg-Williams, S., Noronha, F., Fendley, G., Resasco, D. (2000) CO2 reforming of CH4 over Pt/ZrO2 catalysts promoted with La and Ce oxides. *J. Catal.*, **194** (2), 240–249.

52. Laosiripojana, N., Sangtongkitcharoen, W., Assabumrungrat, S. (2006) Catalytic steam reforming of ethane and propane over CeO_2-doped Ni/Al_2O_3 at SOFC temperature: Improvement of resistance toward carbon formation by the redox property of doping CeO_2. *Fuel*, **85** (3), 323–332.

53. Zhao, C., Lercher, J. (2012) Upgrading pyrolysis oil over Ni/HZSM-5 by cascade reactions. *Angew. Chem. Int. Ed.*, **51** (24), 5935–5940.

54. Wang, L., Zhang, J., Yi, X., Zheng, A., Deng, F., Chen, C., Ji, Y., Liu, F., Meng, X., Xiao, F.-S. (2015) Mesoporous ZSM-5 zeolite-supported Ru nanoparticles as highly efficient catalysts for upgrading phenolic biomolecules. *ACS Catal.*, **5** (5), 2727–2734.

55. Srivastava, R., Iwasa, N., Fujita, S., Arai, M. (2008) Synthesis of nanocrystalline MFI-zeolites with intracrystal mesopores and their application in fine chemical synthesis involving large molecules. *Chem. Eur. J.*, **14** (31), 9507–9511.

56. Chal, R., Gerardin, C., Bulut, M., van Donk, S. (2011) Overview and industrial assessment of synthesis strategies towards zeolites with mesopores. *ChemCatChem*, **3** (1), 67–81.

57. Liu, J., Wang, J., Li, N., Zhao, H., Zhou, H., Sun, P., Chen, T. (2012) Polyelectrolyte-surfactant complex as a template for the synthesis of zeolites with intracrystalline mesopores. *Langmuir*, **28** (23), 8600–8607.

58. Xue, Z., Ma, J., Zhang, T., Miao, H., Li, R. (2012) Synthesis of nanosized ZSM-5 zeolite with intracrystalline mesopores. *Mater. Lett.*, **68**, 1–3.

59. Vu, H., Schneider, M., Bentrup, U., Dang, T., Phan, B., Nguyen, D., Armbruster, U., Martin, A. (2015) Hierarchical ZSM-5 materials for an enhanced formation of gasoline-range hydrocarbons and light olefins in catalytic cracking of triglyceride-rich biomass. *Ind. Eng. Chem. Res.*, **54** (6), 1773–1782.

60. Wang, D., Ma, B., Wang, B., Zhao, C., Wu, P. (2015) One-pot synthesized hierarchical zeolite supported metal nanoparticles for highly efficient biomass conversion. *Chem. Commun.*, **51** (82), 15102–15105.

61. Stocker, M. (2008) Biofuels and biomass-to-liquid fuels in the biorefinery: catalytic conversion of lignocellulosic biomass using porous materials. *Angew. Chem. Int. Ed.*, **47** (48), 9200–9211.

62. Metzger, J. (2006) Production of liquid hydrocarbons from biomass. *Angew. Chem. Int. Ed.*, **45** (5), 696–698.

63. Huber, G., Corma, A. (2007) Synergies between bio- and oil refineries for the production of fuels from biomass. *Angew. Chem. Int. Ed.*, **46** (38), 7184–7201.

64. Liu, Y., Sotelo-Boyas, R., Murata, K., Minowa, T., Sakanishi, K. (2009) Hydrotreatment of Jatropha oil to produce green diesel over trifunctional Ni-Mo/SiO$_2$-Al$_2$O$_3$ catalyst. *Chem. Lett.*, **38** (6), 552–553.

65. Mascarelli, A. (2009) Gold rush for algae. *Nature*, **461** (7263), 460–461.

66. Verma, D., Kumar, R., Rana, B., Sinha, A. (2011) Aviation fuel production from lipids by a single-step route using hierarchical mesoporous zeolites. *Energy Environ. Sci.*, **4** (5), 1667–1671.

67. Ma, B., Zhao, C. (2015) High-grade diesel production by hydrodeoxygenation of palm oil over a hierarchically structured Ni/HBEA catalyst. *Green Chem.*, **17** (3), 1692–1701.

68. Schweyer, F., Braunstein, P., Estournes, C., Guille, J., Kessler, H., Paillaud, J., Rose, J. (2000) Metallic nanoparticles from heterometallic Co-Ru carbonyl clusters in mesoporous silica xerogels and MCM-41-type materials. *Chem. Commun.*, **14**, 1271–1272.

69. Sun, Z., Sun, B., Qiao, M., Wei, J., Yue, Q., Wang, C., Deng, Y., Kaliaguine, S., Zhao, D. (2012) A general chelate-assisted co-assembly to metallic nanoparticles-incorporated ordered mesoporous carbon catalysts for Fischer–Tropsch synthesis. *J. Am. Chem. Soc.*, **134** (42), 17653–17660.

70. Cortial, G., Siutkowski, M., Goettmann, F., Moores, A., Boissiere, C., Grosso, D., Le Floch, P., Sanchez, C. (2006) Metallic nanoparticles hosted in mesoporous oxide thin films for catalytic applications. *Small*, **2** (8-9), 1042–1045.

71. Yamamoto, K., Sunagawa, Y., Takahashi, H., Muramatsu, A. (2005) Metallic Ni nanoparticles confined in hexagonally ordered mesoporous silica material. *Chem. Commun.*, **3**, 348–350.

72. Campesi, R., Paul-Boncour, V., Cuevas, F., Leroy, E., Gadiou, R., Vix-Guterl, C., Latroche, M. (2009) Structural and magnetic properties of Pd$_x$Ni$_{1-x}$ ($x=0$ and 0.54) metallic nanoparticles in an ordered mesoporous carbon template. *J. Phys. Chem. C*, **113** (39), 16921–16926.

73. Overbury, S., Ortiz-Soto, L., Zhu, H., Lee, B., Amiridis, M., Dai, S. (2004) Comparison of Au catalysts supported on mesoporous titania and silica: investigation of Au particle size effects and metal-support interactions. *Catal. Lett.*, **95** (3-4), 99–106.

74. Hampsey, J., Arsenault, S., Hu, Q., Lu, Y. (2005) One-step synthesis of mesoporous metal-SiO$_2$ particles by an aerosol-assisted self-assembly process. *Chem. Mater.*, **17** (9), 2475–2480.

75. Asghari, F., Yoshida, H. (2006) Acid-catalyzed production of 5-hydroxymethyl furfural from D-fructose in subcritical water. *Ind. Eng. Chem. Res.*, **45** (7), 2163–2173.

76. Chheda, J., Roman-Leshkov, Y., Dumesic, J. (2007) Production of 5-hydroxymethylfurfural and furfural by dehydration of biomass-derived mono- and poly-saccharides. *Green Chem.*, **9** (4), 342–350.

77. Nandiwale, K., Pande, A., Bokade, V. (2015) One step synthesis of ethyl levulinate biofuel by ethanolysis of renewable furfuryl alcohol over hierarchical zeolite catalyst. *RSC Adv.*, **5** (97), 79224–79231.

78. Joshi, H., Moser, B., Toler, J., Smith, W., Walker, T. (2011) Ethyl levulinate: A potential bio-based diluent for biodiesel which improves cold flow properties. *Biomass Bioenerg.*, **35** (7), 3262–3266.

79. Tang, X., Hu, L., Sun, Y., Zhao, G., Hao, W., Lin, L. (2013) Conversion of biomass-derived ethyl levulinate into gamma-valerolactone via hydrogen transfer from supercritical ethanol over a ZrO_2 catalyst. *RSC Adv.*, **3** (26), 10277–10284.

80. Tang, X., Chen, H., Hu, L., Hao, W., Sun, Y., Zeng, X., Lin, L., Liu, S. (2014) Conversion of biomass to gamma-valerolactone by catalytic transfer hydrogenation of ethyl levulinate over metal hydroxides. *Appl. Catal. B. Environ.*, **147**, 827–834.

81. Zhou, H., Song, J., Fan, H., Zhang, B., Yang, Y., Hu, J., Zhu, Q., Han, B. (2014) Cobalt catalysts: very efficient for hydrogenation of biomass-derived ethyl levulinate to gamma-valerolactone under mild conditions. *Green Chem.*, **16** (8), 3870–3875.

82. Song, J., Wu, L., Zhou, B., Zhou, H., Fan, H., Yang, Y., Meng, Q., Han, B. (2015) A new porous Zr-containing catalyst with a phenate group: an efficient catalyst for the catalytic transfer hydrogenation of ethyl levulinate to gamma-valerolactone. *Green Chem.*, **17** (3), 1626–1632.

83. Nandiwale, K., Galande, N., Bokade, V. (2015) Process optimization by response surface methodology for transesterification of renewable ethyl acetate to butyl acetate biofuel additive over borated USY zeolite. *RSC Adv.*, **5** (22), 17109–17116.

84. Panagiotopoulou, P., Martin, N., Vlachos, D. (2015) Liquid-phase catalytic transfer hydrogenation of furfural over homogeneous Lewis acid-Ru/C catalysts. *ChemSusChem*, **8** (12), 2046–2054.

85. Cao, Q., Guan, J., Peng, G., Hou, T., Zhou, J., Mu, X. (2015) Solid acid-catalyzed conversion of furfuryl alcohol to alkyl tetrahydrofurfuryl ether. *Catal. Commun.*, **58**, 76–79.

86. Ott, L., Bicker, M., Vogel, H. (2006) Catalytic dehydration of glycerol in sub- and supercritical water: a new chemical process for acrolein production. *Green Chem.*, **8** (2), 214–220.

87. Chai, S., Wang, H., Liang, Y., Xu, B. (2007) Sustainable production of acrolein: investigation of solid acid-base catalysts for gas-phase dehydration of glycerol. *Green Chem.*, **9** (10), 1130–1136.

88. Chai, S., Wang, H., Liang, Y., Xu, B. (2007) Sustainable production of acrolein: Gas-phase dehydration of glycerol over Nb_2O_5 catalyst. *J. Catal.*, **250**, 342–349.

89. Atia, H., Armbruster, U., Martin, A. (2008) Dehydration of glycerol in gas phase using heteropolyacid catalysts as active compounds. *J. Catal.*, **258** (1), 71–82.

90. Zhang, H., Hu, Z., Huang, L., Zhang, H., Song, K., Wang, L., Shi, Z., Ma, J., Zhuang, Y., Shen, W., Zhang, Y., Xu, H., Tang, Y. (2015) Dehydration of glycerol to acrolein over hierarchical ZSM-5 zeolites: effects of mesoporosity and acidity. *ACS Catal.*, **5** (4), 2548–2558.

91. Jia, C., Liu, Y., Schmidt, W., Lu, A., Schueth, F. (2010) Small-sized HZSM-5 zeolite as highly active catalyst for gas phase dehydration of glycerol to acrolein. *J. Catal.*, **269** (1), 71–79.

92. Nimlos, M., Blanksby, S., Qian, X., Himmel, M., Johnson, D. (2006) Mechanisms of glycerol dehydration. *J. Phys. Chem. A*, **110** (18), 6145–6156.

93. Chai, S.-H., Wang, H., Liang, Y., Xu, B. (2008) Sustainable production of acrolein: gas-phase dehydration of glycerol over 12-tungstophosphoric acid supported on ZrO_2 and SiO_2. *Green Chem.*, **10** (10), 1087–1093.

94. Deleplanque, J., Dubois, J., Devaux, J., Ueda, W. (2010) Production of acrolein and acrylic acid through dehydration and oxydehydration of glycerol with mixed oxide catalysts. *Catal. Today*, **157** (1-4), 351–358.

95. Possato, L., Diniz, R., Garetto, T., Pulcinelli, S., Santilli, C., Martins, L. (2013) A comparative study of glycerol dehydration catalyzed by micro/mesoporous MFI zeolites. *J. Catal.*, **300**, 102–112.

96. Aramburo, L., Karwacki, L., Cubillas, P., Asahina, S., de Winter, D., Drury, M., Buurmans, I., Stavitski, E., Mores, D., Daturi, M., Bazin, P., Dumas, P., Thibault-Starzyk, F., Post, J., Anderson, M., Terasaki, O., Weckhuysen, B. (2011) The porosity, acidity, and deactivity of dealuminated zeolite ZSM-5 at the single particle level: The influence of the zeolite architecture. *Chem. Eur. J.*, **17** (49), 13773–13781.

97. Wang, J., Tu, X., Hua, W., Yue, Y., Gao, Z. (2011) Role of the acidity and porosity of MWW-type zeolites in liquid-phase reaction. *Micropor. Mesopor. Mater.*, **142** (1), 82–90.

98. Gonzalez, M., Cesteros, Y., Salagre, P. (2013) Establishing the role of Brønsted acidity and porosity for the catalytic etherification of glycerol with *tert*-butanol by modifying zeolites. *Appl. Catal. A. Gen.*, **450**, 178–188.

99. Melero, J., Vicente, G., Paniagua, M., Morales, G., Munoz, P. (2012) Etherification of biodiesel-derived glycerol with ethanol for fuel formulation over sulfonic modified catalysts. *Biores. Technol.*, **103** (1), 142–151.

100. Melero, J., Vicente, G., Morales, G., Paniagua, M., Moreno, J., Roldan, R., Ezquerro, A., Perez, C. (2008) Acid-catalyzed etherification of bio-glycerol and isobutylene over sulfonic mesostructured silicas. *Appl. Catal. A. Gen.*, **346** (1-2), 44–51.

101. Gu, Y., Azzouzi, A., Pouilloux, Y., Jerome, F., Barrault, J. (2008) Heterogeneously catalyzed etherification of glycerol: new pathways for transformation of glycerol to more valuable chemicals. *Green Chem.*, **10** (2), 164–167.

102. Clacens, J., Pouilloux, Y., Barrault, J. (2002) Selective etherification of glycerol to polyglycerols over impregnated basic MCM-41 type mesoporous catalysts. *Appl. Catal. A. Gen.*, **227** (1-2), 181–190.

103. Saxena, S., Al-Muhtaseb, A., Viswanadham, N. (2015) Enhanced production of high octane oxygenates from glycerol etherification using the desilicated BEA zeolite. *Fuel*, **159**, 837–844.

104. Gonzalez-Arellano, C., Grau-Atienza, A., Serrano, E., Romero, A., Garcia-Martinez, J., Luque, R. (2015) The role of mesoporosity and Si/Al ratio in the catalytic etherification of glycerol with benzyl alcohol using ZSM-5 zeolites. *J. Mol. Catal. A. Chem.*, **406**, 40–45.

105. Li, T., Cheng, J., Huang, R., Zhou, J., Cen, K. (2015) Conversion of waste cooking oil to jet biofuel with nickel-based mesoporous zeolite Y catalyst. *Biores. Technol.*, **197**, 289–294.

106. Lee, Y., Kim, E., Kim, T., Kim, C., Jeong, K., Lee, C., Jeong, S. (2015) Ring opening of naphthenic molecules over metal containing mesoporous Y zeolite catalyst. *J. Nanosci. Nanotechnol.*, **15** (7), 5334–5337.

107. Ivanova, I., Knyazeva, E., Maerle, A., Kasyanov, I. (2015) Design of micro/mesoporous zeolite-based catalysts for petrochemical and organic syntheses. *Kinet. Catal.*, **56** (4), 549–561.

108. Hook, J., Mander, L. (1986) Recent developments in the birch reduction of aromatic-compounds – applications to the synthesis of natural-products. *Nat. Prod. Rep.*, **3** (1), 35–85.

109. Wu, C., Dong, L., Huang, J., Williams, P. (2013) Optimising the sustainability of crude bio-oil via reforming to hydrogen and valuable by-product carbon nanotubes. *RSC Adv.*, **3** (42), 19239–19242.

110. Xiao, W., Wang, F., Xiao, G. (2015) Performance of hierarchical HZSM-5 zeolites prepared by NaOH treatments in the aromatization of glycerol. *RSC Adv.*, **5** (78), 63697–63704.

111. Jang, H., Bae, K., Shin, M., Kim, S., Kim, C., Suh, Y. (2014) Aromatization of glycerol/alcohol mixtures over zeolite H-ZSM-5. *Fuel*, **134**, 439–447.

10

Lignin Depolymerization Over Porous Copper-Based Mixed-Oxide Catalysts in Supercritical Ethanol

Xiaoming Huang, Tamás I. Korányi, and Emiel J. M. Hensen

Laboratory of Inorganic Materials Chemistry, Schuit Institute of Catalysis, Eindhoven University of Technology, Netherlands

10.1 Introduction

10.1.1 Hydrotalcites

Hydrotalcites (HTCs), also known as hydrotalcite-like anionic clays (HTs) and layered double hydroxides (LDHs), are synthetic or natural crystalline materials consisting of positively charged two-dimensional sheets with water and exchangeable charge-compensating anions in the interlayer region [1]. Their anionic clay structure is closely related to the mineral hydrotalcite, with the general formula $Mg_6Al_2(OH)_{16}CO_3 \cdot 4H_2O$ (Figure 10.1) [2]. They may contain divalent (M^{2+}) and trivalent (M^{3+}) ions, with a ratio M^{2+}/M^{3+} of between 1.5 and 4 [3]. HTs are increasingly regarded as a good alternative to traditional homogeneous base catalysts such as NaOH and KOH [2].

HTCs can be prepared with several reducible bivalent (Ni, Cu, Co) and trivalent (Fe, Cr) cations in the structure, together with the classical cations (Mg, Zn, Al) serving as precursors for the preparation of different mixed oxides [2]. They are generally synthesized either by conventional coprecipitation [4] or by sol–gel methods [5]. In the coprecipitation or titration method, a homogeneously mixed solution of Mg^{2+} and Al^{3+} ions is precipitated by an

Nanoporous Catalysts for Biomass Conversion, First Edition. Edited by Feng-Shou Xiao and Liang Wang.
© 2018 John Wiley & Sons Ltd. Published 2018 by John Wiley & Sons Ltd.

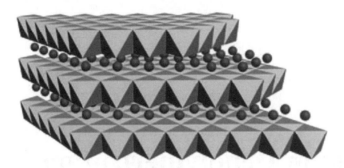

Figure 10.1   *Schematic view of the structure of hydrotalcites. Bivalent and trivalent cations (e.g., Mg^{2+} and Al^{3+}) are sixfold-coordinated to form octahedrals that share edges to constitute infinite layers. Small spheres drawn in the interlayer region represent the compensating anions (e.g., CO_3^{2-}) [2].*

alkali source such as NaOH at a defined pH value close to 10 in order to avoid $Al(OH)_3$ and $Mg(OH)_2$ formation [4]. An aqueous Na_2CO_3 or K_2CO_3 solution is also added as the carbonate source. After aging of the slurry, the resulting material is filtered and washed with deionized water and dried [2]. Sol–gel hydrotalcite-like compounds are prepared using magnesium ethoxide or magnesium acetylacetonate and aluminum acetylacetonate precursors [5]. Microwave irradiation may be used during aging in both preparation methods to reduce the particle sizes, to increase the specific surface areas, and to increase basicity [2]. Hydrotalcites based on other metals (Zn, Ni, Cu, etc.) can be synthesized in the same way by introducing the appropriate nitrate salt [2].

HTCs can be used in the as-synthesized form as anion-exchangers, as solid bases, and they can be designed as bifunctional redox-base catalysts. Their acid–base properties can be controlled. Specific metals can be incorporated in their octahedral layers (Figure 10.1), and small (NO_3^- or Cl^-) exchanged anions can be replaced completely with larger anions and pillared anionic clays can also be obtained [1]. HTCs usually exhibit poor basic properties, porosities, and surface areas (<100 m^2 g^{-1}), because of the adsorbed water, and thus usually need to be activated by thermal decomposition [2]. Typically, calcination is performed in air between 673 and 773K for a prolonged time (several hours to 48 h) [2]. The thermal decomposition of Mg-Al HTCs leads to a well-dispersed mixture of magnesium and aluminum oxides ($MgAlO_x$). These mixed oxides exhibit much higher porosity and surface area ($100{\sim}300$ m^2 g^{-1}) [6]. The mixed oxides show a memory effect, a property by which they can recover the original lamellar structure if they come into contact with water vapor or are immersed in water under a decarbonated atmosphere [7]. Mixed oxides obtained by the thermal decomposition of LDH possess both acid and basic sites. The nature, strength, and relative amounts of these sites depend on the cation and content/composition and the calcination temperature. The basic strength of $MgAlO_x$, calcined at 723K lies between those of MgO and Al_2O_3 [3]. On pure MgO, strong basic sites consisted predominantly of O^{2-} anions. Calcined hydrotalcites contained surface sites of low (OH^- groups), medium (Mg^{2+}-O^{2-} pairs), and strong (O^{2-} anions) basicity. The relative abundance of low- and medium-strength basic sites increased with the Al content. The addition of small amounts of Al to MgO led to a drastic diminution in the density of surface basic sites because of a significant Al surface enrichment [6]. The basic strength, and the ratios of the numbers of basic and acid sites, increases generally with increasing calcination temperature [3]. However, a high calcination temperature leads to reduced basicity due to phase segregation (e.g., $MgAl_2O_4$ spinel formation), and this also lead to a loss of the memory effect [2].

HTCs, in their non-calcined form, are rarely used for biomass conversion, whereas calcined HTC-derived catalysts are quite often used for biodiesel production [8], 5-hydroxymethylfurfural (5-HMF) hydrogenation [9,10], steam gasification [11], steam reforming [12–14], and aldol condensation [15–17] of biomass-derived molecules. A commercial calcined Mg/Al HTC proved to be an active, slowly deactivating, and capable of regeneration catalyst, producing biodiesel [8]. 5-HMF was converted over a Ni-Co-Al mixed oxide [9] and Ru-containing HT catalysts [10]. HT-derived Co-Fe/Mg/Al catalysts were applied in the steam gasification of biomass-derived tar [11]. Ni/Mg/Al [12,14] and Ni/Ca/Al [13] catalysts were used for the steam reforming of tar and different tar model compounds. Mg-Al and Mg-Zr mixed oxides were used for aldol condensation of biomass-derived furfurals and ketones [15–17].

HTCs and their derived mixed-oxide materials were also found to be useful for lignin upgrading. Ford and coworkers applied Cu metal-doped porous metal oxides (Cu-PMO or $CuMgAlO_x$) derived from hydrotalcites by calcination for lignin depolymerization in supercritical methanol [18–21]. The Cu-doped PMO serves multiple purposes, catalyzing substrate hydrogenolysis and hydrogenation, as well as the methanol reforming and shift reactions [20]. The reducing agents are derived by the reforming of MeOH (to H_2 and CO), with water generated by deoxygenation processes being removed by the water–gas shift reaction [20]. Lignin and even crude lignocellulose (various wood chips) were fully depolymerized and markedly deoxygenated under reaction conditions of 300 °C and ~200 bar) [22]. This one-pot process did not result in the production of char [20].

Beckham *et al.* reported a supported LDH containing Ni as an active component for the depolymerization of lignin model compounds, as well as organosolv and ball-milled lignins, into alkylaromatics at 270 °C [23]. The study results indicated that nickel oxide on a solid base support can function as an effective lignin depolymerization catalyst without the need for external hydrogen and reduced metal, and suggested that LDHs offer a novel, active support in multifunctional catalyst applications. Efficient depolymerization by the copper- and vanadium-catalyzed oxidative cleavage of lignin is possible using transition metal-containing HTCs with molecular oxygen as oxidant [24].

10.1.2   Lignin Depolymerization

Over the past decades, numerous scientific reports have appeared related to lignin valorization and lignin model compound conversion. Lignin may find application as a biomaterial, such as a dispersant, wood panels, emulsifier, polyurethane foam, automotive brakes and epoxy resins [25,26]. Apart from these uses, another promising approach is to depolymerize lignin into aromatic compounds such as benzene, toluene, xylenes, or phenols. Several reviews have been published, summarizing the recent advances of lignin depolymerization techniques [24,25,27–32], including gasification, pyrolysis, acid- and base-catalyzed, oxidative and reductive depolymerization. Hydrogenolysis in the presence of hydrogen or hydrogen-donating solvents is promising, because higher monomer yields can be obtained in this way and less char is formed [30]. Solvents such as sub- and supercritical water [33–35], methanol [18,19,36], ethanol [36–38], 2-propanol [36,39], ethanol/water [40–42] and methanol/water [43] have been investigated for the solvolysis and hydrogenolysis of lignin. Despite significant advances made in recent years [20,44–46], the selective conversion of lignin into small molecules remains a challenge [47]. This is mainly due to the recalcitrant nature and heterogeneous structure of lignin, and also because the intermediates formed during its depolymerization are highly reactive, which leads to rapid repolymerization into heavier products than the starting lignin [25,48]. Mechanistic studies have proven that repolymerization usually involves the highly reactive oxygenated species such as phenolic OH groups [49,50], formaldehyde [50,51], aldehyde side chains [48], and ketones [48]. Other factors such as unsaturated double bonds [52,53] and radicals [52,53] also play an important role. Some efforts have been made in order to mitigate the recondensation problem. For example, in the base-catalyzed depolymerization process Roberts *et al.* showed that the use of boric acid can protect the phenolic OH groups and suppress repolymerization [34], and this was confirmed in another study [50]. Barta *et al.* [54] demonstrated a novel strategy to stabilize these reactive aldehyde fragments (C_2-aldehyde) by reacting them with ethylene glycol to form acetals. The formation of higher-molecular-weight side products was markedly suppressed in this way. Other research groups have reported that the addition of phenols for trapping formaldehyde also helps in reducing repolymerization [50,51,55]. Alternatively, hydrogenation of the unsaturated double bonds helps to reduce the negative effect of radical reactions [56]. The reduction of oxygen functionalities also results in less-reactive aromatic products such as benzene, toluene, and xylenes [25,57,58].

The one-step depolymerization of Protobind lignin in supercritical ethanol (critical point: $T_c = 240.8$ °C, $P_c = 61.4$ bar) was investigated over a non-noble Cu-Mg-Al mixed oxide (obtained by calcination of a hydrotalcites precursor) catalyst. Ethanol was the preferred solvent over other alcohols, water and apolar hydrocarbons for three reasons: (i) it is a reasonably good solvent for lignin; (ii) it is a source of hydrogen for removing oxygen functionalities and breaking ether linkages by hydrogenolysis; and (iii) it is unique in protecting the useful lignin fragments against undesired repolymerization [37,59,60]. In this chapter, the roles of the catalyst and solvent are discussed, with the aim of understanding why ethanol is so much more effective than methanol. The influence of reaction time and temperature will also be discussed. Finally, a discussion is provided of how the nature, composition and active site types and distribution in the multi-component catalyst affect catalytic performance. A series of mixed oxides with varying Cu contents and (Cu+Mg)/Al ratios were prepared and tested in lignin depolymerization and phenol alkylation reactions. Both the lignin monomers and ethanol conversion products were analyzed and linked to the relevant chemical reactions. The active sites for Guerbet, esterification and alkylation reactions were revealed.

10.2   Lignin Depolymerization by CuMgAl Mixed-Oxide Catalysts in Supercritical Ethanol

A 20 wt% Cu-containing MgAl mixed-oxide catalyst with a fixed M^{2+}/M^{3+} atomic ratio of 2 was prepared using a coprecipitation method. The catalyst samples are denoted by $Cu_xMgAl(y)$, where x corresponds to the Cu content (by weight) and y is the atomic ratio of (Cu+Mg)/Al. Stirred, high-pressure autoclaves of 50 and 100 ml capacity were used to study the (catalytic) conversion of lignin in (m)ethanol at 300–420 °C temperature and 0–20 h reaction time ranges. Protobind P1000 soda, organosolv Alcell and Kraft lignins were used as feedstocks. Conventional gas chromatography-mass spectrometry (GC-MS), as well as comprehensive GC×GC-MS were employed to analyze the product mixture. Figure 10.2 shows a representative GC×GC-MS chromatogram of the lignin oil sample

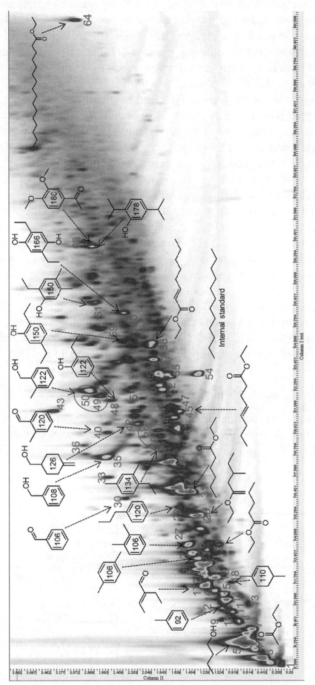

Figure 10.2 GC×GC-MS chromatogram of the product mixture obtained from the catalytic reaction of lignin at 300 °C for 8 h using the CuMgAl mixed-oxide catalyst. (See color plate section for the color representation of this figure.)

obtained from catalytic reaction of the starting lignin in ethanol at 300 °C for 8 h using the $Cu_{20}MgAl(2)$ catalyst. A variety of cyclic as well as linear products were formed. The cyclic products were mainly substituted aromatics, together with a small amount of ring-hydrogenated cyclic products such as cyclohexenes. The aromatics were mostly (partially) deoxygenated products with different degrees of ring alkylation with methyl and/or ethyl groups. Apart from these lignin-derived cyclic products, a wide range of linear products (mainly higher alkyl alcohols and esters) was formed. These products are the result of ethanol conversion reaction of the Guerbet type, and oxidative esterification reactions catalyzed by the CuMgAl mixed-oxide catalyst. Ethanol conversion into such products was confirmed by carrying out the same reaction without lignin under similar conditions. In this reaction, similar linear products were produced and no cyclic products were formed. This confirmed that all of the cyclic products were formed from lignin. In order to rule out any interference from these ethanol-derived products during product analysis, a work-up procedure was developed to distinguish smaller (light residue) and larger (heavy residue) lignin fragments and char. Further experimental details are provided in the original reports [37,59,60]. The light residue represents fragments of lignin that have been depolymerized to a lower molecular weight than that of the original lignin, and this residue is soluble in tetrahydrofuran (THF). The heavy residue was a fraction that was strongly adsorbed onto the solid catalyst, and which is not soluble in THF. After digesting the solid catalyst with HNO_3, it became THF-soluble. This fraction had a higher molecular weight than that of the starting lignin, due to the repolymerization of lignin fragments. Char is characterized by the fraction that is strongly adsorbed to the solid catalyst which could not be washed out by THF after digesting the solid catalyst. It had the lowest H/C ratio of the different solid fractions. The amount of char was determined by thermogravimetric analysis (TGA) [59].

10.2.1   Effect of Catalyst and Ethanol Solvent

P1000 lignin was only partially depolymerized in ethanol solvent at 300 °C, the main product was char and the monomers yield was only 5 wt% without using a catalyst (Table 10.1; entry 1) [37]. Among the monomeric products, guaiacol- and syringol-type

Table 10.1   *Yields of monomers, lignin residues and char following lignin depolymerization in supercritical ethanol at 300 °C for 4 h [37,59].*

			Yield of products (wt%)				
Entry	Catalyst	Solvent	Monomers	Light residue	Heavy residue	Char	Total yield (wt%)
1	Blank	EtOH	5	23	–	40	68
2	Blank	MeOH	4	24	–	46	74
3	$Cu_{20}MgAl(2)$	EtOH	17	73	18	0	108
4	$Cu_{20}MgAl(2)$	MeOH	6	57	39	1	103
5	$Cu_{20}MgAl(2)$	50% MeOH/EtOH	9	77	18	0	104
6	MgAl(3)	EtOH	4	45	22	0	71
7	$Pt_1MgAl(3)$	EtOH	6	43	22	0	71
8	$Pt_3MgAl(3)$	EtOH	7	45	19	0	71
9	$Pt_5MgAl(3)$	EtOH	8	67	9	0	84
10	$Ni_{20}MgAl(2)$	EtOH	4	51	11	17	79

molecules were predominant, and no deoxygenated aromatics or hydrogenated cyclics were formed. A similar product distribution was obtained without catalyst in methanol solvent (Table 10.1; entry 2). In the presence of the $Cu_{20}MgAl(2)$ catalyst, more monomers (17 wt%) were obtained after reaction in ethanol for 4 h (entry 3). A significant amount of deoxygenated aromatic products such as benzene, toluene, xylenes, and ethyl benzene was observed. These results suggested that the catalyst exhibited excellent deoxygenation and low ring-hydrogenation activities [37]. The use of Cu-free MgAl, PtMgAl and NiMgAl mixed-oxide catalysts resulted in very low monomers and total yields (entries 6–10 in Table 10.1), without the formation of deoxygenated aromatics. A combination of Cu and ethanol is essential for obtaining higher monomer yields and the formation of deoxygenated aromatics during lignin conversion [37].

The use of ethanol as a solvent was found to be significantly more effective in producing monomers and avoiding heavy lignin residue and char formation than using methanol (entries 3 and 4 in Table 10.1). In order to understand how the solvent affects the repolymerization of monomers, the conversion of phenol into high-molecular-weight products was compared in methanol and ethanol. The reaction mixtures were analyzed using GC-MS (Figure 10.3), gel permeation chromatography (GPC) (Figure 10.4a) and ^{1}H-^{13}C heteronuclear single quantum coherence (HSQC) NMR spectrometry (Figure 10.5). When phenol was reacted in methanol at 300 °C for 1 h in the presence of the catalyst, a white resin-like

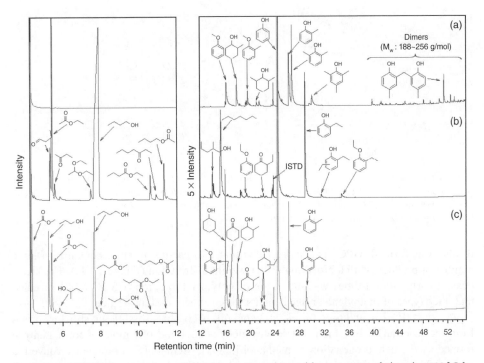

Figure 10.3 *GC-MS chromatograms of reaction mixtures obtained from reaction of phenol at 300 °C for 1 h over the $Cu_{20}MgAl(2)$ catalyst in (a) methanol, (b) ethanol, and (c) 50%/50% (v/v) methanol/ethanol solvents. The GC-MS chromatograms were normalized to the internal standard, ISTD [59].*

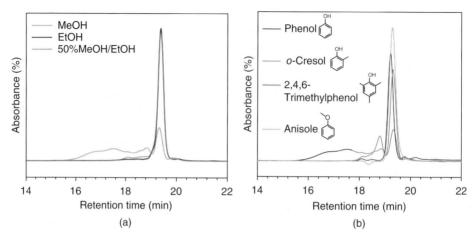

Figure 10.4 *GPC chromatograms of reaction mixtures obtained from the reaction at 300 °C for 1 h over the Cu$_{20}$MgAl(2) catalyst. (a) Using phenol in different solvents; (b) Using different reactants in methanol solvent (the depicted chromatograms have been normalized by the sum of the peak area) [59]. (See color plate section for the color representation of this figure.)*

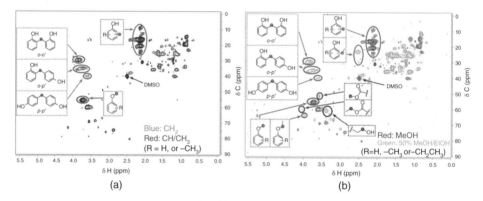

Figure 10.5 *The side-chain region of the ^1H–^{13}C HSQC NMR spectra of the reaction products of phenol conversion (300 °C; 1 h; CuMgAlO$_x$ catalyst). (a) Spectrum for methanol solvent; (b) Combined spectra of methanol (red) and 50%/50% (v/v/) methanol/ethanol solvent (green). The combined spectra have been normalized by the total peak volume [59].*

residue was formed. GPC analysis revealed that the reaction mixture contained a large number of products with a broad molecular weight distribution (Figure 10.4a). The broad peak at early elution times was indicative of polymer formation. HSQC NMR revealed that three types of methylene bridge isomers were formed: o–o' (δ_C/δ_H at 29.3/3.86), o–p' (δ_C/δ_H at 34.4/3.76), and p–p' (δ_C/δ_H at 39.6/3.60) [61]. This crosslinking structure arises from the reaction of two moles of (methylated) phenol and one mole of formaldehyde formed by the dehydrogenation of methanol [30,31]. Methanol can be readily converted to formaldehyde in the presence of metal catalysts [62,63]. The reaction of phenol with formaldehyde for the production of phenolic resins (e.g., resoles and novolaks) is well known in the polymer industry [64,65].

The results were very different when the reaction was performed in ethanol. No resin-like polymer was observed in the reactor and, notably, no products with molecular masses exceeding 150 g mol^{-1} were present, as confirmed by GC-MS (Figure 10.3b). GPC analysis also confirmed the absence of oligomers and polymers after reaction in ethanol (Figure 10.4b). HSQC NMR analysis (Figure 10.5) also showed that no crosslinking structure was formed in ethanol. Instead, many other cross-peaks assigned to higher alcohols, alkyl esters, as well as the C-ethylated and O-ethylated aromatic products, were formed. These products were also observed in GC-MS (Figure 10.3b), indicating that Guerbet reactions of ethanol as well as esterification and C- and O-ethylation reactions of phenol had dominated in ethanol. Notably, when a 50%/50% (v/v) methanol/ethanol mixture was used as solvent, no phenol polymerization reactions were observed, which was also supported by GC-MS, GPC and HSQC NMR analyses (Figures 10.3c, 10.4a and 10.5b). Therefore, it was concluded that ethanol can efficiently scavenge the formaldehyde formed by methanol dehydrogenation. Such scavenging of formaldehyde by ethanol is important, because methanol and formaldehyde can be formed from methoxy groups [35,40,66] and alkyl side-chains during the lignin depolymerization process [30,55,67,68]. During lignin depolymerization in a methanol/ethanol mixture (entry 5, Table 10.1), a significant decrease was observed in the yield of heavy lignin residue (18 wt%) compared to the yield obtained in methanol solvent (39 wt%; entry 3, Table 10.1). GPC analysis of this residue further indicated a lower rate of repolymerization (Figure 10.4a). These results were consistent with those achieved in phenol model compound reactions in the methanol/ethanol mixture.

In order to verify whether the alkylation reaction also contributes to protect the reactive phenolic intermediates from polymerization, similar reactions were performed using *o*-cresol, 2,4,6-trimethylphenol, and anisole as reactants in methanol. Figure 10.4b shows the GPC analysis results for the product mixtures. A good correlation was observed between the extent of polymerization and the degree of C-alkylation. With one methyl group present in the *ortho*-position (*o*-cresol), polymerization occurred at a much lower rate compared with phenol as the reactant. When the phenolic ring contained three methyl groups at the *ortho*- and *para*-positions, almost no polymerized product was observed. Furthermore, polymerization was found to be completely suppressed when anisole was the reactant. These results highlighted the important role of the phenolic hydroxyl group in repolymerization processes during lignin depolymerization. Both, C-alkylation and O-alkylation contribute to suppress repolymerization, which provides solid evidence for the importance of alkylation during lignin conversion [37].

The most important aspects of lignin depolymerization in supercritical ethanol are summarized in Scheme 10.1. Ethanol has three important functions. First, it serves as a source of hydrogen to facilitate the lignin depolymerization and deoxygenation reactions by hydrogenolysis. Hydrogen is observed among the gas-phase products of ethanol-mediated lignin depolymerization. Second, ethanol acts as a scavenger for the formaldehyde formed by removal of methoxy groups from the lignin, thereby suppressing repolymerization reactions involving formaldehyde. Third, ethanol serves as a capping agent to stabilize the reactive phenolic intermediates by O-alkylating the phenolic hydroxyl groups and C-alkylating the aromatic rings. The latter two roles of ethanol cause the rate of repolymerization of phenolic products derived from lignin disassembly to be low, which explains the absence of char formation in ethanol solvent.

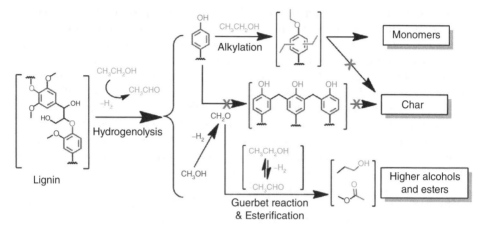

Scheme 10.1   *The roles of alkylation, the Guerbet reaction, and esterification in suppressing char forma-tion during lignin depolymerization over the $Cu_{20}MgAl(2)$ catalyst in supercritical ethanol [59].*

10.2.2   Influence of Reaction Parameters and Lignin Source

The effect of reaction temperature and time has been evaluated. Without catalyst, and with increasing the reaction temperature, both the yield of lignin monomers and char were seen to increase, while the yield of light lignin residue decreased (entries 1-4, Table 10.2). This suggests that both lignin depolymerization and repolymerization were enhanced at a higher temperature. However, without catalyst the reaction is dominated by char-forming reactions. The total yield is low (<80 wt%), which is due to the formation of large amounts of char, which sticks to the reactor and its internal parts and cannot be recovered completely. In the presence of a $Cu_{20}MgAl(2)$ catalyst, at lower reaction temperatures (200–250 °C, entries 5 and 6 in Table 10.2), the dominant product was heavy lignin residue and this decreased at moderate temperatures (300–340 °C, entries 7, 8, 13, and 14). At high temperatures (380–420 °C, entries 15 and 16), almost no heavy lignin residue could be recovered. Instead, considerable amounts (6–12 wt%) of char were obtained. The reaction time also has a profound influence on the reaction outcome. Significant amounts of heavy (49 wt%) and light (40 wt%) lignin residues were obtained when the reaction mixture was only heated to 300 °C, followed by immediate cooling (entry 9). By increasing the reaction time at 300 °C (entries 7 and 9–12), the monomer yield was seen to increase up to 23 wt% at 8 h reaction time, although significant amounts of light and heavy lignin residues were obtained, but char was not formed. It was observed that the heavy lignin residue was primarily formed at the early stages of the reaction and decreased for longer reaction times (entries 7 and 9–12).

Based on the change in the yields of heavy residue and char, it was concluded that condensation is dominant at low reaction temperatures (200–250 °C), while char-forming reactions become significant at high reaction temperatures (380–420 °C). At low temperature, the lignin depolymerization reactions are catalyzed by hydrogenolysis, with thermolysis being largely absent. Rather, condensation reactions that involve reactive side-chains such as C=C double bonds [53] and reactive species such as formaldehyde [59] are important in this temperature regime. These condensation reactions are known to take place at low temperature. Another aspect is that the rates of reactions that can limit char formation, such as

Table 10.2   *Yields of monomers, lignin residues, and char and the total yields following lignin depolymerization under varying conditions.*

Entry	Catalyst/ Lignin	Temp. (°C)	Time (h)	Monomers	Light residue	Heavy residue	Char	Total yield (wt%)
				Yield of products (wt%)				
Reactions in 50-ml autoclave[a]								
1	Blank	200	4	2	49	0	26	77
2	Blank	250	4	4	39	0	35	78
3	Blank	300	4	6	30	0	41	78
4	Blank	340	4	8	15	0	36	58
5	$Cu_{20}MgAl(2)$	200	4	1	21	41	1	64
6	$Cu_{20}MgAl(2)$	250	4	3	35	59	1	98
7	$Cu_{20}MgAl(2)$	300	4	17	73	18	0	108
8	$Cu_{20}MgAl(2)$	340	4	20	69	9	3	101
9	$Cu_{20}MgAl(2)$	300	0	5	40	49	0	94
10	$Cu_{20}MgAl(2)$	300	2	8	52	31	0	91
11	$Cu_{20}MgAl(2)$	300	8	23	63	16	0	102
12	$Cu_{20}MgAl(2)$	300	20	21	78	11	0	110
Reactions in 100-ml autoclave[b]								
13	$Cu_{20}MgAl(2)$	300	4	19	67	11	0	97
14	$Cu_{20}MgAl(2)$	340	4	30	72	8	1	111
15	$Cu_{20}MgAl(2)$	380	4	42	56	1	6	105
16	$Cu_{20}MgAl(2)$	420	4	49	55	0	12	116
17	$Cu_{20}MgAl(2)$	380	8	60	52	1	10	123
18	$Cu_{20}MgAl(2)$	380	20	49	47	1	18	115
19	Alcell	380	8	62	47	1	6	116
20	Kraft	380	8	86	26	3	31	146

[a]50-ml autoclave conditions: 1 g lignin; 0.5 g catalyst; 20 ml solvent.
[b]100-ml autoclave conditions: 1 g lignin; 0.5 g catalyst; 40 ml solvent.

alkylation, Guerbet and esterification reactions, are low at the low end of the reaction temperatures explored. The lower activity in depolymerization and char-hindering reactions shifts the reaction balance towards condensation, which explains the high yield of heavy lignin residue at low temperature. At moderate reaction temperatures (300–340 °C), depolymerization reactions are enhanced and more recalcitrant bonds in the lignin can be cleaved, generating more lignin fragments and reactive phenolic intermediates. These reactive phenolic intermediates can be better protected by the enhanced alkylation and Guerbet and esterification reactions. This shifts the reaction balance towards depolymerization, which explains the decreased yields of heavy residue and increased yields of light residue and monomers.

Under optimized conditions (380 °C, 8 h), a 60 wt% yield of monomers could be obtained from P1000 lignin (Table 10.2; entry 17). The lignin monomeric product distribution under this condition is shown in Figure 10.6. Using organosolv lignin (Alcell, entry 19 in Table 10.2) gave a similar product yield to soda lignin, with Kraft lignin (entry 20), the monomers yield was 86 wt% [59]. This result showed that the catalyst can also convert sulfur-containing lignins. The excess mass balances shown in Table 10.1 are due to the extensive alkylation and Guerbet-type coupling reactions of the lignin products by the ethanol solvent.

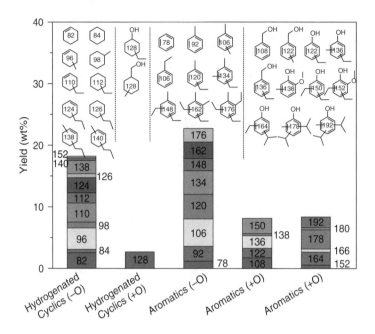

Figure 10.6 *Monomeric product distribution following lignin reaction at 380 °C for 8h over the Cu₂₀MgAl(2) catalyst in ethanol solvent [59].*

10.2.3   Effect of Catalyst Composition

An investigation was made into the role of Cu-Mg-Al mixed oxides in the depolymerization of soda lignin in supercritical ethanol. A series of mixed oxides ($Cu_xMgAl(y)$) with varying $(Cu+Mg)/Al$ ratios ($y = 2, 3, 4,$ and 6) and Cu content ($x = 0, 10, 20, 40$ wt%) were prepared. Figure 10.7a shows the X-ray diffraction (XRD) patterns of the $Cu_{20}MgAl$ LDH precursors.

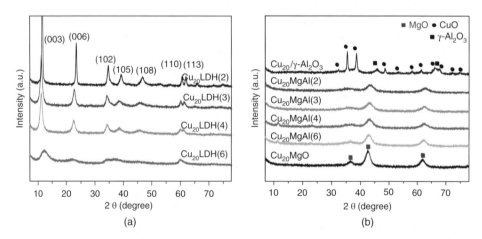

Figure 10.7 *XRD patterns of the (a) Cu₂₀LDH(y) precursors and (b) mixed-oxide catalysts after calcination at 460 °C for 6 h [60].*

Table 10.3   *Textural properties and chemical composition of mixed oxide catalysts with different M^{2+}/M^{3+} ratios [60].*

Catalyst	S_{BET} ($m^2\,g^{-1}$)	V^a ($cm^3\,g^{-1}$)	d^b (nm)	Cu (wt%)	Mg (wt%)	Al (wt%)	(Cu+Mg)/Al (atomic ratio)
Cu_{20}/γ-Al_2O_3	150	0.41	7.6	17.8	–	38.0	0.2
Cu_{20}MgAl(2)	189	0.53	15.6	16.1	21.9	14.2	2.2
Cu_{20}MgAl(3)	183	0.80	13.8	16.2	26.2	11.1	3.3
Cu_{20}MgAl(4)	206	0.72	14.9	16.3	28.3	9.1	4.2
Cu_{20}MgAl(6)	170	0.71	14.9	16.4	31.5	6.5	6.5
Cu_{20}MgO	173	0.49	10.9	20.2	33.9	–	–

^{a} Average pore volume.
^{b} Average pore diameter.

The patterns contain the characteristic diffraction peaks of the hydrotalcite structure [69]. The peaks become broader and less intense with increasing M^{2+}/M^{3+} ratios, which indicates that the crystallinity decreases with decreasing Al^{3+} content. The Cu_{20}LDH(6) hydrotalcite shows the lowest crystallinity and, likely, contains the largest amount of amorphous material. Figure 10.7b shows the XRD patterns of the mixed oxides obtained by calcination of the Cu_{20}LDH(y) precursors at 460 °C for 6 h. All of the Cu_{20}MgAl(y) samples have the MgO structure with Cu^{2+} and Al^{3+} cations dissolved in the lattice [70].

The textural properties and elemental compositions of the mixed-oxide catalyst samples are summarized in Table 10.3. The Cu_{20}MgAl(y) samples have surface areas in the range of 170– 206 $m^2\,g^{-1}$ and large pore volumes (0.53–0.80 $cm^3\,g^{-1}$). Elemental analysis shows that the actual Cu content is slightly lower than the intended content, which is likely due to the absorption of atmospheric CO_2 and H_2O after calcination. The (Cu+Mg)/Al ratios were slightly higher than expected.

CO_2-temperature-programmed desorption (TPD) was used to determine the number and strength of basic sites in mixed oxides obtained by calcination of hydrotalcites [6,71]. The CO_2-TPD profiles of the catalysts are shown in Figure 10.8. In all cases, a broad desorption band is observed between 100 °C and 460 °C, which can be deconvoluted into three contributions at about 165 °C (weak basic strength), 200 °C (medium basic strength) and 255 °C (high basic strength) [52]. The low-temperature desorption peak corresponds to basic surface OH^- groups, while the medium-temperature peak can be ascribed to Mg^{2+}-O^{2-}, Al^{3+}-O^{2-} and Cu^{2+}-O^{2-} acid–base pairs. The high-temperature peak is usually attributed to the strong basic sites associated with low-coordinated O^{2-} anions [6,71]. A small shift of the desorption peak towards a higher temperature was observed when the M^{2+}/M^{3+} ratio was increased from 2 to 4 (Figure 10.8). The total amount of basic sites of these catalysts was also increased. When the M^{2+}/M^{3+} ratio was further increased from 4 to 6, both the number of basic sites and their strength decreased. This was attributed to a partial collapse of the hydrotalcite structure during synthesis of the sample with the highest M^{2+}/M^{3+} ratio, in keeping with the lowest crystallinity of this sample, as followed with XRD (see Figure 10.7). Among all of the samples, Cu_{20}MgAl(4) contained the largest amount of basic sites (0.35 mmol g^{-1}) of which weak, medium, and strong basic sites accounted for 29%, 38%, and 33%, respectively. The Cu_{20}/γ-Al_2O_3 sample contained the least basic sites (0.03 mmol g^{-1}), while the Cu_{20}MgO contained an intermediate amount of basic sites (0.25 mmol g^{-1}).

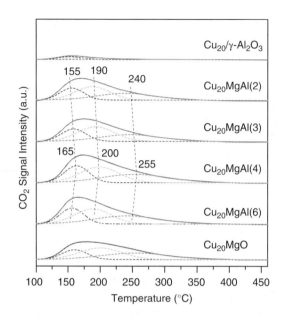

Figure 10.8 *CO_2-TPD profiles of the mixed-oxide catalysts with different M^{2+}/M^{3+} ratios [60].*

These samples were evaluated as catalysts in the conversion of soda lignin in super-critical ethanol at a temperature of 340 °C and a reaction time of 4 h. Comparisons with Cu/MgO and Cu/γ-Al$_2$O$_3$ catalysts emphasized the excellent performance of the CuMgAl mixed oxides (Table 10.4). A 20 wt% Cu content and (Cu+Mg)/Al = 4 molar ratio gave the best performance (highest lignin monomer yields without char formation) (entry 4 in Table 10.4). A comparison of the ethanol conversion products (Table 10.4) with the catalyst properties (Figure 10.7) showed a strong correlation between the basic sites and the overall yield of alcohols and esters. The higher the basicity of the catalyst, the higher the amount of alcohols and esters formed via Guerbet-type and esterification reactions, and the less repolymerized products are formed [72,73]. In order to clarify the role of alkylation, instead of lignin, phenol model compound reactions were carried out under similar conditions (at

Table 10.4 *Product distribution for the reaction of lignin in ethanol at 340 °C for 4 h over mixed oxide catalysts with different M^{2+}/M^{3+} ratios [60].*

		Yield lignin products (wt%)					Yield ethanol products (C_{4+}, mg)				
Entry	Catalyst	Mono-mers	Light residue	Heavy residue	Char	Total yield	Alcohols	Esters	Alde-hydes	Ethers	Hydro-carbons
1	Cu$_{20}$/γ-Al$_2$O$_3$	26	40	4	23	93	234	130	1	2490	83
2	Cu$_{20}$MgAl(2)	30	72	8	0	110	2276	776	32	103	193
3	Cu$_{20}$MgAl(3)	31	75	8	0	114	2348	663	40	125	317
4	Cu$_{20}$MgAl(4)	36	69	6	0	110	3214	1069	67	105	336
5	Cu$_{20}$MgAl(6)	31	70	7	0	108	2996	695	31	153	300
6	Cu$_{20}$MgO	20	47	15	0	82	2722	371	96	51	155

Table 10.5 *Product distribution for the reaction of lignin in ethanol at 340 °C for 4 h over mixed oxide catalysts as a function of Cu content [60].*

Entry	Catalyst	Yield lignin products (wt%)					Yield ethanol products (C_{4+}, mg)				
		Mono-mers	Light residue	Heavy residue	Char	Total yield	Alcohols	Esters	Alde-hydes	Ethers	Hydro-carbons
1	MgAl(4)	9	42	15	0	66	1343	150	10	247	102
2	$Cu_{10}MgAl(4)$	21	61	13	0	95	2471	506	36	123	127
3	$Cu_{20}MgAl(4)$	36	69	6	0	110	3214	1069	67	105	336
4	$Cu_{40}MgAl(4)$	26	79	8	0	113	2372	593	80	116	254

340 °C for 4 h) over $Cu_{20}MgAl(y)$ catalysts. The alkylation activity strongly correlated with the Al content, and although Cu_{20}/γ-Al_2O_3 had the highest alkylation activity its performance in lignin conversion was worse than that of $Cu_{20}MgAl(4)$. The most basic $Cu_{20}MgAl(4)$ sample gave the highest yield of ethanol-derived products and the lowest yield of heavy residue (6 wt%, entry 4 in Table 10.4). It was concluded that Guerbet and esterification reactions are more important than alkylation in suppressing repolymerization and char formation [60].

In order to better understand the role of Cu, the Cu content was further varied (0 wt%, 10 wt%, 20 wt%, and 40 wt%) in the MgAl(4) precursor and mixed-oxide catalysts were obtained by calcination. The characteristic peaks corresponding to hydrotalcite were preserved below 40 wt% Cu content according to the XRD patterns of the $Cu_xLDH(4)$ precursors. All of the samples below 40 wt% Cu content had the MgO structure in the mixed oxides obtained by calcination of the corresponding hydrotalcite precursors.

Using the Cu-free MgAl(4) sample as the catalyst, the monomers yield was only 9 wt%, with the major products being light residue (42 wt%) and heavy residue (15 wt%) (entry 1 in Table 10.5). The low total yield was due to the low rates of Guerbet, esterification, and alkylation reactions in the absence of Cu. Consistent with this, the yield of light residue and the total yield increased when the rates of Guerbet, esterification, and alkylation reactions catalyzed by Cu were enhanced by an increased Cu content. Cu and Al sites are both key to catalyzing alkylation reactions, and it is reasonable to state that these sites constitute Lewis acid sites for aromatics alkylation. Compared with the $Cu_{40}MgAl(4)$, $Cu_{10}MgAl(4)$ exhibited very similar activity in Guerbet and esterification reactions, but gave a higher yield of heavy residue (Table 10.5, entries 2 and 4). This can be related to the lower alkylation activity of $Cu_{10}MgAl(4)$. From phenol model reactions, it was observed that the alkylation degree increased with the Cu content. Accordingly, it could be inferred that Cu is pivotal to the high alkylation activity of the hydrotalcites-derived catalysts.

All data considered, it can be firmly concluded that lignin depolymerization strongly benefits from Guerbet, esterification, and alkylation reactions, because these reactions suppress the repolymerization of lignin fragments. Among these reactions, the Guerbet and esterification reactions catalyzed by basic sites (Cu, Mg, O) are more important in suppressing repolymerization than the alkylation reaction catalyzed by Lewis acid sites (Cu, Al, O) (Scheme 10.2).

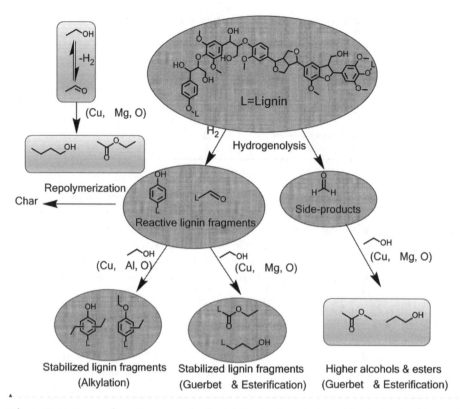

Scheme 10.2 *Proposed reaction network of catalytic depolymerization of lignin in ethanol over the Cu$_x$MgAl(y) catalysts [60].*

10.3   Conclusions

Lignocellulosic biomass is a low-cost, renewable feedstock that is uniquely suited for the production of sustainable liquid fuels. The aim of these studies was to systematically investigate different thermocatalytic approaches to valorize lignins into fuel components and high-value chemicals. A one-step process to do so was developed that comprises the use of supercritical ethanol and a cheap Cu-based catalyst to upgrade technical lignin into a mixture of alkylated cycloalkanes, aromatics, and phenolic compounds. It was found that Cu-containing hydrotalcite-derived mixed-oxide catalysts gave much higher lignin monomer yields than MgAl, NiMgAl, and Pt-promoted MgAl mixed-oxide catalysts. Ethanol is significantly more effective in producing monomers and avoiding char than methanol. The conversion of phenol into high-molecular-weight products in methanol and ethanol solvents was compared under similar conditions. Consistently, it was found that the polymerization of phenol took place in methanol, but not in ethanol, because methanol can be readily dehydrogenated to form formaldehyde, which easily reacts with phenol to form bonds that are also common in phenol-formaldehyde resins. Furthermore, it was proved that ethanol also acts as a formaldehyde scavenger as it reacts with the formaldehyde formed from methoxy group cleavage or by elimination of the γ-CH$_2$OH groups of the lignin structure. This explains the absence of char formation in ethanol. Explorations were also made in more detail to determine how alkylation suppresses the repolymerization of model monomers, using different alkylphenols as reactants. The results further confirmed that both C-alkylation and O-alkylation contribute to suppressing repolymerization. The influence of reaction time and temperature also has a profound influence on the reaction outcome. Condensation reactions are dominant at low temperatures (200–250 °C), while char-forming reactions become significant at very high reaction temperatures (380–420 °C). At low temperatures, lignin depolymerization involves hydrogenolysis reactions, and thermal cracking is almost absent under these conditions. As the rates of condensation reactions involving reactive side-chains such as C=C double bonds and species such as formaldehyde are higher than that of hydrogenolysis, heavy residue is formed in this temperature regime. Moreover, the rates of reactions that can limit char formation such as alkylation, Guerbet and esterification reactions, are low below 300 °C, and this shifts the balance towards condensation rather than depolymerization. At moderate temperatures (300–340 °C), depolymerization reactions are enhanced, mostly because thermal cracking starts, and the more recalcitrant bonds in the lignin can now also be cleaved, including these of heavy residues formed during heating. This generates more lignin fragments and reactive phenolic intermediates. The reactive phenolic intermediates are obtained in a higher proportion, because they are protected by alkylation and Guerbet and esterification reactions, and this shifts the reaction balance towards depolymerization. At the highest temperatures used in this study (380–420 °C), char-forming reactions due to carbonization become dominant, but these carbonization reactions cannot be suppressed. Three types of lignin (soda, Alcell, and Kraft) were used as feedstocks under optimized conditions. High monomer yields (typically ~60 wt% for soda and organosolv lignins and 86 wt% for Kraft lignin) can be obtained after the reaction of lignin in ethanol at 380 °C for 8 h, with little char formation.

Further variations were made in the (Cu+Mg)/Al ratios (M^{2+}/M^{3+} ratios = 2, 3, 4, and 6) and the Cu contents (0 wt%, 10 wt%, 20 wt%, 40 wt%) of the catalysts, and the influence of these compositional changes on lignin conversion were studied at 340 °C. Comparison with Cu/MgO and Cu/γ-Al$_2$O$_3$ catalysts emphasized the excellent performance of Cu-MgAl mixed oxides. In addition to lignin products, the side products derived from ethanol conversion reactions were also analyzed and linked to the chemical reactions (Guerbet and esterification). Phenol was used as a model reactant to understand the role of acid and base sites in alkylation reactions. The combination of Cu and basic sites facilitates hydrogen production by the dehydrogenation of ethanol. Cu and basic sites also catalyze Guerbet and esterification reactions, producing higher alcohols and esters from ethanol. The higher the activity of the catalyst in Guerbet and esterification reactions is, the less repolymerization products it produces. This can be linked to the scavenging of reactive species such as formaldehyde and aldehyde side-chains. The Lewis acid sites due to the presence of Cu and Al catalyze C- and O-alkylation reactions. Guerbet and esterification reactions are more important in suppressing repolymerization and char formation than alkylation reactions.

In conclusion, lignin, being the second most abundant but least valorized biomass component, has attracted increasing research interest during recent years. These studies have demonstrated the various possibilities of valorizing different technical lignins, but foremost have led to new insights into the mechanism by which these materials can be effectively deconstructed. The current studies have also helped to identify the technical barriers that need to be overcome, and have provided direction to make lignin usage economically viable.

References

1. Sels, B. F., De Vos, D. E., Jacobs, P. A. (2001) Hydrotalcite-like anionic clays in catalytic organic reactions. *Catal. Rev.*, **43** (4), 443–488.
2. Debecker, D. P., Gaigneaux E. M., Busca, G. (2009) Exploring, tuning, and exploiting the basicity of hydrotalcites for applications in heterogeneous catalysis. *Chem. Eur. J.*, **15** (16), 3920–3935.
3. Tichit, D., Coq, B. (2003) Catalysis by hydrotalcites and related materials. *CATTECH*, **7** (6), 206–217.
4. Nishimura, S., Takagaki A., Ebitani, K. (2013) Characterization, synthesis and catalysis of hydrotalcite-related materials for highly efficient materials transformations. *Green Chem.*, **15** (8), 2026–2042.
5. Paredes, S. P., Fetter, G., Bosch, P., Bulbulian, S. (2006) Sol-gel synthesis of hydrotalcite- like compounds. *J. Mater. Sci.*, **41** (11), 3377–3382.
6. Di Cosimo, J. I., Diez, V. K., Xu, M., Iglesia, E., Apesteguia, C. R. (1998) Structure and surface and catalytic properties of Mg-Al basic oxides. *J. Catal.*, **178** (2), 499–510.
7. Abello, S., Medina, F., Tichit, D., Perez-Ramirez, J., Groen, J. C., Sueiras, J. E., Salagre, P., Cesteros, Y. (2005) Aldol condensations over reconstructed Mg-Al hydrotalcites: Structure-activity relationships related to the rehydration method. *Chem. Eur. J.*, **11** (2), 728–739.
8. Di Serio, M., Mallardo, S., Carotenuto, G., Tesser R., Santacesaria, E. (2012) Mg/Al hydrotalcite catalyst for biodiesel production in continuous packed bed reactors. *Catal. Today*, **195** (1), 54–58.
9. Yao, S., Wang, X., Jiang, Y., Wu, F., Chen X., Mu, X. (2014) One-step conversion of biomass-derived 5-hydroxymethylfurfural to 1,2,6-hexanetriol over Ni-Co-Al mixed-oxide catalysts under mild conditions. *ACS Sustainable Chem. Eng.*, **2** (2), 173–180.
10. Nagpure, A. S., Venugopal, A. K., Lucas, N., Manikandan, M., Thirumalaiswamy, R., Chilukuri, S. (2015) Renewable fuels from biomass-derived compounds: Ru-containing hydrotalcites as catalysts for conversion of HMF to 2,5-dimethylfuran. *Catal. Sci. Technol.*, **5** (3), 1463–1472.
11. Wang, L., Chen, J. H., Watanabe, H., Xu, Y., Tamura, M., Nakagawa, Y., Tomishige, K. (2014) Catalytic performance and characterization of Co-Fe bcc alloy nanoparticles prepared from hydrotalcite-like precursors in the steam gasification of biomass-derived tar. *Appl. Catal. B*, **160**, 701–715.
12. Li, D., Wang, L., Koike, M., Nakagawa Y., Tomishige, K. (2011) Steam reforming of tar from pyrolysis of biomass over Ni/Mg/Al catalysts prepared from hydrotalcite-like precursors. *Appl. Catal. B*, **102** (3-4), 528–538.
13. Ashok, J., Kawi, S. (2014) Nickel-iron alloy supported over iron-alumina catalysts for steam reforming of biomass tar model compound. *ACS Catal.*, **4** (1), 289–301.
14. Josuinkas, F. M., Quitete, C. P. B., Ribeiro, N. F. P., Souza, M. M. V. M. (2014) Steam reforming of model gasification tar compounds over nickel catalysts prepared from hydrotalcite precursors. *Fuel Process. Technol.*, **121**, 76–82.
15. Chheda, J. N., Dumesic, J. A. (2007) An overview of dehydration, aldol-condensation and hydrogenation processes for production of liquid alkanes from biomass-derived carbohydrates. *Catal. Today*, **123** (1-4), 59–70.
16. Sadaba, I., Ojeda, M., Mariscal R., Fierro, J. L. G., Granados, M. L. (2011) Catalytic and structural properties of co-precipitated Mg-Zr mixed oxides for furfural valorization via aqueous aldol condensation with acetone. *Appl. Catal. B*, **101** (3-4), 638–648.
17. Liu, H. H., Xu, W. J., Liu, X. H., Guo, Y., Guo, Y. L., Lu, G. Z., Wang, Y. Q. (2010) Aldol condensation of furfural and acetone on layered double hydroxides. *Kinet. Catal.*, **51** (1), 75–80.
18. Barta, K., Matson, T. D., Fettig, M. L., Scott, S. L., Iretskii, A. V., Ford, P. C. (2010) Catalytic disassembly of an organosolv lignin via hydrogen transfer from supercritical methanol. *Green Chem.*, **12** (9), 1640–1647.
19. Matson, T. D., Barta, K., Iretskii, A. V., Ford, P. C. (2011) One-pot catalytic conversion of cellulose and of woody biomass solids to liquid fuels. *J. Am. Chem. Soc.*, **133** (35), 14090–14097.

20. Barta, K., Ford, P. C. (2014) Catalytic conversion of nonfood woody biomass solids to organic liquids. *Acc. Chem. Res.*, **47** (5), 1503–1512.

21. Barta, K., Warner, G. R., Beach, E. S., Anastas, P. T. (2014) Depolymerization of organosolv lignin to aromatic compounds over Cu-doped porous metal oxides. *Green Chem.*, **16** (1), 191–196.

22. Deuss, P. J., Barta, K. (2016) From models to lignin: Transition metal catalysis for selective bond cleavage reactions. *Coord. Chem. Rev.*, **306**, 510–532.

23. Sturgeon, M. R., O'Brien, M. H., Ciesielski, P. N., Katahira, R., Kruger, J. S., Chmely, S. C., Hamlin, J., Lawrence, K., Hunsinger, G. B., Foust, T. D., Baldwin, R. M., Biddy, M. J., Beckham, G. T. (2014) Lignin depolymerisation by nickel supported layered-double hydroxide catalysts. *Green Chem.*, **16** (2), 824–835.

24. Mottweiler, J., Puche, M., Rauber, C., Schmidt, T., Concepcion, P., Corma A., Bolm, C. (2015) *ChemSusChem*, **8** (12), 2106–2113.

25. Ragauskas, A. J., Beckham, G. T., Biddy, M. J., Chandra, R., Chen, F., Davis, M. F., Davison, B. H., Dixon, R. A., Gilna, P., Keller, M., Langan, P., Naskar, A. K., Saddler, J. N., Tschaplinski, T. J., Tuskan, G. A., Wyman, C. E. (2014) Lignin valorization: improving lignin processing in the biorefinery. *Science*, **344** (6185), 1246843.

26. Holladay, J. E., White, J. F., Bozell, J. J., Johnson, D. (2007) Top Value-Added Chemicals from Biomass – Volume II—Results of Screening for Potential Candidates from Biorefinery Lignin, Report PNNL-16983; 7, Pacific Northwest National Laboratory (PNNL), Richland, WA (US).

27. Azadi, P., Inderwildi, O. R., Farnood, R., King, D. A. (2013) Liquid fuels, hydrogen and chemicals from lignin: A critical review. *Renew. Sustain. Energy Rev.*, **21**, 506–523.

28. Li, C. Z., Zhao, X. C., Wang, A. Q., Huber, G. W., Zhang, T. (2015) Catalytic transformation of lignin for the production of chemicals and fuels. *Chem. Rev.*, **115** (21), 11559–11624.

29. Joffres, B., Laurenti, D., Charon, N., Daudin, A., Quignard A., Geantet, C. (2013) Thermochemical conversion of lignin for fuels and chemicals: a review. *Oil Gas Sci. Technol.*, **68** (4), 753–763.

30. Pandey, M. P., Kim, C. S. (2011) Lignin depolymerization and conversion: a review of thermochemical methods. *Chem. Eng. Technol.*, **34** (1), 29–41.

31. Deuss, P. J., Barta, K. (2016) From models to lignin: Transition metal catalysis for selective bond cleavage reactions. *Coord. Chem. Rev.*, **306**, 510–532.

32. Rinaldi, R., Jastrzebski, R., Clough, M. T., Ralph, J., Kennema, M., Bruijnincx, P. C. A., Weckhuysen, B. M. (2016) Paving the way for lignin valorisation: recent advances in bioengineering, biorefining and catalysis. *Angew. Chem. Int. Ed.*, **55** (29), 8164–8215.

33. Miller, J. E., Evans, L., Littlewolf, A., Trudell, D. E. (1999) Batch microreactor studies of lignin and lignin model compound depolymerization by bases in alcohol solvents. *Fuel*, **78** (11), 1363–1366.

34. Roberts, V. M., Stein, V., Reiner, T., Lemonidou, A., Li, X. B., Lercher, J. A. (2011) Towards quantitative catalytic lignin depolymerization. *Chem. Eur. J.*, **17** (21), 5939–5948.

35. Onwudili, J. A., Williams, P. T. (2014) Catalytic depolymerization of alkali lignin in subcritical water: influence of formic acid and Pd/C catalyst on the yields of liquid monomeric aromatic products. *Green Chem.*, **16** (11), 4740–4748.

36. Song, Q., Wang, F., Cai, J. Y., Wang, Y. H., Zhang, J. J., Yu, W. Q., Xu, J. (2013) Lignin depolymerization (LDP) in alcohol over nickel-based catalysts via a fragmentation-hydrogenolysis process. *Energy Environ. Sci.*, **6** (3), 994–1007.

37. Huang, X., Korányi, T. I., Boot, M. D., Hensen, E. J. M. (2014) Catalytic depolymerization of lignin in supercritical ethanol. *ChemSusChem*, **7** (8), 2276–2288.

38. Ma, R., Hao, W. Y., Ma, X. L., Tian, Y., Li, Y. D. (2014) Catalytic ethanolysis of Kraft lignin into high-value small-molecular chemicals over a nanostructured alpha-molybdenum carbide catalyst. *Angew. Chem. Int. Ed.*, **53** (28), 7310–7315.

39. Wang, X. Y., Rinaldi, R. (2012) Solvent effects on the hydrogenolysis of diphenyl ether with Raney nickel and their implications for the conversion of lignin. *ChemSusChem*, **5** (8), 1455–1466.

40. Zakzeski, J., Weckhuysen, B. M. (2011) Lignin solubilization and aqueous phase reforming for the production of aromatic chemicals and hydrogen. *ChemSusChem*, **4** (3), 369–378.

41. Jongerius, A. L., Bruijnincx, P. C. A., Weckhuysen, B. M. (2013) Liquid-phase reforming and hydrodeoxygenation as a two-step route to aromatics from lignin. *Green Chem.*, **15** (11), 3049–3056.

42. Cheng, S. N., Wilks, C., Yuan, Z. S., Leitch, M., Xu, C. B. (2012) Hydrothermal degradation of alkali lignin to bio-phenolic compounds in sub/supercritical ethanol and water-ethanol co-solvent. *Polym. Degrad. Stab.*, **97** (6), 839–848.

43. Deepa, A. K., Dhepe, P. L. (2014) Lignin depolymerization into aromatic monomers over solid acid catalysts. *ACS Catal.*, **5** (1), 365–379.

44. Rahimi, A., Ulbrich, A., Coon, J. J., Stahl, S. S. Formic-acid-induced depolymerization of oxidized lignin to aromatics. (2014) *Nature*, **515** (7526), 249–252.

45. Ferrini, P., Rinaldi, R. (2014) Catalytic biorefining of plant biomass to non-pyrolytic lignin bio-oil and carbohydrates through hydrogen transfer reactions. *Angew. Chem. Int. Ed.*, **53** (33), 8634–8639.

46. Parsell, T., Yohe, S., Degenstein, J., Jarrell, T., Klein, I., Gencer, E., Hewetson, B., Hurt, M., Kim, J. I., Choudhari, H., Saha, B., Meilan, R., Mosier, N., Ribeiro, F., Delgass, W. N., Chapple, C., Kenttamaa, H. I., Agrawal R., Abu-Omar, M. M. (2015) A synergistic biorefinery based on catalytic conversion of lignin prior to cellulose starting from lignocellulosic biomass. *Green Chem.*, **17** (3), 1492–1499.

47. Lancefield, C. S., Westwood, N. J. (2015) The synthesis and analysis of advanced lignin model polymers. *Green Chem.*, **17** (11), 4980–4990.

48. Deuss, P. J., Scott, M., Tran, F., Westwood, N. J., de Vries, J. G., Barta, K. (2015) Aromatic monomers by in situ conversion of reactive intermediates in the acid-catalyzed depolymerization of lignin. *J. Am. Chem. Soc.*, **137** (23), 7456–7467.

49. Roberts, V., Fendt, S., Lemonidou, A. A., Li, X. B., Lercher, J. A. (2010) Influence of alkali carbonates on benzyl phenyl ether cleavage pathways in superheated water. *Appl. Catal. B*, **95** (1-2), 71–77.

50. Toledano, A., Serrano, L., Labidi, J. (2014) Improving base-catalyzed lignin depolymerization by avoiding lignin repolymerization. *Fuel*, **116**, 617–624.

51. Okuda, K., Umetsu, M., Takami S., Adschiri, T. (2004) Disassembly of lignin and chemical recovery – rapid depolymerization of lignin without char formation in water-phenol mixtures. *Fuel Process. Technol.*, **85** (8-10), 803–813.

52. Hosoya, T., Kawamoto, H., Saka, S. (2008) Secondary reactions of lignin-derived primary tar components. *J. Anal. Appl. Pyrolysis*, **83** (1), 78–87.

53. Nakamura, T., Kawamoto, H., Saka, S. (2008) Pyrolysis behavior of Japanese cedar wood lignin studied with various model dimers. *J. Anal. Appl. Pyrolysis*, **81** (2), 173–182.

54. Deuss, P. J., Scott, M., Tran, F., Westwood, N. J., de Vries, J. G., Barta, K. (2015) Aromatic monomers by in situ conversion of reactive intermediates in the acid-catalyzed depolymerization of lignin. *J. Am. Chem. Soc.*, **137** (23), 7456–7467.

55. Saisu, M., Sato, T., Watanabe, M., Adschiri, T., Arai, K. (2003) Conversion of lignin with supercritical water-phenol mixtures. *Energy Fuels*, **17** (4), 922–928.

56. Roberts, D. V. M. (2008) PhD Thesis, Technology University of Munich.

57. Zhang, J. G., Asakura, H., van Rijn, J., Yang, J., Duchesne, P., Zhang, B., Chen, X., Zhang, P., Saeys, M., Yan, N. (2014) Highly efficient, NiAu-catalyzed hydrogenolysis of lignin into phenolic chemicals. *Green Chem.*, **16** (5), 2432–2437.

58. Xin, J. N., Zhang, P., Wolcott, M. P., Zhang, X., Zhang, J. W. (2014) Partial depolymerization of enzymolysis lignin via mild hydrogenolysis over Raney nickel. *Bioresource Technol.*, **155**, 422–426.

59. Huang, X., Korányi, T. I., Boot, M. D., Hensen, E. J. M. (2015) Ethanol as capping agent and formaldehyde scavenger for efficient depolymerization of lignin to aromatics. *Green Chem.*, **17** (11), 4941–4950.

60. Huang, X., Atay, C., Korányi, T. I., Boot, M. D., Hensen, E. J. M. (2015) Role of Cu-Mg-Al mixed-oxide catalysts in lignin depolymerization in supercritical ethanol. *ACS Catal.*, **5** (12), 7359–7370.

61. Werstler, D. D. (1986) Quantitative C-13 NMR characterization of aqueous formaldehyde resins. 1. Phenol-formaldehyde resins. *Polymer*, **27** (5), 750–756.

62. Bravo-Suarez, J. J., Subramaniam, B., Chaudhari, R. V. (2013) Vapor-phase methanol and ethanol coupling reactions on CuMgAl mixed metal oxides. *Appl. Catal. A*, **455**, 234–246.
63. Carlini, C., Marchionna, M., Noviello, M., Galletti, A. M. R., Sbrana, G., Basile, F., Vaccari, A. (2005) Guerbet condensation of methanol with n-propanol to isobutyl alcohol over heterogeneous bifunctional catalysts based on Mg-Al mixed oxides partially substituted by different metal components. *J. Mol. Catal. A: Chem.*, **232** (1-2), 13–20.
64. Pilato, L. (2010) *Phenolic Resins: A Century of Progress*, Springer, Berlin Heidelberg, pp. 1–151.
65. NPSC Board of Consultants & Engineers (2008) *Phenolic Resins Technology Handbook*, NIIR Project Consultancy Services, Delhi.
66. Bui, V. N., Laurenti, D., Delichere, P., Geantet, C. (2011) Hydrodeoxygenation of guaiacol Part II: Support effect for CoMoS catalysts on HDO activity and selectivity. *Appl. Catal. B*, **101** (3-4), 246–255.
67. Evans, R. J., Milne, T. A., Soltys, M. N. (1986) Direct mass-spectrometric studies of the pyrolysis of carbonaceous fuels. 3. Primary pyrolysis of lignin. *J. Anal. Appl. Pyrolysis*, **9** (3), 207–236.
68. Lundquis. K., Ericsson, L. (1970) Acid degradation of lignin. III. Formation of formaldehyde. *Acta Chem. Scand.*, **24**, 3681–3686.
69. Cavani, F., Trifirò, F., Vaccari, A. (1991) Hydrotalcite-type anionic clays: preparation, properties and applications. *Catal. Today*, **11** (2), 173–301.
70. Millange, F., Walton, R. I., O'Hare, D. (2000) Time-resolved in situ X-ray diffraction study of the liquid-phase reconstruction of Mg-Al-carbonate hydrotalcite-like compounds. *J. Mater. Chem.*, **10** (7), 1713–1720.
71. Liu, P., Derchi, M., Hensen, E. J. M. (2013) Synthesis of glycerol carbonate by transesterification of glycerol with dimethyl carbonate over MgAl mixed-oxide catalysts. *Appl. Catal. A*, **467**, 124–131.
72. Sad, M. E., Neurock, M., Iglesia, E. (2011) Synthesis of glycerol carbonate by transesterification of glycerol with dimethyl carbonate over MgAl mixed-oxide catalysts. *J. Am. Chem. Soc.*, **133** (50), 20384–20398.
73. Kozlowski, J. T., Davis, R. J. (2013) Heterogeneous catalysts for the Guerbet coupling of alcohols. *ACS Catal.*, **3** (7), 1588–1600.

11

Niobium-Based Catalysts for Biomass Conversion

Qineng Xia and Yanqin Wang

School of Chemistry & Molecular Engineering, East China University of Science and Technology, China

11.1 Introduction

The sustainable conversion of biomass to produce fuels, chemicals and materials is one of the promising solutions to tackle the global issues caused by the excessive consumption of fossil fuel reserves, which are projected to be depleted within a few decades [1–3]. Catalysts with unique structures and properties that are active and selective for biomass conversion play a pivotal role in biomass utilization. Thus, effective and practical catalysts are desperately needed for efficient biomass transformations from both academic and industrial points of view.

Niobium-based materials are extremely promising catalysts, and are capable of catalyzing a wide variety of reactions, including oxidation, esterification, hydrolysis, dehydration, hydration, hydrodeoxygenation, alkylation, condensation, and photocatalysis [4,5]. Niobium compounds and materials have unique properties, with their catalytic performance being somewhat different from those of their neighbors in the Periodic Table (V, Zr, Mo, Ti, W) [5]. Despite the distinctive behaviors of niobium-based catalysts for a number of reactions, however, very few reviews have been prepared on niobium chemistry except for some earlier instances, before 2003 [4–9]. Moreover, although the use of niobium materials as catalysts for the catalytic conversion of biomass or biomass-derivatives to transportation fuels and fine chemicals has attracted much attention during recent years, no summaries have been prepared covering these aspects of niobium chemistry.

Nanoporous Catalysts for Biomass Conversion, First Edition. Edited by Feng-Shou Xiao and Liang Wang.
© 2018 John Wiley & Sons Ltd. Published 2018 by John Wiley & Sons Ltd.

In this chapter, a brief review is provided of niobium-based catalysts (e.g., niobium phosphates, niobium oxides, niobic acid, niobium-containing composite materials such as niobia-silica, niobia-carbon and layered niobium molybdates) for biomass conversion, demonstrating their structures, catalytic properties, and related mechanisms. Generally, the catalytic behaviors of niobium materials mainly include their redox properties, acidic properties, and photocatalytic properties. Here, three main features are discussed with regards to biomass conversion: (i) their unique acid properties to cleave and/or cooperatively cleave the C–O bond in biomass-derived chemicals; (ii) their mesoporous structures and large surface areas that can be used as supports for metal

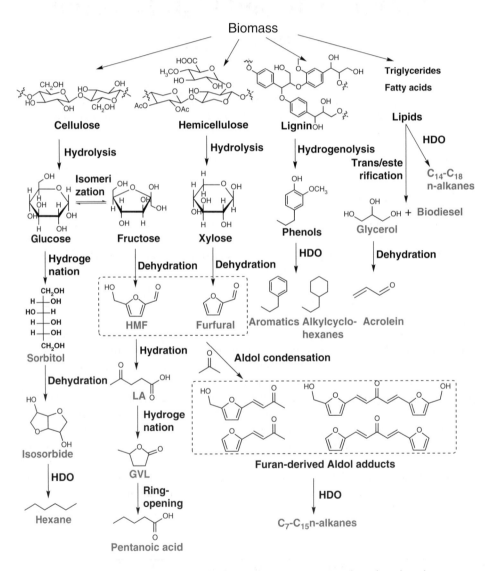

Figure 11.1   *The reactions involved in biomass conversion over niobium-based catalysts.*

loading; and (iii) their synergistic effects with metals to allow their use as bifunctional or multifunctional catalysts for cooperative catalysis (e.g., hydrolysis/hydrogenation, dehydration/hydrogenation, hydrodeoxygenation). The reactions involved in biomass conversion which can be catalyzed by niobium-based catalysts are summarized in Figure 11.1. Although some homogeneous niobium compounds may also have catalytic applications in biomass conversion, this chapter is limited to heterogeneous catalysts with solid niobium materials, taking into consideration the theme of this book. It should be noted that some typical acid-catalyzed reactions such as alkylation, etherification, alcohol dehydration that are not related to biomass conversion are not included in this chapter. For a clear description of recent studies, this chapter is structured according to the different types of reaction involved in biomass conversion.

11.2   Hydrolysis

The hydrolysis of cellulose or hemicellulose to monosaccharides is of central importance for lignocellulosic biomass valorization. Niobium-based solid acids are water-tolerant catalysts with strong acidity, which makes them ideal candidates for hydrolysis reactions in the aqueous phase [10]. Carniti *et al.* [11] studied the hydrolysis of sucrose, maltose and cellobiose at 50–80 °C over different solid-acid catalysts, including three niobium-based solid acids (i.e., niobium phosphate, silica-niobia, and niobic acid) in the aqueous phase. Among the Nb-based solid acids studied, niobium phosphate provided the best catalytic activity towards disaccharides hydrolysis with 60% conversion of sucrose, while only 10% and minimal conversion were observed over Nb_2O_5 and SiO_2-Nb_2O_5, respectively [11].

Niobium phosphate ($NbOPO_4$) is a typical water-tolerant solid acid with its acid strength ($H_0 \leq -8.2$) equivalent to the acid strength of 90% H_2SO_4 [5,10]. Although the acid properties of materials are largely subject to the preparation methods employed, $NbOPO_4$ usually possesses large amounts of both Brønsted acid sites from P-OH and Nb-OH, and Lewis acid sites from the Nb^{5+} cation, which is coordinatively unsaturated [12,13]. The representative structures of these materials are presented in Figure 11.2. The acid characteristics of niobium phosphate make it a perfect candidate for hydrolysis reactions in hot water.

The hydrolysis of polysaccharides to monosaccharides is often accompanied by side reactions such as isomerization, dehydration, or oligomerization of monosaccharides, which are also acid-catalyzed reactions. Therefore, in order to achieve a high selectivity of products, the hydrolysis of cellulose to glucose is usually combined with subsequent purposeful reactions such as hydrogenation to sorbitol [14–16], dehydration to 5-hydroxymethylfurfural (HMF) [17], dehydration/hydration to levulinic acid [18,19], and conversion to ethylene glycol [20]. Niobium phosphate has a high specific surface

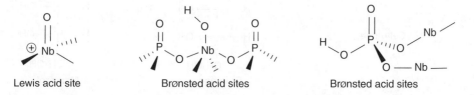

Figure 11.2   *Representative structures of Lewis and Brønsted acid sites in NbOPO₄.*

area and mesoporous structure with a narrow pore size distribution when synthesized by hydrothermal methods with a cationic or anionic surfactant as template [13,21,22]. These features of niobium phosphate enable the material to be an excellent support for metal loading. The resultant metal-loaded niobium phosphate can act as a bifunctional catalyst for the hydrolysis of cellulose and the subsequent hydrogenation [16,20].

A novel Ru-loaded mesoporous $NbOPO_4$ ($Ru/NbOPO_4$) bifunctional catalyst was prepared by Wang *et al.* [16] for the selective production of sorbitol from cellulose via hydrolysis and subsequent dehydration. A high yield (59–69%) of sorbitol with >90% conversion of cellulose was obtained at 170 °C, as a result of cooperative catalysis between the Ru metal sites and acid sites in the $NbOPO_4$ support. In this process, $NbOPO_4$ played a role in the efficient hydrolysis of cellulose to glucose, and Ru nanoparticles catalyzed the subsequent hydrogenation of the *in-situ*-generated glucose to sorbitol, suppressing the side reactions of glucose (see Figure 11.3). By incorporating this process with the subsequent dehydration of sorbitol, isosorbide could be produced from cellulose via a one-pot method [23,24], as be discussed in the following section.

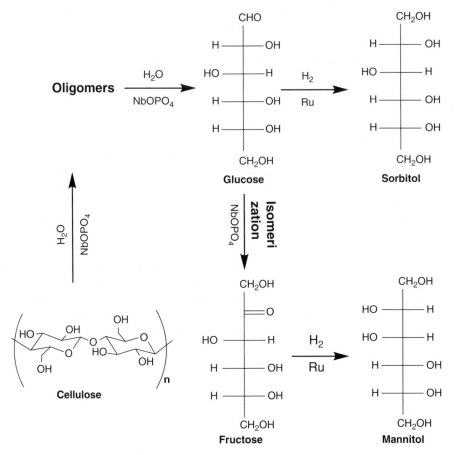

Figure 11.3 *Production of sorbitol from cellulose by successive hydrolysis and hydrogenation over Ru/NbOPO₄ bifunctional catalyst. Reproduced from Ref. [16] with permission from Elsevier.*

11.3 Dehydration

11.3.1 Sorbitol Dehydration

The dehydration product of sorbitol, isosorbide, is a useful platform chemical for the synthesis of renewable biomass-based products as a monomer or a building block for new polymers and functional materials [25]. Both sorbitol and isosorbide have high solubility in water, which makes the aqueous-phase sorbitol dehydration an attractive and necessary step. Niobium-based solid acids are suitable catalysts for aqueous-phase sorbitol dehydration because of their strong acidity and water-tolerant properties. The use of niobium materials for sorbitol dehydration was first reported by Huang *et al.* [26], who used Nb_2O_5 and H_3PO_4-modified Nb_2O_5 as catalysts. As a result, up to 100.0% conversion of sorbitol with 62.5% selectivity for isosorbide were obtained under optimal conditions over the phosphated Nb_2O_5 catalyst. The modification of H_3PO_4 to Nb_2O_5 prevented the crystallization of Nb_2O_5 and retained the relatively large surface area and stable acidity, thereby dramatically enhancing the selectivity to isosorbide [26]. In addition to niobium, tantalum materials (with similar acid properties) were also employed as solid acids for the dehydration of sorbitol [27].

As mentioned above, the feedstock sorbitol can be produced *in situ* from the hydrolysis/hydrogenation of cellulose over metal-loaded solid-acid bifunctional catalysts [16,23,24]. Wang *et al.* [23] reported a two-step sequential process for the one-pot production of isosorbide from cellulose, combining the first hydrolysis/hydrogenation of cellulose to sorbitol with the second dehydration to isosorbide over a bifunctional $Ru/NbOPO_4$ catalyst (see Figure 11.4). The reaction was first conducted at a lower temperature (170 °C) in the presence of H_2 for the hydrolysis/hydrogenation step, after which the temperature was raised 230 °C for the dehydration step. A 32.9% yield of isosorbide was obtained when 5 wt% $Ru/NbOPO_4$ was used as the only catalyst, whereas when 5 wt% $Ru/NbOPO_4$ and $NbOPO_4$ was used as a co-catalyst system the yield of isosorbide could be remarkably enhanced to 56.7% [23]. Very importantly, the as-prepared mesoporous $NbOPO_4$ catalyst has a long-term stability and can last for up to 200 h when dehydrating sorbitol to isosobide

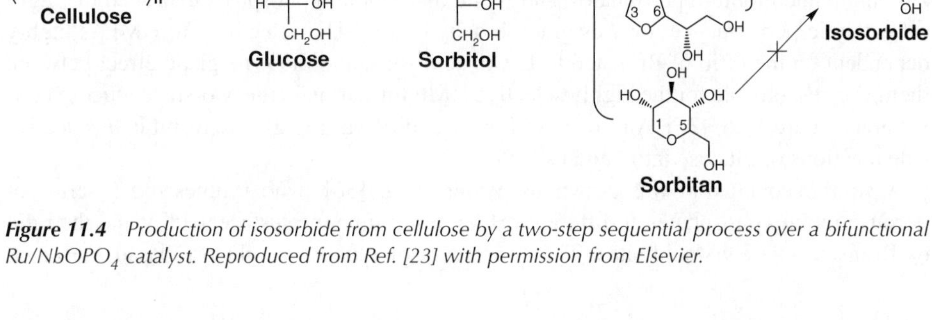

Figure 11.4 *Production of isosorbide from cellulose by a two-step sequential process over a bifunctional $Ru/NbOPO_4$ catalyst. Reproduced from Ref. [23] with permission from Elsevier.*

at 220–230 °C. A similar study was conducted by Li *et al.* [24], with over 50% yield of isosorbide obtained from cellulose over a mesoporous Ru/NbOPO$_4$ with an Ru particle size centered at 5.5 nm.

Layered niobium molybdate (HNbMoO$_6$) was also reported to be selective for the aqueous-phase dehydration of sorbitol to monomolecular-dehydrated intermediate 1,4-sorbitan [28]. The sorbitol can be intercalated into the interlayers of the layered compounds and thereby selectively converted to 1,4-sorbitan because of the steric effect. The results indicated that HNbMoO$_6$, Nb$_2$O$_5$·nH$_2$O, and HZSM-5 were more active than sulfuric acid, and HNbMoO$_6$ exhibited a higher selectivity towards 1,4-sorbitan than Nb$_2$O$_5$·nH$_2$O and HZSM-5 at similar conversions.

11.3.2 Carbohydrate Dehydration

The use of niobium-based catalysts for carbohydrate dehydration to obtain HMF has long been documented [12,29–31]. In 1999, Carlini *et al.* [29] reported a comprehensive study on the dehydration of fructose, sucrose and inulin to HMF over niobium catalysts in water. Soon after, the same group [12] spectroscopically characterized the Brønsted and Lewis acidity of niobic acid and various niobium phosphate catalysts, and employed these catalysts for fructose dehydration in an aqueous medium. No levulinic acid (LA) or formic acid (from the subsequent hydration of HMF and C–C bond cleavage on the aldehyde) were obtained, and >90% selectivity to HMF was achieved at a lower (<30%) conversion [12]. Carniti *et al.* [30] studied the catalytic activity of niobic acid and niobium phosphate for fructose dehydration in a continuous reactor, and observed that niobium phosphate had a superior activity and selectivity towards HMF in comparison to niobic acid.

These earlier studies opened a new chapter for the efficient production of platform chemicals from renewable biomass-derived saccharides over niobic acid and niobium phosphate catalysts. Nb atoms in niobic acid and niobium phosphate are mainly in the form of distorted NbO$_6$ octahedra and NbO$_4$ tetrahedra [32,33]. The unsaturated NbO$_4$ tetrahedra can act as strong Lewis acid sites and still function even in aqueous solution after forming NbO$_4$-H$_2$O adducts [34]. Therefore, niobic acid and niobium phosphate are effective catalysts for HMF production from glucose or glucose-based carbohydrates in aqueous phase [34–36], which requires the isomerization of glucose to fructose catalyzed by Lewis acid sites and the dehydration of fructose to HMF catalyzed by Brønsted or Lewis acid sites. However, over-strong acidity will lead to severe undesired side reactions, such as polymerization into humins and rehydration into LA, with simultaneous formation of a molecule of formic acid (see Figure 11.5).

For example, Nijhuis *et al.* [35] studied the glucose dehydration over aluminum, titanium, zirconium and niobium phosphates, and found that niobium phosphate exhibited the highest catalytic performance towards glucose dehydration. The selectivity to HMF is highly dependent on the ratio of Brønsted to Lewis acid sites, and the synergistic effect between them was responsible for the highly selective HMF formation in the two-stage glucose conversion (Figure 11.6) [35]. An excess of Lewis acidity was reported to result in unselective side reactions of glucose into humans [35].

A similar conclusion was drawn by Wang *et al.* [36], who synthesized a series of porous niobium phosphates at different pH-values to tune the surface acidity and the ratio of Brønsted to Lewis acid sites. The conversion of glucose or glucose-based feedstock

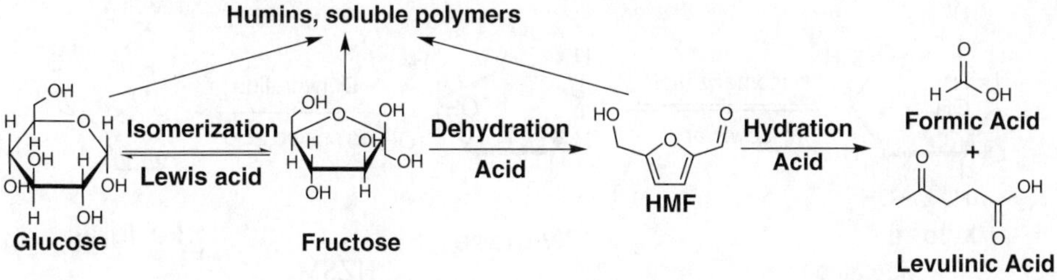

Figure 11.5 *Schematic reaction pathway of glucose dehydration to HMF over Nb-based catalyst containing both Lewis and Brønsted acid sites.*

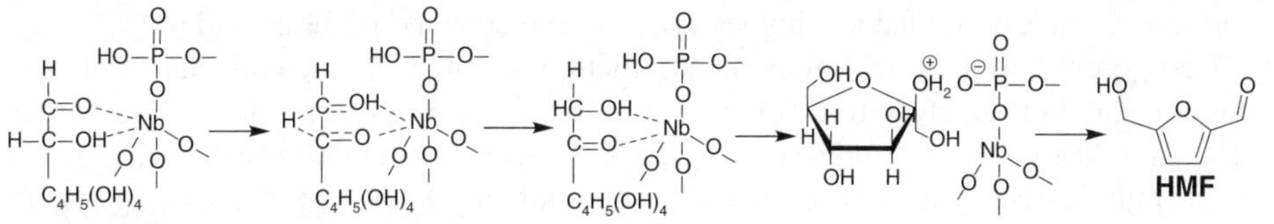

Figure 11.6 *Schematic reaction mechanism for the production of HMF from glucose over NbOPO₄. Reproduced from Ref. [35], with permission from Elsevier.*

and the selectivity to HMF were greatly influenced by the acidic properties and the ratio of Brønsted/Lewis acid sites in the aqueous phase. It was found that the excessive Brønsted acid sites could inhibit the isomerization of glucose to fructose, which is a Lewis acid-promoted process. On the other hand, the excessive Lewis acid had a negative influence on the dehydration of fructose, which would lead to the undesired side reactions into humans [36]. The NbOPO$_4$ prepared at pH 7 had the best catalytic performance, and up to 33.2% yield of HMF was achieved from glucose in aqueous phase under optimal conditions. More importantly, this catalyst exhibited an excellent stability with almost no decrease in activity or selectivity after seven successive runs, and could be fully regenerated [36]. In addition, the use of phosphoric acid- or sulfuric acid-treated niobic acid was also attempted for the dehydration of a wide range of sugar substrates, with medium to high yields of HMF being achieved [37,38].

Niobium-based catalysts were reported to be effective for the dehydration of xylose to furfural [39–43], another pivotal platform chemical in biomass conversion. Wang *et al.* [39] synthesized a mesoporous NbOPO$_4$ containing both Lewis and Brønsted acid sites via a hydrothermal method, and used this for the one-pot efficient conversion of xylose to furfural, including the isomerization of xylose to xylulose and subsequent dehydration of xylulose to furfural (Figure 11.7). Owing to the large surface area (>200 m^2 g^{-1}), the ideal pore size (3.5 nm) and relatively strong Lewis and Brønsted acidity, a very high yield of furfural (49.8%) with 96.5% conversion of xylose was achieved in aqueous phase under the optimal reaction conditions. By employing methyl isobutyl ketone (MIBK)/water (NaCl) as a biphasic solvent, the yield of furfural could be further enhanced to 68.4% [39]. Li *et al.* [40] investigated phosphoric acid-treated niobic acid for the dehydration of xylose to furfural in an organic/aqueous biphasic system, and found the catalyst to exhibit the best performance among those catalysts tested (H-β, H-Y, sulfated zirconia, zirconium phosphate, niobium

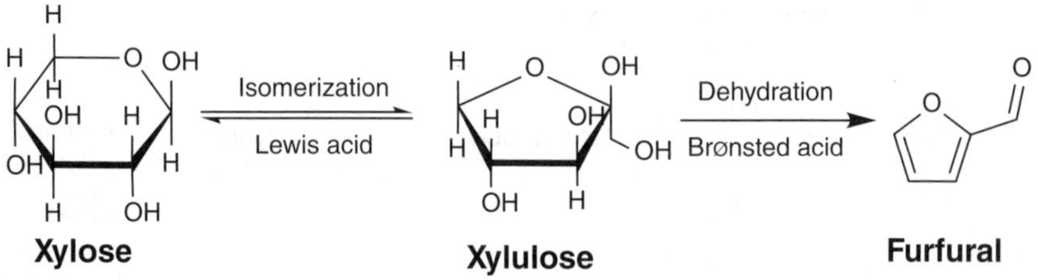

Figure 11.7   *Production of furfural from xylose via isomerization and dehydration over niobium phosphate.*

phosphate), with excellent stability. The synergistic effect between the Brønsted and Lewis acid sites of the catalyst led to a higher xylose conversion and furfural yield [40].

Mesoporous Nb_2O_5 with an amorphous structure prepared via a neutral templating route was efficient for the dehydration of D-xylose to furfural in a biphasic water–toluene system [41]. Over 90% conversion of D-xylose and a 50% yield of furfural were achieved at 170 °C and 90 min. The catalytic performance of mesoporous Nb_2O_5 was superior in comparison to commercial Nb_2O_5, due to a relatively larger amount of acid and the presence of mesopores. Nb_2O_5 nanowires were synthesized via a hydrothermal method, using polyethylenimine as a soft template at pH <7 and employed for the dehydration of D-xylose to furfural [44]. In this case, a 90.3% conversion of D-xylose and a 73.7% yield of furfural were obtained with Nb_2O_5 nanowires prepared at pH 1–2 and calcined at 500 °C as the catalyst.

In addition to niobic acid, niobium phosphate and niobia, other niobium materials (e.g., niobia-silica [31,45], niobia-ceria [46], exfoliated niobate [47,48]) were also employed as catalysts for sugar dehydration. When Valente *et al.* [31] studied microporous AM-11 crystalline niobium silicates for the dehydration of D-xylose in a water–toluene biphase at 140–180 °C, a 90% conversion of xylose and a 50% yield of furfural were achieved after 6 h at 160 °C. Moreover, the catalysts exhibited good reusability, without any loss in catalytic activity between the first and third runs. A niobia phase dispersing in/on a silica matrix was reported to have a superior stability in comparison to niobic acid [45]. Ceria-niobia mixed oxides with different Nb/Ce ratios were prepared via a coprecipitation method, and used as catalysts for the dehydration of fructose to HMF [46]. The mixed oxides contained both Lewis and Brønsted acid sites on the surface, and the conversion of fructose and selectivity to HMF were highly dependent on the content of niobia in the samples.

Layered H^+-exchanged niobates or titanoniobates (HNb_3O_8, $HTiNbO_5$, $H_4Nb_6O_{17}$) are potentially strong solid acids. Exfoliation of the layered niobates in aqueous solution will generate colloidal anionic nanosheets that can be precipitated under acidic conditions [49,50]. These nanosheets possess much larger surface areas that enable the bulky substrates to become accessible to the surface and react with exposed H^+ ions. Valente and coworkers found that $HTiNbO_5$, HTi_2NbO_7, HNb_3O_8, $H_4Nb_6O_{17}$, and $H_2Ti_3O_7$ nanosheets obtained by exfoliation had much higher catalytic activities than the nonexfoliated, acid-exchanged layered materials for the dehydration of xylose into furfural [47]. The exfoliation of layered HNb_3O_8 with the aid of microwave irradiation was reported to be efficient for HMF production in aqueous phase [48]. The rapid in-situ exfoliation of layered HNb_3O_8 leads to a quasi-homogeneous catalytic behavior, which is responsible for the high performance of the catalytic system. The catalytic system is also applicable for HMF production from di- and/or polysaccharides via consecutive hydrolysis/dehydration.

Supported niobium catalysts usually have specific properties and improved hydrother-mal stabilities, and therefore have been studied extensively during recent years for sugar dehydration [43,51,52]. Xiong *et al.* [51] used niobia/carbon composites to catalyze glucose dehydration in a biphasic system, where the functionalized carbon black allowed a selec-tive location of the catalyst either in the organic phase, aqueous phase, or at the interface. MCM-41-supported niobia catalysts were reported to be efficient and highly stable for the dehydration of D-xylose to furfural [43]. The xylose conversion was shown to increase with increasing Nb_2O_5 content, with a 59.9% yield of furfural being achieved at 170 °C after 3 h reaction over 16 wt% Nb_2O_5/MCM-41 catalyst in a water–toluene biphasic system. The catalyst showed excellent stability, with no loss of catalytic activity for at least three runs, and without significant niobium leaching [43]. When other niobium oxide-supported cat-alysts (using silica, γ-Al_2O_3, SBA-15) were also studied and tested for the dehydration of D-xylose to furfural, the best results (84% conversion and 93% selectivity) were achieved with the Nb/SBA-15 catalyst [52].

The catalytic performance of various catalysts for the dehydration of different carbohy-drates, under their respective optimal conditions, are summarized in Table 11.1.

11.3.3 Glycerol Dehydration

Glycerol is a main byproduct of biodiesel synthesis from natural triglycerides (vegetable oil and animal fat) through transesterification with methanol or ethanol. Although the amount of glycerol produced from per tonne of biodiesel is small (ca. 1:10 weight ratio), the huge-amount and increasing production of biodiesel has resulted in an excess of glycerol production, which makes glycerol a attractive building block for the production of value-added chemicals [54–56]. One of the important routes for glycerol valorization is converting it to acrolein, which is currently based on petroleum-derived propylene, *via* a double-dehydration reaction (Figure 11.8) [57–59]. Varies of niobium-based solid acid catalysts, including Nb_2O_5 [60], supported Nb_2O_5 [61–64], $NbOPO_4$ or phosphated Nb_2O_5[65–67], heteropolyacids supported on Nb_2O_5 [68–70], and Nb-Zr mixed oxides [71,72] have been tested for the dehydration of glycerol in either gaseous or liquid phases.

Xu and coworkers [60] studied the gas-phase dehydration of glycerol over Nb_2O_5 with different catalyst textures, crystal structures, and acidities by calcining at different tem-peratures. The study results showed the medium acid sites to be effective for the selective dehydration of glycerol to acrolein. The Nb_2O_5 amorphous catalyst prepared by calcina-tion at 400 °C gave the highest amount of medium acid sites, and thus showed the highest catalytic performance towards acrolein formation. The other samples which had a higher fraction of either stronger or weaker acid sites were less effective and selective for glycerol dehydration [60].

Silica- [61], alumina- [62], titania- [62], zirconia- [63], and zirconium-doped silica- [64] supported niobia were also employed for the gas-phase dehydration of glycerol to acrolein. The acid strength can be adjusted by varying the niobia content and catalyst calcination temperature [61]. The results showed the glycerol conversion and acrolein selectivity to depend heavily on the Brønsted acidity and its acid strength [61–63]. Supported tungsten

Table 11.1   *Summary of the catalytic performance of various catalysts for the dehydration of different carbohydrates under their respective optimal conditions.*

Catalyst	Substrate	Reaction conditions	Conv. (%)	Sel. (%)	Reference
P-niobic acid [a]	Fructose	6 wt%, 100 °C, 0.5 h, water	28.5	97.8	[29]
P-niobic acid	Glucose	1 wt%, 120 °C, 3 h, water	92	52.1	[34]
P-niobic acid	Fructose	6 wt%, 160 °C, 50 min, water/2-butanol	90	99	[37]
P-niobic acid	Inulin	6 wt%, 160 °C, 140 min, water/2-butanol	86	63	[37]
S-niobic acid [a]	Fructose	9 wt%, 120 °C, 5 h, DMSO	–	88.1 [b]	[38]
$NbOPO_4$	Fructose	6 wt%, 100 °C, 1 h, water	75.8	97.8	[29]
$NbOPO_4$	Inulin	6 wt%, 100 C, 1.5 h, water	76.3	72.0	[29]
$NbOPO_4$	Fructose	5 wt%, 90–110 °C, 2–6 bar, water, fixed-bed	70–80	31–33	[30]
$NbOPO_4$, pH = 2	Fructose	8 wt%, 130 °C, 0.5 h, water	57.6	78.2	[22]
$NbOPO_4$, pH = 7	Fructose	8 wt%, 130 °C, 0.5 h, water	67.7	49.6	[22]
$NbOPO_4$, pH = 10	Fructose	8 wt%, 130 °C, 0.5 h, water	22.3	9.8	[22]
$NbOPO_4$, pH = 7	Glucose	5 wt%, 140 °C, 60 min, water	68.1	49.3	[36]
$NbOPO_4$, pH = 7	Glucose	5 wt%, 140 °C, 60 min, methyl isobutyl ketone/water	–	39.3 [b]	[36]
$NbOPO_4$	Xylose	5 wt%, 160 °C, methyl isobutyl ketone /water	99.0	68.4	[39]
$NbOPO_4$	Xylose	10 wt%, 210 °C, 1 h, water/toluene	96	46.6	[40]
Nb_2O_5	Fructose	2 wt%, 120 °C, 2 h, DMSO	100	86.2	[53]
Nb_2O_5	Xylose	10 wt%, 170 °C, 1.5 h, water/toluene	90	55.6	[41]
Nb_2O_5	Xylose	0.5 wt%, 120 °C, 2 h, DMSO	90.3	81.6	[44]
E-HNb_3O_8 [c]	Fructose	10 wt%, 155 °C, 18 min, water, microwave-assisted	85.1	65.7	[48]
E-$HTiNbO_5$-MgO [c]	Xylose	10 wt%, 160 °C, 4 h, water/toluene,	92	60	[47]
Nb_2O_5/C	Glucose	5 wt%, 170 °C, 2 h, water/sec-butyl phenol	78	26	[51]
AM-11 Nb-Si	Xylose	10 wt%, 160 °C, 6 h, water/toluene,	90	50	[31]
Nb/MCM-41	Xylose	10 wt%, 170 °C, 3 h, water/toluene	84	73	[43]
Nb/SBA-15	Xylose	2 wt%, 160 °C, 24 h, water/toluene	84	93	[52]

[a] 'P-' or 'S-' represents phosphoric acid or sulfuric acid-treated.
[b] The yield of the product.
[c] 'E-' represents exfoliated.
DMSO, dimethylsulfoxide.

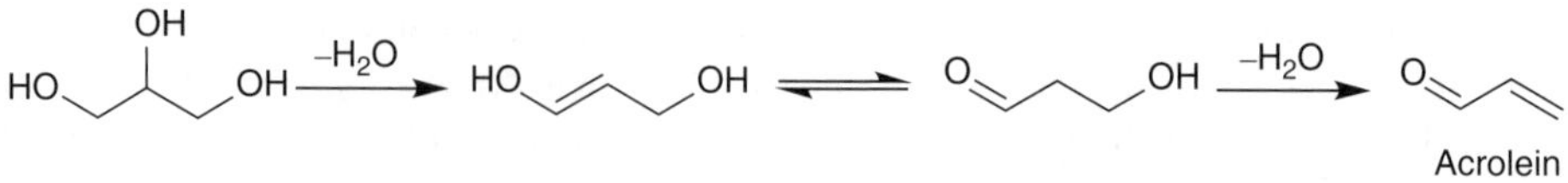

Figure 11.8   *Production of acrolein from glycerol via double dehydration.*

oxides were slightly better than supported niobia, with approximately 80% selectivity for acrolein at high glycerol conversions [62,63]. Catalyst deactivation by coking occurred with time on stream over supported niobia catalysts, though the catalysts could be regenerated by simple oxidative treatment [61–64]; alternatively, the stability could be improved by adding oxygen to the feed in air [62,63].

In view of the high dependence of the glycerol conversion and acrolein selectivity on the Brønsted acidity, the addition of phosphorus into niobium catalysts (niobium phosphates) would greatly enhance the yield of acrolein, since the catalysts possess a large amount of P-OH with medium acid strength. The mesoporous siliconiobium phosphate (NbPSi) composite is a promising solid acid for glycerol dehydration, as it contains almost pure Brønsted acid sites, such that its acidity is weaker than that of niobium phosphate and HZSM-5 zeolite [65]. The NbPSi catalyst exhibited a high catalytic performance for the dehydration of glycerol and with excellent stability that was attributed to the nearly pure Brønsted acidity and large mesopores, which suppressed any side reactions and reduced the pore blocking that could lead to coke formation [65]. Loading H_3PO_4 onto the Nb_2O_5 support significantly increased the amount of acid sites and strength of the surface acidity such that the activity for glycerol dehydration was largely improved, although the stability of H_3PO_4/Nb_2O_5 was less effective due to severe leaching of H_3PO_4 during the reaction [66].

Chary *et al.* [67] found that amorphous niobium phosphate calcined at 550 °C contains higher fractions of moderate acid sites, with the majority being Brønsted acid sites, which is responsible for the higher selectivity for glycerol dehydration to acrolein. A high selectivity to acrolein on the Brønsted acid sites could be attributed to a selective activation of glycerol through the secondary –OH group and the kinetically favored subsequent consecutive steps (see Figure 11.9). The presence of strong Lewis acid sites will activate the primary –OH and lead to hydroxyacetone (acetol) formation, which can be detected in large amounts (>9%) at lower reaction temperatures (270 °C) and would decompose to acetaldehyde and acetic acid at higher temperatures [67].

Heteropolyacids supported on Nb_2O_5 are very efficient and selective catalysts for glycerol dehydration to acrolein owing to their large Brønsted acid density. Wang and coworkers investigated a $Cs_{2.5}H_{0.5}PW_{12}O_{40}$ (CsPW) -supported Nb_2O_5 catalyst for the gas-phase glycerol dehydration [68]. The CsPW/Nb_2O_5 catalyst with 20 wt% loading calcined at 500 °C gave up to 96% conversion of glycerol and 80% selectivity for acrolein. The co-feeding of oxygen at an appropriate ratio significantly enhanced the selectivity

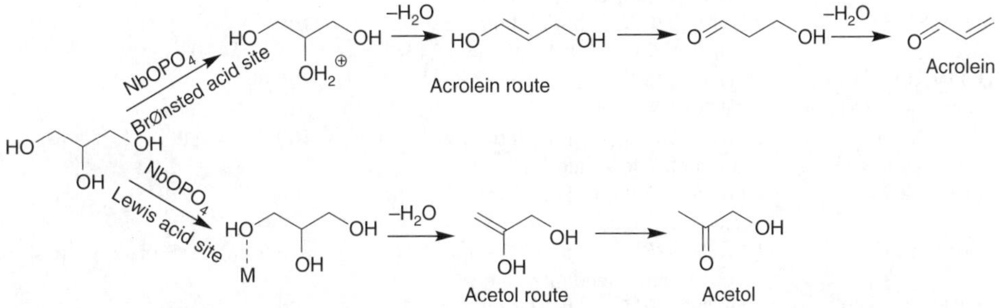

Figure 11.9 *Schematic reaction pathways in the dehydration of glycerol over NbOPO₄. Reproduced from Ref. [67], with permission from Wiley.*

of the acrolein dehydration route by suppressing the oligomerization of glycerol [73]. Other heteropolyacid (e.g., phosphotungstic acid) -supported Nb_2O_5 catalysts were also investigated for glycerol dehydration [70].

The combination of the double-dehydration of glycerol to acrolein with the oxidation of acrolein to acrylic acid represents a very attractive approach to glycerol utilization. A two-bed system involving the dehydration catalysts of $CsPW/Nb_2O_5$ and oxidation

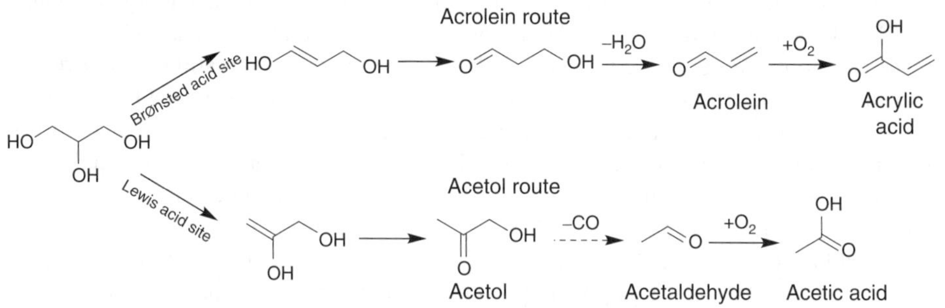

Figure 11.10 *Schematic reaction pathways for the conversion of glycerol in a two-bed system combining $CsPW/Nb_2O_5$ and V-Mo/SiC. Reproduced from Ref. [69], with permission from the American Chemical Society.*

Table 11.2 *The catalytic performance of various Nb catalysts for the dehydration of glycerol to acrolein under their respective optimal conditions.*

Catalyst	Reaction conditions	Conv. (%)	Sel. (%)	Reference
Nb_2O_5	36.2 wt%, 315 °C, 1 atm, 30 ml min^{-1} N_2, GHSV = 80 h^{-1}	88	51	[60]
NbPSi-0.5	10 wt%, 250 °C, 1 atm, 30 ml min^{-1} N_2, GHSV = 14940 ml g^{-1} h^{-1},	100	76.3	[65]
$NbOPO_4$	20 wt%, 300 °C, 1 atm, 30 ml min^{-1} N_2, WHSV = 2.6 h^{-1}	90	86	[67]
$ZrNbO_x$	20 wt%, 300 °C, 1 atm, 75 ml min^{-1} N_2, GHSV = 1930 h^{-1}	100	72	[71]
50 wt%PO_4/Nb_2O_5	Batch reactor, 10 wt%, 240 °C, autogenous pressure, 2 h	68.1	72	[66]
20%$CsPW/Nb_2O_5$	20 wt%, 300 °C, 1 atm, 18 ml min^{-1} N_2:O_2 = 5:1, GHSV = 97.4 h^{-1}	96	80	[68]
$CsPW/Nb_2O_5$	20 wt%, 320 °C, 1 atm, 18 ml min^{-1} N_2:O_2 = 5:1, WHSV = 0.24 h^{-1}	100	76.5	[73]
$H_3PW_{12}O_{40}/Nb_2O_5$	30 wt%, 325 °C, 1 atm, 10 ml min^{-1} N_2, GHSV = 420 h^{-1}	99.8	92	[70]
$Nb_2O_5/WTiO_x$	20 wt%, 305 °C, 1 atm, 15 ml min^{-1} Ar, 3.6 ml h^{-1} feed rate	100	80	[62]
Nb_2O_5/ZrO_2	20 wt%, 305 °C, 1 atm, 15 ml min^{-1} Ar, 3.6 ml h^{-1} feed rate	100	75	[63]
$Nb_2O_5/SiZrO_x$	10 wt%, 325 °C, 1 atm, 30 ml min^{-1} N_2, 0.1 ml min^{-1} feed rate	76.9	45.0	[64]
20%Nb_2O_5/SiO_2	30 wt%, 320 °C, 1 atm, 60 ml min^{-1} N_2, WHSV = 80 h^{-1}	100	67	[61]

GHSV = gas hourly space velocity. WHSV = weight hourly space velocity.

catalyst of V-Mo/SiC was reported for the conversion of glycerol to acrylic acid [69]. Two competitive routes were noted in this reaction: (i) the Brønsted acid sites on the dehydration catalysts catalyze the dehydration of glycerol to acrolein, followed by the consecutive oxidation to acrylic acid; and (ii) the Lewis acid sites catalyzed the dehydration of glycerol to the byproducts (acetol), which was then oxidized to acetic acid (Figure 11.10).

The catalytic performances of various Nb catalysts for the dehydration of glycerol to acrolein under their respective optimal conditions are summarized in Table 11.2.

11.4 HMF Hydration to Levulinic Acid

During the process of HMF production from the acid-catalyzed dehydration of carbohydrates, the rehydration of HMF to form levulinic acid (LA) always exists (Figure 11.11), which is also an acid-catalyzed reaction. LA is a water-soluble, organic compound with a ketone and carboxylic group, giving it a wide range of functionality and reactivity [74]. Moreover, LA is much more stable than HMF towards thermal treatment or light exposure, which makes it easier for the compound's separation, storage, and transportation. Hence, the one-pot production of LA from cellulose or raw biomass materials is particularly attractive, via the hydrolysis–dehydration–rehydration route catalyzed by water-tolerant, solid-acid catalysts.

Several niobium-based solid-acid catalysts, including niobium oxide [19], niobium phosphate [18,19,75,76] and Nb/Al mixed oxide [77] have been employed for the direct conversion of cellulose or raw biomass materials to LA, and/or further upgrading to γ- valerolactone (GVL) via subsequent hydrogenation. Galletti [19] has investigated the acid hydrothermal conversion of giant reed to LA and subsequent hydrogenation to GVL by a bifunctional catalytic system based on Ru/C and niobium oxide or niobium phosphate. A high yield of 16.6 wt% GVL was obtained from the dry raw biomass, with an almost quantitative conversion of the intermediate LA [19].

Aluminum-modified mesoporous niobium phosphate (NbAlPO) was synthesized and used for the direct catalytic conversion of cellulose to LA via a sequential hydrolysis of cellulose to glucose, isomerization, and dehydration to HMF, and rehydration of HMF to LA. The obtained LA was then hydrogenated to GVL over Ru/C in the presence of H_2 (Figure 11.12) [18]. An increase in the amount and the strength of Lewis acid sites was found to significantly enhance the yield of LA, as confirmed by control experiments in which lanthanum trifluoroacetate was added into a catalytic system containing HCl, TfOH, or niobium phosphate to reach a different Lewis/Brønsted acid ratio. The doping of aluminum significantly increased the amount of the strong Lewis and Brønsted acid sites (especially of the strong Lewis acid sites), leading to a remarkable increase in LA yield from cellulose [18]. While for the alcoholysis of cellulose into methyl levulinate (LA ester)

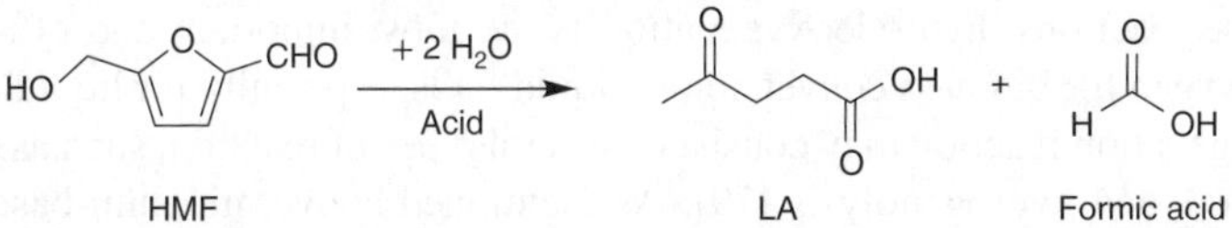

Figure 11.11 *The rehydration of HMF to levulinic and formic acid catalyzed by acid.*

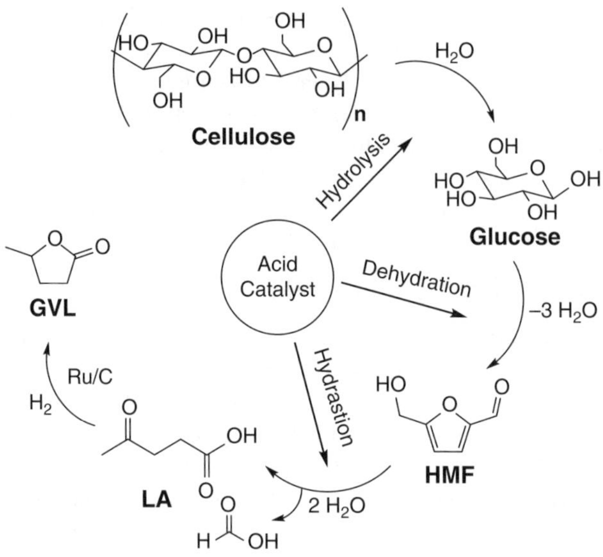

Figure 11.12   *Schematic reaction process of cellulose conversion to GVL via sequential hydrolysis, dehydration, rehydration, and hydrogenation. Reproduced from Ref. [18], with permission from The Royal Society of Chemistry.*

in 95% methanol media, which involves the rehydration of an HMF ether (5-alkoxymethyl furfural) to the LA ester, Brønsted acid sites were the active sites towards the formation of methyl levulinate, and a high yield of methyl levulinate (56%) was obtained over catalyst with a high Brønsted/Lewis acid ratio, niobium phosphate synthesized at pH = 1 [75].

The use of niobium phosphate was also investigated for the hydrothermal conversion of inulin and wheat straw to LA [76], and yields of about 20 wt% and 10 wt% LA were achieved from these feedstocks, respectively. The use of microwave irradiation allowed a significant time saving, while the yield of LA was enhanced from 20 wt% to about 30 wt% for the conversion of inulin. A series of Nb/Al oxides that was prepared by low-cost co-precipitation proved to be efficient for the production of LA from monosaccharides such as fructose, or from kiwifruit waste residue containing monosaccharides. Such efficiency was based on the high acidity and large surface area of the Nb/Al oxide catalysts [77].

11.5   Hydrodeoxygenation

One structural feature of biomass feedstocks is their high oxygen content, which is over 40 mol% in common lignocellulosic feedstocks [78]. The ultimate goal of converting biomass resources to liquid fuels and chemicals is the removal of oxygen through various reactions. Among these reactions, hydrodeoxygenation is the most important and efficient way of removing oxygen for biomass conversion to liquid fuels, especially for liquid alkanes. The hydrodeoxygenation reaction may consist of several types of reaction, such as dehydration, hydrogenation, and hydrogenolysis [79]. As mentioned above, niobium-based solid-acid catalysts are efficient catalysts for dehydration and effective supports for metal loadings that can then be used for hydrogenation and hydrogenolysis reactions. Thus, metal-loaded, niobium solid-acid catalysts are predicted to be efficient for hydrodeoxygenation reactions.

The use of a metal-loaded, niobium-based catalyst for total hydrodeoxygenation was first reported by Dumesic *et al.* in 2008 [80]. These authors used Pt/NbOPO$_4$ to catalyze the hydrodeoxygenation of aldol adducts of furanic compounds to liquid alkanes in a fixed-bed reactor, and found the catalytic performance of Pt/NbOPO$_4$ to be superior in comparison to that of Pt/SiO$_2$-Al$_2$O$_3$ and Pt/zeolites. The group then compared the catalytic performances of niobium-based solid-acid catalysts with SiO$_2$-Al$_2$O$_3$ based catalysts for the aqueous-phase hydrodeoxygenation of sorbitol to alkanes at 530K and 54 bar [81]. The study results showed that niobium-based catalysts had a superior catalytic activity in comparison to the SiO$_2$-Al$_2$O$_3$ catalyst. The authors speculated that the high catalytic activity of niobium-based catalysts for hydrodeoxygenation reactions was not relevant to the density of acid sites but rather was attributed to the coordination environment of the niobium acid center, although a detailed mechanism or verification was not given.

Wang *et al.* [82] carried out the dehydration/hydrogenation of C$_8$-ols (obtained from the selective ring-opening/hydrogenation of 4-(2-furyl)-3-buten-2-one) to octane over Pt/NbOPO$_4$ in a fixed-bed reactor. They found that Pt/NbOPO$_4$ is able not only to convert octanediols to octane via dehydration/hydrogenation, but can also convert 4-(2-tetrahydrofuryl)-butan-2-ol into octane via dehydration, ring-opening and hydrogenation under very mild conditions (165–175 °C, 25 bar) (Figure 11.13). The suspicion was that the ring-opening catalytic activity of Pt/NbOPO$_4$ may be related to its Lewis acid sites, in consideration of the fact that the ring-opening cannot take place over a Pt/ZSM-5 catalyst containing strong Brønsted acid sites.

The same research group then prepared catalysts with different active metals (Pt, Pd, Ru) and supports (NbOPO$_4$, Nb$_2$O$_5$, Al$_2$O$_3$, H-ZSM-5), and compared their catalytic activity

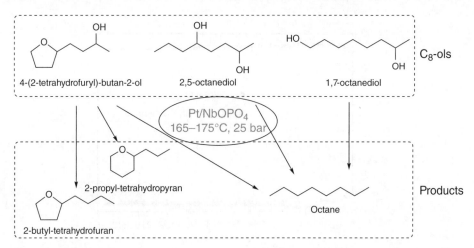

Figure 11.13 *Schematic reactions of dehydration/hydrogenation of C$_8$-ols over Pt/NbOPO$_4$ catalyst. Reproduced from Ref. [82], with permission from Wiley.*

for the direct hydrodeoxygenation of aldol adducts such as furfural-acetone [83]. Through control experiments, the NbO$_x$ species in NbOPO$_4$ was found to have a significant promotional effect for C–O bond cleavage, which was then confirmed by density functional theory (DFT) calculations. Thus, the Pd/NbOPO$_4$ catalyst actually acts as a multifunctional catalyst that plays three roles: (i) the active metal (Pd) plays a role in hydrogen dissociation and hydrogenation; (ii) the NbO$_x$ species plays a role in C–O bond activation, especially of the tetrahydrofuran ring; and (iii) the acid sites in NbOPO$_4$ catalyze the dehydration. Hence, the biomass-derived furan adducts can be directly hydrodeoxygenated to liquid alkanes at very mild reaction conditions (170 °C, 2 MPa), with almost no catalyst deactivation throughout a 256-h time-on-stream test (Figure 11.14). The catalyst was also efficient for the direct hydrodeoxygenation of vegetable oils to diesel-range alkanes [84]. However, the same catalyst (Pd/NbOPO$_4$) showed a completely different selectivity from furfural-acetone hydrodeoxygenation in the aqueous phase (to 1-octanol rather than octane), due to the high hydrophilicity of the catalyst, which allows a spontaneous separation of the catalyst with water-insoluble products (such as 1-octanol) [85].

The unique feature of NbO$_x$ for C–O bond activation makes the hydrodeoxygenation (C–O bond cleavage) a facile process over niobium-based catalysts. Wang and coworkers [86] employed Pt/NbOPO$_4$ as a multifunctional catalyst for the direct total hydrodeoxygenation of raw wood biomass (containing unseparated cellulose, hemicellulose and lignin component), a much more inert and robust chemical than furanic aldol adducts, to their respective liquid alkanes. Owing to the outstanding activity of the catalyst, not only cellulose and hemicellulose components can be directly converted into hexane and pentane, but also can the more complex lignin fractions be totally hydrodeoxygenated into alkylcyclohexanes (Figure 11.15). In addition to the DFT calculations, these authors used a new technique called Inelastic Neutron Scattering (INS) to directly visualize the interaction between a model compound of the reaction (THF) and the catalyst surface, and to study the molecular details of binding, ring-opening, and C–O cleavage of THF. A combination

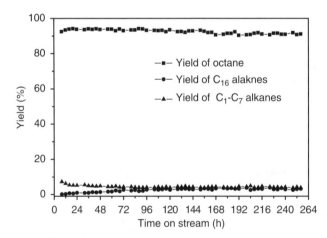

Figure 11.14 *Catalytic performance of direct hydrodeoxygenation of furfural-acetone over Pd/NbOPO$_4$ in a fixed-bed reactor. Reprinted from Ref. [83], with permission from Wiley.*

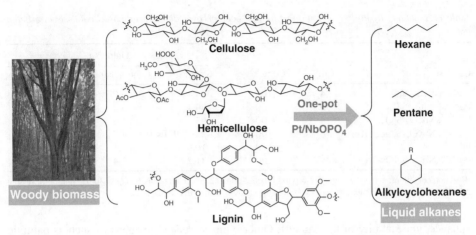

Figure 11.15   *One-pot conversion of raw woody biomass to liquid alkanes via direct hydrodeoxygenation over multifunctional Pt/NbOPO₄. Reprinted from Ref. [86], with permission from Nature.*

of the *in-situ* INS study and theoretical modeling with experimental data showed that the high efficiency for the hydrodeoxygenation reaction originated from the synergistic effect between Pt, NbO_x species, and acidic sites of the catalyst.

Wang and colleagues also studied the total hydrodeoxygenation of sorbitol to hexane over Pt/NbOPO₄ in aqueous phase [87]. In these studies, the Pt/NbOPO₄ catalyst showed the best performance among a series of catalysts towards sorbitol hydrodeoxygenation in aqueous phase, with 60% total yield of liquid alkanes (C_5 and C_6 alkanes) at optimal reaction conditions. The same group also studied the reaction pathways and kinetics, and found isosorbide to be a very important intermediate during this process, and was generated from the dehydration of sorbitol and underwent hydrogenolysis to hexane [87]. The activation energies of sorbitol dehydration and the subsequent hydrogenolysis of isosorbide over Pt/NbOPO₄ catalyst (72.7 kJ mol^{-1} and 147.6 kJ mol^{-1}, respectively) were much lower than those over Pt/ZrP and Pt/H-Beta catalysts, owing to the significant promotional effect of NbO_x for C–O bond cleavage. More significantly, the aqueous phase hydrodeoxygenation of sorbitol was conducted in a fixed-bed reactor, and the Pt/NbOPO₄ catalyst showed excellent stability with almost no decrease in catalytic performance through a 100-h time-on-stream test [87].

The promotional effect of NbO_x on C–O bond cleavage was further verified by using Pd-loaded Nb_2O_5-modified SiO_2 as catalyst for the hydrodeoxygenation of biomass-derived furfural-acetone to octane under mild conditions of 170 °C, 2.5 MPa [88]. With Pd/SiO₂ as catalyst, the catalytic activity was low and only a 16% yield of octane was obtained. However, after adding 10% Nb_2O_5 to SiO_2 (by the sol–gel method) as a support, Pd/10%Nb_2O_5/SiO₂ showed a much higher activity and an octane yield of up to 95.5% was achieved from furfural-acetone, which was comparable with that over Pd/NbOPO₄ [83]. The direct impregnation of Nb_2O_5 in SiO_2 also enhanced the yield of octane from 16.0% to 31.2% (Table 11.3), further confirming that NbO_x species could promote the hydrodeoxygenation process. The significant activation of NbO_x species on the C–O bond and the mild reaction conditions restrained the C–C bond cleavage and

Table 11.3   *Direct hydrodeoxygenation of furfural-acetone over different Pd-based catalysts in a batch reactor.*[a]

Catalyst	T (h)	Conv. (%)	Yield (%)		
			Octane	Other alkanes	Oxygenates
Pd/5%Nb$_2$O$_5$/SiO$_2$	24	100	84.1	6.0	9.9
Pd/5%Nb$_2$O$_5$/SiO$_2$	30	100	90.1	6.3	3.6
Pd/10%Nb$_2$O$_5$/SiO$_2$	24	100	95.5	3.6	0.9
Pd/20%Nb$_2$O$_5$/SiO$_2$	16	100	95.3	4.7	0
Pd/SiO$_2$	24	100	16.0	1.1	82.9
Pd/10%Nb$_2$O$_5$/SiO$_2$-im	24	100	31.2	1.4	67.4

[a]Reaction conditions: catalyst (0.2 g); 4-(2-furyl)-3-buten-2-one (0.2 g); cyclohexane (6.46 g); 170 °C; initial H$_2$ pressure 2.5 MPa. Pd loading in all catalysts was 4 wt% and the catalysts were reduced in situ.

enhanced the stability of the catalyst. Other biomass-related oxygenates such as palmitic acid, tristearin and diphenyl ether could also be totally hydrodeoxygenated to alkanes over Pd/10%Nb$_2$O$_5$/SiO$_2$ with high yields (>94%) and minimal C–C cleavage [88].

Suh *et al.* [89] used Pt-supported silica-niobia aerogels for the hydrodeoxygenation of 1-propanol in aqueous phase at 230 °C and 35 bar. The Pt supported on amorphous silica-niobia aerogel catalysts exhibited superior activity in comparison to Pt supported on the alumina-niobia aerogel, or crystalline niobic acid. The Nb$_2$O$_5$-SiO$_2$ aerogels were found to maintain their amorphous and porous structures, while the other supports suffered from structural crystallization after hydrodeoxygenation in the aqueous phase. Nb$_2$O$_5$-supported Pt nanoparticles were reported to be efficient for the selective hydrodeoxygenation of various fatty acids (lauric, capric, palmitic, myristic, oleic, and stearic acids) to long-chain alkanes [90]. The authors concluded that the activity of hydrodeoxygenation depends on the active metal, the support materials, and the pre-reduction temperature. Pt supported on Nb$_2$O$_5$ reduced at 300 °C gave the highest activity towards fatty acids hydrodeoxygenation, with high alkane yields (88–100%). The conversion of tristearin was also attempted over Pt/Nb$_2$O$_5$, which gave a 93% yield of n-octadecane.

Layered niobium compounds (LiNbMoO$_6$, HNbMoO$_6$, LiNbWO$_6$) combined with Ru/C and phosphoric acid were also employed for the total hydrodeoxygenation of cellulose to hexane in aqueous [91]. In this catalytic system, layered niobium compounds played an important role in the selective production of hexane from cellulose by suppressing the formation of isosorbide, which is difficult to undergo further hydrodeoxygenation, and attributes to the steric restrictions of sorbitol dehydration within the interlayers. Particularly, by using layered LiNbMoO$_6$ as a co-catalyst, a 72% yield of hexane was produced directly from cellulose hydrodeoxygenation [91]. Shortly thereafter, the same group used a layered Ta catalyst (LiTaMoO$_6$) as a co-catalyst, combined with Ru/C and phosphoric acid, to produce gasoline alkanes and chemicals from one-pot hydrodeoxygenation of raw lignocelluloses [92].

The catalytic results of the hydrodeoxygenation of various biomass-derived feedstocks over Nb-based bifunctional catalysts are summarized in Table 11.4.

Table 11.4 Summary of hydrodeoxygenation of various biomass-derived oxygenates over Nb-based bifunctional catalysts.

Catalyst	Feedstock	Products	Reaction conditions	Results	Reference
4 wt% Pt/NbPO$_5$	Hydrogenated C$_8$–C$_{15}$ adducts	n-alkanes Oxygenates	31–94 wt% aq. 255–275 °C, 54–60 bar, 0.15–0.79 h^{-1}	Aq. layer: 3.0–3.5%, org. layer: 82–87%, gas species: 10–15%	[80]
4 wt% Pt/NbOPO$_4$	Sorbitol	C$_1$–C$_6$ oxygenates, C$_1$–C$_6$ alkanes	25 wt% aq. 255 °C, 54 bar, 0.37 h^{-1}	63% yield alkanes, 7% monofunctional species	[81]
4 wt% Pt/niobic acid	Sorbitol	C$_1$–C$_6$ oxygenates, C$_1$–C$_6$ alkanes	25 wt% aq., 255 °C, 54 bar, 0.4 h^{-1}	50% alkanes, 18% monofunctional species	[81]
4 wt% Pt/P-niobic acid	Sorbitol	C$_1$–C$_6$ oxygenates, C$_1$–C$_6$ alkanes	25 wt% aq., 255 °C, 54 bar, 0.5 h^{-1}	61% alkanes, 21% monofunctional species	[81]
5 wt% Pt/NbOPO$_4$	C$_8$-alcohols	Octane	5 wt% in THF, 175 °C, 25 bar, 0.6 h^{-1}	100% conv. 84.2% yield octane	[82]
1 mol% Pt/Nb$_2$O$_5$	Fatty acids	Linear alkanes	solvent-free, 8 bar, 180–250 °C, 24 h	100% conv. 88–99% yield n-alkanes	[90]
1 mol% Pt/Nb$_2$O$_5$	Tristearin	n-Octadecane	solvent-free, 260 °C, 50 bar, 24 h	100% conv. 93% yield	[90]
5 wt% Pd/NbOPO$_4$	C$_8$–C$_{15}$ adducts	C$_8$–C$_{15}$ alkanes	3 wt% in cyclohexane, 170 °C, 20 bar, 24 h	100% conv. >90% yield	[83]
4 wt% Pd/Nb$_2$O$_5$/SiO$_2$	C$_8$ adduct	Octane	3 wt% in cyclohexane, 170 °C, 25 bar, 24 h	100% conv. 95.5% yield	[88]
4 wt% Pd/Nb$_2$O$_5$/SiO$_2$	Palmitic acid	n-C$_{16}$ alkane	3 wt% in cyclohexane, 170 °C, 25 bar, 24 h	100% conv. 95.3% yield	[88]
4 wt% Pd/Nb$_2$O$_5$/SiO$_2$	Tristearin	n-C$_{18}$ alkane	3 wt% in cyclohexane, 170 °C, 25 bar, 24 h	100% conv. 94.2% yield	[88]
4 wt% Pd/Nb$_2$O$_5$/SiO$_2$	Diphenyl ether	Cyclohexane	3 wt% in dodecane, 170 °C, 25 bar, 24 h	100% conv. 98.2% yield	[88]
Pt/NbOPO$_4$ (0.5 mol% Pt)	Furanyl adducts	C$_8$–C$_{19}$ alkanes	0.125 M in octane, 250 °C, 34 bar, 4 h	100% conv. 75–78% yield	[93]
4 wt% Pt/NbOPO$_4$	Sorbitol	C$_5$–C$_6$ alkanes	5 wt% aq., 250 °C, 40 bar, 10 h	100% conv. 60% total yield	[87]
5 wt% Pd/NbOPO$_4$	Trioleate, Stearic acid	C$_{17}$–C$_{18}$ alkanes	2 wt% in cyclohexane, 180 °C, 30 bar H$_2$, 24 h	100% conv. >97% total yield	[84]
5 wt% Pd/NbOPO$_4$	Palm oil, Soybean oil	C$_{14}$–C$_{18}$ alkanes	2 wt% in cyclohexane, 180 °C, 30 bar H$_2$, 24 h	100% conv. 84–85 wt% yield	[84]
5 wt% Pt/NbOPO$_4$	Raw woody biomass	C$_5$–C$_6$ alkanes, alkylcyclohexanes	3 wt% in cyclohexane, 190 °C, 50 bar, 20 h	Up to 28.1 wt% total yield of liquid alkanes	[86]

11.6 C–C Coupling Reactions

C–C coupling reactions (aldol condensation, ketonization, alkylation) are very important reactions in biomass conversion, particularly in the field of high-grade liquid fuel synthesis from biomass-derived carbohydrates or platform chemicals, which have only six carbons in maximum. The strong acidity of Nb catalysts meets the requirement for catalyzing many of the C–C coupling reactions, though only a few cases have been recorded.

A Pd/Nb_2O_5 bifunctional catalyst was reported to be effective for the one-pot synthesis of methyl isobutyl ketone (MIBK) from acetone via the self-aldol condensation of acetone and subsequent selective hydrogenation of the self-aldol adduct [94]. Montes *et al.* [95] prepared a niobium oxide solid acid with a large surface area and tested the activity for the gas-phase aldol condensation of acetone, with the main products being mesitylene and acetic acid.

Pt/Nb_2O_5 was reported as an effective bifunctional catalyst for the aqueous phase processing of lactic acid to produce C_4–C_7 ketones, involving *in-situ* ketonization and aldol condensation reactions [96,97]. Reaction pathways studies showed that acetaldehyde and propanoic acid were first formed over Pt/Nb_2O_5, as the key intermediates in the overall process. The strong acidity and water-tolerant property of Nb_2O_5 allowed the propanoic acid to undergo ketonization reactions to form pentanones, and allowed the acetaldehyde to undergo successive aldol condensations to form C_4–C_7 ketones finally. In this process, Pt was used for hydrogenation and the niobia support played a key role for dehydration and C–C coupling reactions [96,97].

Alkylation is a common strategy for C–C coupling to increase the carbon chain and control the final molecular weight of liquid hydrocarbons, such as the alkylation/hydroxyalkylation of 2-methylfuran (2-MF) with furfural to produce C_{15} fuel precursors [98,99]. The use of Nb-based solid acids for alkylation reactions has been reported in many cases [100–102], though no reports have been made on the alkylation reactions, particularly for biomass valorization. Considering that the alkylation/hydroxyalkylation reactions are catalyzed by strong acids, it is conceivable that the Nb-based solid acids will also be efficient for the alkylation/hydroxyalkylation of 2-MF with various ketones/aldehydes for biomass-derived fuel synthesis.

11.7 Esterification/Transesterification

Biodiesel can be produced either by the esterification of fatty acids or by the transesterification of triglycerides with small alcohols such as methanol and ethanol. The transesterification reaction of triglycerides with small alcohols, which can be catalyzed by either base or acid catalyst, is shown in Figure 11.16. Base catalysts such as sodium and potassium hydroxide are the most common catalysts used for transesterification reactions because of their higher reaction rates and lesser equipment corrosion. However, the basic catalysis of transesterification requires strict anhydrous conditions and the absence of free fatty acids, which would otherwise undergo saponification. Another obstacle is the intractable separation of the products with a base catalyst, and this is a common challenge in homogeneous processes. The use of solid-acid catalysts for biodiesel production can solve these drawbacks, but will inevitably suffer from low activity.

Niobium catalysts have been used for the esterification of fatty acids [103–107] and the transesterification of triglycerides [107,108] with methanol to produce biodiesel. Taft

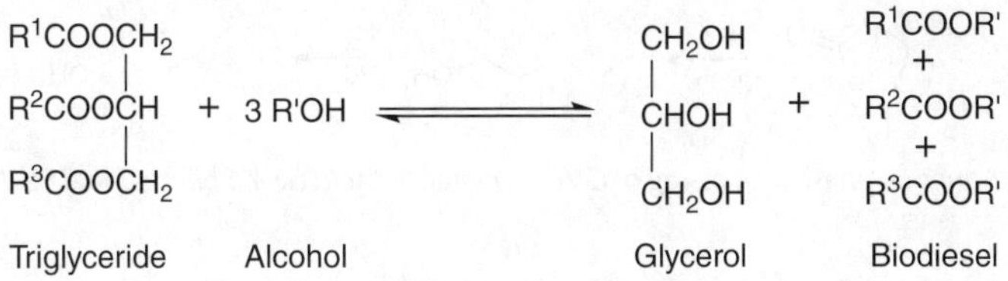

R^{1}COOCH$_2$ CH$_2$OH   R^{1}COOR'

|

R^{2}COOCH   + 3 R'OH ⇌ CHOH   +   R^{2}COOR'

|

R^{3}COOCH$_2$ CH$_2$OH   R^{3}COOR'

Triglyceride   Alcohol   Glycerol   Biodiesel

Figure 11.16   *Glycerol as a byproduct from biodiesel production by transesterification of triglycerides with small alcohols.*

et al. reported that Nb_2O_5 exhibited a better performance than zeolite catalysts towards the esterification of palm fatty acids with methanol [103]. The same authors investigated the catalytic activity of niobic acid towards the esterification of lauric, palmitic, stearic, oleic, and linoleic fatty acids with methanol, both experimentally and theoretically [104]. Lachter and coworkers [106] compared the activity of niobic acid and niobium phosphate in the esterification of a series of fatty acids (C_{12}–C_{18}) with small alcohols (methanol, ethanol, butanol), and found niobium phosphate to show superior activity in the esterification of lauric acid with butanol.

Phosphoric acid- or sulfuric acid-treated niobic acid was also used for the esterification and transesterification reactions [107]. In this case, 78% and 40% yields of targeted methyl esters were achieved from the esterification of oleic acid and transesterification of soybean oil with methanol, respectively, over sulfuric acid-treated niobic acid. Tesser *et al.* reported that supported niobia was efficient for both esterification and transesterification reactions in a continuous-flow reactor at 220 °C and 60 bar, with 90–95% conversion of fatty acid and 80–90% yield of fatty acid methyl ester (FAME) [108]. The catalyst showed an improved stability that maintained a constant activity for more than 100 h times-on-stream, although a non-negligible leaching phenomenon was detected.

11.8 Other Reactions in Biomass Conversion

11.8.1 Delignification

The goal of delignification is to reduce the lignin in lignocellulose that might influence the subsequent hydrolysis step and alter the crystalline structure of cellulose, making it accessible for hydrolysis [109]. Typically, the pretreatment or delignification was catalyzed using dilute H_2SO_4, but solid acids were also used in some cases. Niobium oxide was employed as a pretreatment agent by Kolar *et al.* for the delignification of *Alamo* switchgrass [110]. The effects of temperature, catalyst loading and pretreatment time on delignification over niobium oxide were evaluated. The results showed that niobium oxide was able to reduce total lignin by up to 44.6%, with good activity [110].

11.8.2 Ring-Opening of GVL

Niobium-based catalysts were also used for the ring-opening of GVL to pentanoic acid by loading Pd nanoparticles onto Nb-based supports as bifunctional catalysts (Figure 11.17) [111,112]. The addition of silica to Pd/Nb_2O_5 was found to improve its hydrothermal stability and maintain a small particle size of Pd, leading to an enhanced catalytic activity in

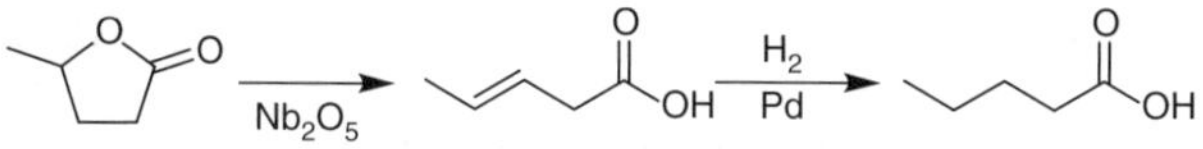

Figure 11.17   *Aqueous-phase processing of GVL to pentanoic acid over a bifunctional Pd/Nb$_2$O$_5$ catalyst.*

the aqueous-phase processing of GVL to pentanoic acid [111]. Other methods such as the atomic layer deposition of niobia within a mesoporous material can also used to prepare catalysts with excellent hydrothermal stability and catalytic activity [112]. Pd supported on SBA-15 coated with 19 cycles of niobia showed a better stability in comparison to conventional Pd/HY340 for the aqueous-phase processing of GVL to pentanoic acid.

11.8.3   Steam Reforming Reaction

Steam reforming of ethanol, which is considered as an environmentally fuel that can be produced from lignocellulosic biomass either by fermentation or catalysis, is a good alternative for H$_2$ production. Although the steam reforming reaction is mainly conducted on active metal sites, the nature of the support is also of significant importance for efficient catalysis. It has been reported that Nb$_2$O$_5$-supported Cu is a low-cost and efficient catalyst for the steam reforming of ethanol [113–115]. For example, when Schmal *et al.* [115] compared the catalytic performance of Cu/Nb$_2$O$_5$ and Ni/Al$_2$O$_3$ for ethanol steam reforming and partial oxidation, the Cu/Nb$_2$O$_5$ catalyst presented a much higher activity in comparison to the Ni/Al$_2$O$_3$ catalyst. For the same level of H$_2$ production, the reaction temperature for Cu/Nb$_2$O$_5$ catalyst was 200 °C lower than that for the Ni/Al$_2$O$_3$ catalyst. Moreover, very little CO formation occurred over the Cu/Nb$_2$O$_5$ catalyst, owing to the strong interaction between Cu and Nb$_2$O$_5$. A detailed mechanism of the Cu/Nb$_2$O$_5$ for ethanol reforming and oxidation was determined by conducting ethanol-TPD and surface reactions of ethanol [115].

11.8.4   Ketalization

Ketalization, catalyzed by acid, is a process in which a ketone or aldehyde reacts with alcohols to form ketal. Lachter *et al.* [116] investigated the ketalization of ethyl levulinate with ethylene glycol and 1,2-dodecanediol over a series of solid acids, including NbOPO$_4$. In comparison to other acid catalysts (e.g., *p*-toluenesulfonic acid, Amberlyst 70, H-ZSM-5), NbOPO$_4$ achieved the best performance, with excellent activity and recyclability in the ketalization of ethyl levulinate with ethylene glycol. Even better results were obtained in the ketalization of ethyl levulinate with longer-chain 1,2-dodecanediol, with a higher conversion and selectivity to targeted ketal. The resultant ketal, with a long alkyl chain, may serve as a new bio-surfactant by the subsequent alkaline hydrolysis.

11.9   Summary and Outlook

The essential chemistry process in biomass conversion is cleavage of the C–O bond by means of a variety of reaction types (e.g., hydrolysis, hydration, hydrogenolysis, hydrodeoxygenation etc.), since raw biomass feedstock contains a large volume of oxygen. Such C–O bond cleavage processes are mostly catalyzed by acid catalysts. In other words, acid catalysts play a significant role in the valorization of biomass to transportation fuels and fine chemicals. On the other hand, duo to the high oxygen content and strong hydrophilic properties of biomass feedstocks, most of the reactions in biomass conversion take place in the aqueous phase. Niobium-based materials are water-tolerant solid acids with controllable compositions, acid properties (acid types, amount and strength) and surface states (specific surface areas, pore size and volume), with the capability of efficiently catalyzing a variety of reactions. Hence, they have extensive application prospects in biomass conversion.

Since niobium-based catalysts have great potential in biomass conversion, it is necessary to develop new methods for their large-scale preparation, with low cost and simple operation. Besides, when compared to homogeneous acid catalysts (sulfuric acid, nitric acid), the efficiency of heterogeneous Nb solid acids remains to be improved. Therefore, the design and preparation of novel Nb catalysts with high efficiencies are the ultimate goal. Finally, although Nb materials are water-tolerant, Nb leaching can still occur during long-term operations in the aqueous phase. Consequently, the stability of these catalysts must be further improved from the point of view of industrial applications.

References

1. Huber, G.W., Iborra, S., Corma, A. (2006) Synthesis of transportation fuels from biomass: Chemistry, catalysts, and engineering. *Chem. Rev.*, **106** (9), 4044–4098.
2. Corma, A., Iborra, S., Velty, A. (2007) Chemical routes for the transformation of biomass into chemicals. *Chem. Rev.*, **107** (6), 2411–2502.
3. Cho, A. (2010) Energy's tricky tradeoffs. *Science*, **329** (5993), 786–787.
4. Nowak, I., Ziolek, M. (1999) Niobium compounds: Preparation, characterization, and application in heterogeneous catalysis. *Chem. Rev.*, **99** (12), 3603–3624.
5. Tanabe, K. (2003) Catalytic application of niobium compounds. *Catal. Today*, **78** (1-4), 65–77.
6. Tanabe, K. (1990) Application of niobium oxides as catalysts. *Catal. Today*, **8** (1), 1–11.
7. Tanabe, K., Okazaki, S. (1995) Various reactions catalyzed by niobium compounds and materials. *Appl. Catal. A. Gen.*, **133** (2), 191–218.
8. Ushikubo, T. (2000) Recent topics of research and development of catalysis by niobium and tantalum oxides. *Catal. Today*, **57** (3-4), 331–338.
9. Ziolek, M. (2003) Niobium-containing catalysts – the state of the art. *Catal. Today*, **78** (1-4), 47–64.
10. Okuhara, T. (2002) Water-tolerant solid acid catalysts. *Chem. Rev.*, **102** (10), 3641–3665.
11. Marzo, M., Gervasini, A., Carniti, P. (2012) Hydrolysis of disaccharides over solid acid catalysts under green conditions. *Carbohydr. Res.*, **347** (1), 23–31.
12. Armaroli, T., Busca, G., Carlini, C., Giuttari, M., Galletti, A.M.R., Sbrana, G. (2000) Acid sites characterization of niobium phosphate catalysts and their activity in fructose dehydration to 5-hydroxymethyl-2-furaldehyde. *J. Mol. Catal. A. Chem.*, **151** (1-2), 233–243.
13. Sarkar, A., Pramanik, P. (2009) Synthesis of mesoporous niobium oxophosphate using niobium tartrate precursor by soft templating method. *Micropor. Mesopor. Mater.*, **117** (3), 580–585.
14. Ding, L.-N., Wang, A.-Q., Zheng, M.-Y., Zhang, T. (2010) Selective transformation of cellulose into sorbitol by using a bifunctional nickel phosphide catalyst. *ChemSusChem*, **3** (7), 818–821.
15. Liu, M., Deng, W., Zhang, Q., Wang, Y., Wang, Y. (2011) Polyoxometalate-supported ruthenium nanoparticles as bifunctional heterogeneous catalysts for the conversions of cellobiose and cellulose into sorbitol under mild conditions. *Chem. Commun.*, **47** (34), 9717–9719.

16. Xi, J., Zhang, Y., Xia, Q., Liu, X., Ren, J., Lu, G., Wang, Y. (2013) Direct conversion of cellulose into sorbitol with high yield by a novel mesoporous niobium phosphate supported ruthenium bifunctional catalyst. *Appl. Catal. A. Gen.*, **459**, 52–58.

17. Zhao, S., Cheng, M., Li, J., Tian, J., Wang, X. (2011) One pot production of 5-hydroxymethylfurfural with high yield from cellulose by a Brønsted-Lewis-surfactant-combined heteropolyacid catalyst. *Chem. Commun.*, **47** (7), 2176–2178.

18. Ding, D., Wang, J., Xi, J., Liu, X., Lu, G., Wang, Y. (2014) High-yield production of levulinic acid from cellulose and its upgrading to gamma-valerolactone. *Green Chem.*, **16** (8), 3846–3853.

19. Galletti, A.M.R., Antonetti, C., Ribechini, E., Colombini, M.P., Di Nasso, N.N.O, Bonari, E. (2013) From giant reed to levulinic acid and gamma-valerolactone: a high yield catalytic route to valeric biofuels. *Appl. Energy*, **102**, 157–162.

20. Xi, J., Ding, D., Shao, Y., Liu, X., Lu, G., Wang, Y. (2014) Production of ethylene glycol and its monoether derivative from cellulose. *ACS Sust. Chem. Eng.*, **2** (10), 2355–2362.

21. Mal, N.K., Bhaumik, A., Fujiwara, M., Matsukata, M. (2006) Novel organic-inorganic hybrid and organic-free mesoporous niobium oxophosphate synthesized in the presence of an anionic surfactant. *Micropor. Mesopor. Mater.*, **93** (1-3), 40–45.

22. Zhang, Y., Wang, J., Ren, J., Liu, X., Li, X., Xia, Y., Lu, G., Wang, Y. (2012) Mesoporous niobium phosphate: an excellent solid acid for the dehydration of fructose to 5-hydroxymethylfurfural in water. *Catal. Sci. Technol.*, **2** (12), 2485–2491.

23. Xi, J., Zhang, Y., Ding, D., Xia, Q., Wang, J., Liu, X., Lu, G., Wang, Y. (2014) Catalytic production of isosorbide from cellulose over mesoporous niobium phosphate-based heterogeneous catalysts via a sequential process. *Appl. Catal. A. Gen.*, **469**, 108–115.

24. Sun, P., Long, X., He, H., Xia, C., Li, F. (2013) Conversion of cellulose into isosorbide over bifunctional ruthenium nanoparticles supported on niobium phosphate. *ChemSusChem*, **6** (11), 2190–2197.

25. Fenouillot, F., Rousseau, A., Colomines, G., Saint-Loup, R., Pascault, J.P. (2010) Polymers from renewable 1,4:3,6-dianhydrohexitols (isosorbide, isomannide and isoidide): A review. *Prog. Polym. Sci.*, **35** (5), 578–622.

26. Tang, Z.-C., Yu, D.-H., Sun, P., Li, H., Huang, H. (2010) Phosphoric acid modified Nb_2O_5: A selective and reusable catalyst for dehydration of sorbitol to isosorbide. *Bull. Korean Chem. Soc.*, **31** (12), 3679–3683.

27. Zhang, X., Yu, D., Zhao, J., Zhang, W., Dong, Y., Huang, H. (2014) The effect of P/Ta ratio on sorbitol dehydration over modified tantalum oxide by phosphoric acid. *Catal. Commun.*, **43**, 29–33.

28. Morita, Y., Furusato, S., Takagaki, A., Hayashi, S., Kikuchi, R., Oyama, S.T. (2014) Intercalation-controlled cyclodehydration of sorbitol in water over layered-niobium-molybdate solid acid. *Chem-SusChem*, **7** (3), 748–752.

29. Carlini, C., Giuttari, M., Galletti, A.M.R., Sbrana, G., Armaroli, T., Busca, G. (1999) Selective saccharides dehydration to 5-hydroxymethyl-2-furaldehyde by heterogeneous niobium catalysts. *Appl. Catal. A. Gen.*, **183** (2), 295–302.

30. Carniti, P., Gervasini, A., Biella, S., Auroux, A. (2006) Niobic acid and niobium phosphate as highly acidic viable catalysts in aqueous medium: Fructose dehydration reaction. *Catal. Today*, **118** (3-4), 373–378.

31. Dias, A.S., Lima, S., Brandao, P., Pillinger, M., Rocha, J., Valente, A.A. (2006) Liquid-phase dehydration of D-xylose over microporous and mesoporous niobium silicates. *Catal. Lett.*, **108** (3-4), 179–186.

32. Jehng, J.M., Wachs, I.E. (1991) Molecular structures of supported niobium oxide catalysts under in situ conditions. *J. Phys. Chem.*, **95** (19), 7373–7379.

33. Maurer, S.M., Ko, E.I. (1992) Structural and acidic characterization of niobia aerogels. *J. Catal.*, **135** (1), 125–134.

34. Nakajima, K., Baba, Y., Noma, R., Kitano, M., Kondo, J.N., Hayashi, S., Hara, M. (2011) $Nb_2O_5.nH_2O$ as a heterogeneous catalyst with water-tolerant Lewis acid sites. *J. Am. Chem. Soc.*, **133** (12), 4224–4227.

35. Ordomsky, V.V., Sushkevich, V.L., Schouten, J.C., van der Schaaf, J., Nijhuis, T.A. (2013) Glucose dehydration to 5-hydroxymethylfurfural over phosphate catalysts. *J. Catal.*, **300**, 37–46.

36. Zhang, Y., Wang, J., Li, X., Liu, X., Xia, Y., Hu, B., Lu, G., Wang, Y. (2015) Direct conversion of biomass-derived carbohydrates to 5-hydroxymethylfururual over water-tolerant niobium-based catalysts. *Fuel*, **139**, 301–307.

37. Yang, F., Liu, Q., Bai, X., Du, Y. (2011) Conversion of biomass into 5-hydroxymethylfurfural using solid acid catalyst. *Bioresource Technol.*, **102** (3), 3424–3429.

38. Ngee, E.L.S., Gao, Y., Chen, X., Lee, T.M., Hu, Z., Zhao, D., Yan, N. (2014) Sulfated mesoporous niobium oxide catalyzed 5-hydroxymethylfurfural formation from sugars. *Ind. Eng. Chem. Res.*, **53** (37), 14225–14233.

39. Li, X.-C., Zhang, Y., Xia, Y.-J., Hu, B.-C., Zhong, L., Wang, Y.-Q., Lu, G.-Z. (2012) One-pot catalytic conversion of xylose to furfural on mesoporous niobium phosphate. *Acta Phys. Chim. Sin.*, **28** (10), 2349–2354.

40. Pholjaroen, B., Li, N., Wang, Z., Wang, A., Zhang, T. (2013) Dehydration of xylose to furfural over niobium phosphate catalyst in biphasic solvent system. *J. Energy Chem.*, **22** (6), 826–832.

41. Garcia-Sancho, C., Rubio-Caballero, J.M., Merida-Robles, J.M., Moreno-Tost, R., Santamaria-Gonzalez, J., Maireles-Torres, P. (2014) Mesoporous Nb_2O_5 as solid acid catalyst for dehydration of D-xylose into furfural. *Catal. Today*, **234**, 119–124.

42. Bernal, H.G., Galletti, A.M.R., Garbarino, G., Busca, G., Finocchio, E. (2015) NbP catalyst for furfural production: FT IR studies of surface properties. *Appl. Catal. A. Gen.*, **502**, 388–398.

43. Garcia-Sancho, C., Sadaba, I., Moreno-Tost, R., Merida-Robles, J., Santamaria-Gonzalez, J., Lopez-Granados, M., Maireles-Torres, P. (2013) Dehydration of xylose to furfural over MCM-41-supported niobium-oxide catalysts. *ChemSusChem*, **6** (4), 635–642.

44. Zhang, Z., Zhang, G., He, L., Sun, L., Jiang, X., Yun, Z. (2014) Synthesis of niobium oxide nanowires by polyethylenimine as template at varying pH values. *CrystEngCommun*, **16** (17), 3478–3482.

45. Carniti, P., Gervasini, A., Marzo, M. (2010) Silica-niobia oxides as viable acid catalysts in water: Effective vs. intrinsic acidity. *Catal. Today*, **152** (1-4), 42–47.

46. Stosic, D., Bennici, S., Rakic, V., Auroux, A. (2012) CeO_2-Nb_2O_5 mixed oxide catalysts: Preparation, characterization and catalytic activity in fructose dehydration reaction. *Catal. Today*, **192** (1), 160–168.

47. Dias, A.S., Lima, S., Carriazo, D., Rives, V., Pillinger, M., Valente, A.A. (2006) Exfoliated titanate, niobate and titanoniobate nanosheets as solid acid catalysts for the liquid-phase dehydration of D-xylose into furfural. *J. Catal.*, **244** (2), 230–237.

48. Wu, Q., Yan, Y., Zhang, Q., Lu, J., Yang, Z., Zhang, Y., Tang, Y. (2013) Catalytic dehydration of carbohydrates on in-situ exfoliatable layered niobic acid in an aqueous system under microwave irradiation. *ChemSusChem*, **6** (5), 820–825.

49. Takagaki, A., Sugisawa, M., Lu, D.L., Kondo, J.N., Hara, M., Domen, K., Hayashi, S. (2003) Exfoliated nanosheets as a new strong solid acid catalyst. *J. Am. Chem. Soc.*, **125** (18), 5479–5485.

50. Takagaki, A., Lu, D.L., Kondo, J.N., Hara, M., Hayashi, S., Domen, K. (2005) Exfoliated HNb_3O_8 nanosheets as a strong protonic solid acid. *Chem. Mater.*, **17** (10), 2487–2489.

51. Xiong, H., Wang, T., Shanks, B.H., Datye, A.K. (2013) Tuning the location of niobia/carbon composites in a biphasic reaction: dehydration of D-glucose to 5-hydroxymethylfurfural. *Catal. Lett.*, **143** (6), 509–516.

52. Garcia-Sancho, C., Agirrezabal-Telleria, I., Gueemez, M.B., Maireles-Torres, P. (2014) Dehydration of D-xylose to furfural using different supported niobia catalysts. *Appl. Catal. B. Environ.*, **152**, 1–10.

53. Wang, F., Wu, H.-Z., Liu, C.-L., Yang, R.-Z., Dong, W.-S. (2013) Catalytic dehydration of fructose to 5-hydroxymethylfurfural over Nb_2O_5 catalyst in organic solvent. *Carbohydr. Res.*, **368**, 78–83.

54. Pagliaro, M., Ciriminna, R., Kimura, H., Rossi, M., Della Pina, C. (2007) From glycerol to value-added products. *Angew. Chem. Int. Ed.*, **46** (24), 4434–4440.

55. Behr, A., Eilting, J., Irawadi, K., Leschinski, J., Lindner, F. (2008) Improved utilisation of renewable resources: New important derivatives of glycerol. *Green Chem.*, **10** (1), 13–30.

56. Zhou, C.H.C., Beltramini, J.N., Fan, Y.X., Lu, G.Q.M. (2008) Chemoselective catalytic conversion of glycerol as a biorenewable source to valuable commodity chemicals. *Chem. Soc. Rev.*, **37** (3), 527–549.

57. Chai, S.-H., Wang, H.-P., Liang, Y., Xu, B.-Q. (2007) Sustainable production of acrolein: investigation of solid acid-base catalysts for gas-phase dehydration of glycerol. *Green Chem.*, **9** (10), 1130–1136.

58. Corma, A., Huber, G.W., Sauvanauda, L., O'Connor, P. (2008) Biomass to chemicals: Catalytic conversion of glycerol/water mixtures into acrolein, reaction network. *J. Catal.*, **257** (1), 163–171.

59. Katryniok, B., Paul, S., Belliere-Baca, V., Rey, P., Dumeignil, F. (2010) Glycerol dehydration to acrolein in the context of new uses of glycerol. *Green Chem.*, **12** (12), 2079–2098.

60. Chai, S.-H., Wang, H.-P., Liang, Y., Xu, B.-Q. (2007) Sustainable production of acrolein: Gas-phase dehydration of glycerol over Nb_2O_5 catalyst. *J. Catal.*, **250** (2), 342–349.

61. Shiju, N.R., Brown, D.R., Wilson, K., Rothenberg, G. (2010) Glycerol valorization: dehydration to acrolein over silica-supported niobia catalysts. *Top. Catal.*, **53** (15-18), 1217–1223.

62. Massa, M., Andersson, A., Finocchio, E., Busca, G. (2013) Gas-phase dehydration of glycerol to acrolein over Al_2O_3-, SiO_2-, and TiO_2-supported Nb- and W-oxide catalysts. *J. Catal.*, **307**, 170–184.

63. Massa, M., Andersson, A., Finocchio, E., Busca, G., Lenrick, F., Wallenberg, L.R. (2013) Performance of ZrO_2-supported Nb- and W-oxide in the gas-phase dehydration of glycerol to acrolein. *J. Catal.*, **297**, 93–109.

64. Garcia-Sancho, C., Cecilia, J.A., Moreno-Ruiz, A., Merida-Robles, J.M., Santamaria-Gonzalez, J., Moreno-Tost, R., Maireles-Torres, P. (2015) Influence of the niobium supported species on the catalytic dehydration of glycerol to acrolein. *Appl. Catal. B. Environ.*, **179**, 139–149.

65. Choi, Y., Park, D.S., Yun, H.J., Baek, J., Yun, D., Yi, J. (2012) Mesoporous siliconiobium phosphate as a pure Brønsted acid catalyst with excellent performance for the dehydration of glycerol to acrolein. *ChemSusChem*, **5** (12), 2460–2468.

66. Lee, Y.Y., Lee, K.A., Park, N.C., Kim, Y.C. (2014) The effect of PO_4 to Nb_2O_5 catalyst on the dehydration of glycerol. *Catal. Today*, **232**, 114–118.

67. Rao, G.S., Rajan, N.P., Pavankumar, V., Chary, K.V.R. (2014) Vapour phase dehydration of glycerol to acrolein over $NbOPO_4$ catalysts. *J. Chem. Technol. Biotechnol.*, **89** (12), 1890–1897.

68. Liu, R., Wang, T., Liu, C., Jin, Y. (2013) Highly selective and stable $CsPW/Nb_2O_5$ catalysts for dehydration of glycerol to acrolein. *Chinese J. Catal.*, **34** (12), 2174–2182.

69. Liu, R., Wang, T., Cai, D., Jin, Y. (2014) Highly efficient production of acrylic acid by sequential dehydration and oxidation of glycerol. *Ind. Eng. Chem. Res.*, **53** (21), 8667–8674.

70. Viswanadham, B., Pavankumar, V., Chary, K.V.R. (2014) Vapor phase dehydration of glycerol to acrolein over phosphotungstic acid catalyst supported on niobia. *Catal. Lett.*, **144** (4), 744–755.

71. Lauriol-Garbay, P., Millet, J.M.M., Loridant, S., Belliere-Baca, V., Rey, P. (2011) New efficient and long-life catalyst for gas-phase glycerol dehydration to acrolein. *J. Catal.*, **280** (1), 68–76.

72. Lauriol-Garbey, P., Postole, G., Loridant, S., Auroux, A., Belliere-Baca, V., Rey, P., Millet, J.M.M. (2011) Acid-base properties of niobium-zirconium mixed oxide catalysts for glycerol dehydration by calorimetric and catalytic investigation. *Appl. Catal. B. Environ.*, **106** (1-2), 94–102.

73. Liu, C., Liu, R., Wang, T. (2015) Glycerol dehydration to acrolein: selectivity control over $CsPW/Nb_2O_5$ catalyst. *Can. J. Chem. Eng.*, **93** (12), 2177–2183.

74. Rackemann, D.W., Doherty, W.O.S. (2011) The conversion of lignocellulosics to levulinic acid. *Biofuel. Bioprod. Biores.*, **5** (2), 198–214.

75. Ding, D., Xi, J., Wang, J., Liu, X., Lu, G., Wang, Y. (2015) Production of methyl levulinate from cellulose: selectivity and mechanism study. *Green Chem.*, **17** (7), 4037–4044.

76. Galletti, A.M.R., Antonetti, C., De Luise, V., Licursi, D., Di Nasso, N.N.O. (2012) Levulinic acid production from waste biomass. *Bioresources*, **7** (2), 1824–1835.

77. Wang, R., Xie, X., Liu, Y., Liu, Z., Xie, G., Ji, N., Ma, L., Tang, M. (2015) Facile and low-cost preparation of Nb/Al oxide catalyst with high performance for the conversion of kiwifruit waste residue to levulinic acid. *Catalysts*, **5** (4), 1636–1648.

78. Vassilev, S.V., Baxter, D., Andersen, L.K., Vassileva, C.G. (2010) An overview of the chemical composition of biomass. *Fuel*, **89** (5), 913–933.

79. Nakagawa, Y., Liu, S., Tamura, M., Tomishige, K. (2015) Catalytic total hydrodeoxygenation of biomass-derived polyfunctionalized substrates to alkanes. *ChemSusChem*, **8** (7), 1114–1132.

80. West, R.M., Liu, Z.Y., Peter, M., Dumesic, J.A. (2008) Liquid alkanes with targeted molecular weights from biomass-derived carbohydrates. *ChemSusChem*, **1** (5), 417–424.

81. West, R.M., Tucker, M.H., Braden, D.J., Dumesic, J.A. (2009) Production of alkanes from biomass derived carbohydrates on bi-functional catalysts employing niobium-based supports. *Catal. Commun.*, **10** (13), 1743–1746.

82. Xu, W., Xia, Q., Zhang, Y., Guo, Y., Wang, Y., Lu, G. (2011) Effective production of octane from biomass derivatives under mild conditions. *ChemSusChem*, **4** (12), 1758–1761.

83. Xia, Q.-N., Cuan, Q., Liu, X.-H., Gong, X.-Q., Lu, G.-Z., Wang, Y.-Q. (2014) Pd/NbOPO$_4$ multifunctional catalyst for the direct production of liquid alkanes from aldol adducts of furans. *Angew. Chem. Int. Ed.*, **53** (37), 9755–9760.

84. Xia, Q., Zhuang, X., Li, M.M.-J., Peng, Y.-K., Liu, G., Wu, T.-S., Soo, Y.-L., Gong, X.-Q., Wang, Y., Tsang, S.C.E. (2016) Cooperative catalysis for the direct hydrodeoxygenation of vegetable oils into diesel-range alkanes over Pd/NbOPO$_4$. *Chem. Commun.*, **52** (29), 5160–5163.

85. Xia, Q., Xia, Y., Xi, J., Liu, X., Wang, Y. (2015) Energy-efficient production of 1-octanol from biomass-derived furfural-acetone in water. *Green Chem.*, **17** (8), 4411–4417.

86. Xia, Q.N., Chen, Z.J., Shao, Y., Gong, X.Q., Wang, H.F., Liu, X.H., Parker, S.F., Han, X., Yang, S.H., Wang, Y.Q. (2016) Direct hydrodeoxygenation of raw woody biomass into liquid alkanes. *Nat. Commun.*, **7**, 11162.

87. Xi, J., Xia, Q., Shao, Y., Ding, D., Yang, P., Liu, X., Lu, G., Wang, Y. (2016) Production of hexane from sorbitol in aqueous medium over Pt/NbOPO$_4$ catalyst. *Appl. Catal. B. Environ.*, **181**, 699–706.

88. Shao, Y., Xia, Q., Liu, X., Lu, G., Wang, Y. (2015) Pd/Nb$_2$O$_5$/SiO$_2$ catalyst for the direct hydrodeoxygenation of biomass-related compounds to liquid alkanes under mild conditions. *ChemSusChem*, **8** (10), 1761–1767.

89. Ryu, J., Kim, S.M., Choi, J.-W., Ha, J.-M., Ahn, D.J., Suh, D.J., Suh, Y.-W. (2012) Highly durable Pt-supported niobia-silica aerogel catalysts in the aqueous-phase hydrodeoxygenation of 1-propanol. *Catal. Commun.*, **29**, 40–47.

90. Kon, K., Onodera, W., Takakusagi, S., Shimizu, K.-i. (2014) Hydrodeoxygenation of fatty acids and triglycerides by Pt-loaded Nb$_2$O$_5$ catalysts. *Catal. Sci. Technol.*, **4** (10), 3705–3712.

91. Liu, Y., Chen, L., Wang, T., Zhang, X., Long, J., Zhang, Q., Ma, L. (2015) High yield of renewable hexanes by direct hydrolysis-hydrodeoxygenation of cellulose in an aqueous phase catalytic system. *RSC Adv.*, **5** (15), 11649–11657.

92. Liu, Y., Chen, L., Wang, T., Zhang, Q., Wang, C., Yan, J., Ma, L. (2015) One-pot catalytic conversion of raw lignocellulosic biomass into gasoline alkanes and chemicals over LiTaMoO$_6$ and Ru/C in aqueous phosphoric acid. *ACS Sust. Chem. Eng.*, **3** (8), 1745–1755.

93. Sreekumar, S., Balakrishnan, M., Goulas, K., Gunbas, G., Gokhale, A.A., Louie, L., Grippo, A., Scown, C.D., Bell, A.T., Toste, F.D. (2015) Upgrading lignocellulosic products to drop-In biofuels via dehydrogenative cross-coupling and hydrodeoxygenation sequence. *ChemSusChem*, **8** (16), 2609–2614.

94. Higashio, Y., Nakayama, T. (1996) One-step synthesis of methyl isobutyl ketone catalyzed by palladium supported on niobic acid. *Catal. Today*, **28** (1-2), 127–131.

95. Paulis, M., Martin, M., Soria, D.B., Diaz, A., Odriozola, J.A., Montes, M. (1999) Preparation and characterization of niobium oxide for the catalytic aldol condensation of acetone. *Appl. Catal. A. Gen.*, **180** (1-2), 411–420.

96. Serrano-Ruiz, J. C., Dumesic, J.A. (2009) Catalytic upgrading of lactic acid to fuels and chemicals by dehydration/hydrogenation and C-C coupling reactions. *Green Chem.*, **11** (8), 1101–1104.

97. Serrano-Ruiz, J.C., Dumesic, J.A. (2009) Catalytic processing of lactic acid over Pt/Nb_2O_5. *ChemSusChem*, **2** (6), 581–586.

98. Corma, A., de la Torre, O., Renz, M., Villandier, N. (2011) Production of high-quality diesel from biomass waste products. *Angew. Chem. Int. Ed.*, **50** (10), 2375–2378.

99. Li, G., Li, N., Wang, Z., Li, C., Wang, A., Wang, X., Cong, Y., Zhang, T. (2012) Synthesis of high-quality diesel with furfural and 2-methylfuran from hemicellulose. *ChemSusChem*, **5** (10), 1958–1966.

100. Pereira, C.C.M., Lachter, E.R. (2004) Alkylation of toluene and anisole with 1-octen-3-ol over niobium catalysts. *Appl. Catal. A. Gen.*, **266** (1), 67–72.

101. de la Cruz, M.H.C., Rocha, A.S., Lachter, E.R., Forrester, A.M.S., Reis, M.C., San Gil, R.A.S., Caldarelli, S., Farias, A.D., Gonzalez, W.A. (2010) Investigation of the catalytic activity of niobium phosphates for liquid phase alkylation of anisole with benzyl chloride. *Appl. Catal. A. Gen.*, **386** (1-2), 60–64.

102. Murayama, T., Chen, J., Hirata, J., Matsumoto, K., Ueda, W. (2014) Hydrothermal synthesis of octahedra-based layered niobium oxide and its catalytic activity as a solid acid. *Catal. Sci. Technol.*, **4** (12), 4250–4257.

103. Aranda, D.A.G., Goncalves, J.D., Peres, J.S., Ramos, A.L.D., de Melo, C.A.R., Antunes, O.A.C., Furtado, N.C., Taft, C.A. (2009) The use of acids, niobium oxide, and zeolite catalysts for esterification reactions. *J. Phys. Org. Chem.*, **22** (7), 709–716.

104. Goncalves, J.D., Ramos, A.L.D., Rocha, L.L.L., Domingos, A.K., Monteiro, R.S., Peres, J.S., Furtado, N.C., Taft, C.A., Aranda, D.A.G. (2011) Niobium oxide solid catalyst: esterification of fatty acids and modeling for biodiesel production. *J. Phys. Org. Chem.*, **24** (1), 54–64.

105. Banchero, M., Gozzelino, G. (2015) Nb_2O_5-catalyzed kinetics of fatty acids esterification for reactive distillation process simulation. *Chem. Eng. Res. Des.*, **100**, 292–301.

106. Bassan, I.A.L., Nascimento, D.R., San Gil, R.A.S., da Silva, M.I.P., Moreira, C.R., Gonzalez, W.A., Faro, A.C., Onfroy, T., Lachter, E.R. (2013) Esterification of fatty acids with alcohols over niobium phosphate. *Fuel Process. Technol.*, **106**, 619–624.

107. de Pietre, M.K., Almeida, L.C.P., Landers, R., Vinhas, R.C.G., Luna, F.J. (2010) H_3PO_4- and H_2SO_4-treated niobic acid as heterogeneous catalyst for methyl ester production. *React. Kinet. Mech. Catal.*, **99** (2), 269–280.

108. Tesser, R., Vitiello, R., Carotenuto, G., Garcia Sancho, C., Vergara, A., Maireles Torres, P.J., Li, C., Di Serio, M. (2015) Niobia supported on silica as a catalyst for biodiesel production from waste oil. *Catal. Sust. Energy*, **2** (1), 33–42.

109. Kumar, P., Barrett, D.M., Delwiche, M.J., Stroeve, P. (2009) Methods for pretreatment of lignocellulosic biomass for efficient hydrolysis and biofuel production. *Ind. Eng. Chem. Res.*, **48** (8), 3713–3729.

110. Ansanay, Y., Kolar, P., Sharma-Shivappa, R.R., Cheng, J.J. (2014) Niobium oxide catalyst for delignification of switchgrass for fermentable sugar production. *Ind. Crop. Prod.*, **52**, 790–795.

111. Pham, H.N., Pagan-Torres, Y.J., Serrano-Ruiz, J.C., Wang, D., Dumesic, J.A., Datye, A.K. (2011) Improved hydrothermal stability of niobia-supported Pd catalysts. *Appl. Catal. A. Gen.*, **397** (1-2), 153–162.

112. Pagan-Torres, Y.J., Gallo, J.M.R., Wang, D., Pham, H.N., Libera, J.A., Marshall, C.L., Elam, J.W., Datye, A.K., Dumesic, J.A. (2011) Synthesis of highly ordered hydrothermally stable mesoporous niobia catalysts by atomic layer deposition. *ACS Catal.*, **1** (10), 1234–1245.

113. Alonso, C.G., Furtado, A.C., Cantao, M.P., dos Santos, O.A.A., Fernandes-Machado, N.R.C. (2009) Reactions over Cu/Nb_2O_5 catalysts promoted with Pd and Ru during hydrogen production from ethanol. *Int. J. Hydrog. Energy*, **34** (8), 3333–3341.

114. Dancini-Pontes, I., DeSouza, M., Silva, F.A., Scaliante, M.H.N.O., Alonso, C.G., Bianchi, G.S., Neto, A.M., Pereira, G.M., Fernandes-Machado, N.R.C. (2015) Influence of the CeO_2 and Nb_2O_5 supports and the inert gas in ethanol steam reforming for H-2 production. *Chem. Eng. J.*, **273**, 66–74.

115. Guarido, C.E.M., Cesar, D.V., Souza, M.M.V.M., Schmal, M. (2009) Ethanol reforming and partial oxidation with Cu/Nb_2O_5 catalyst. *Catal. Today*, **142** (3-4), 252–257.

116. Freitas, F.A., Licursi, D., Lachter, E.R., Galletti, A.M.R., Antonetti, C., Brito, T.C., Nascimento, R.S.V. (2016) Heterogeneous catalysis for the ketalisation of ethyl levulinate with 1,2-dodecanediol: Opening the way to a new class of bio-degradable surfactants. *Catal. Commun.*, **73**, 84–87.

12

Towards More Sustainable Chemical Synthesis, Using Formic Acid as a Renewable Feedstock

Shu-Shuang Li, Lei Tao, Yong-Mei Liu, and Yong Cao

Department of Chemistry, Shanghai Key Laboratory of Molecular Catalysis and Innovative Materials, Fudan University, China

12.1 Introduction

In this era of diminishing supplies of fossil fuels, biomass – being the only renewable source of carbon – has the potential to serve as a sustainable platform for the production of energy and chemicals [1]. Until now, however, the direct use of biomass has generally been considered unsuitable, primarily due to the inherent recalcitrant and complex nature of the raw biomass, and this has underscored the need to develop alternative pathways for converting biomass into useful chemicals [2–6]. In this context, the cascaded valorization of biomass involving the intermediate processing of a set of defined platform compounds has emerged as a viable strategy for converting crude biomass into diverse, value-added chemical structures [7–11]. Well-practiced examples of such advanced biorefinery schemes include the use of various biogenic platform molecules, including 5-hydroxymethylfurfural (HMF), furfural (FAL), and levulinic acid (LA), as the starting points for subsequent bio-processing [12–16]. This concept (Figure 12.1) has recently been expanded to establish formic acid (FA), one of the major byproducts from biomass processing, as an alternative biomass-derived feedstock resource [17].

Formic acid is a high-demand commodity chemical that is widely used in the chemical, agricultural, textile, leather, pharmaceutical, and rubber industries [18]. Historically, FA was usually produced as a byproduct of acetic acid production, via the liquid-phase oxidation of hydrocarbons [19], but current FA production is mainly based on the

Nanoporous Catalysts for Biomass Conversion, First Edition. Edited by Feng-Shou Xiao and Liang Wang.
© 2018 John Wiley & Sons Ltd. Published 2018 by John Wiley & Sons Ltd.

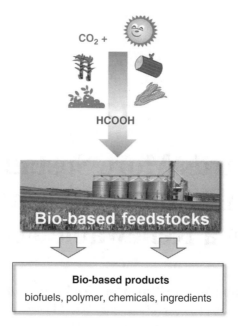

Figure 12.1   *The concept of formic acid-based biorefinery.* (See color plate section for the color representation of this figure.)

carbonylation of methanol [20–22]. The attraction of FA as a feedstock for a bio-based economy has increased with continuing advances in the direct and selective synthesis of FA in high yields (up to 85%) from bio-derived carbohydrates [23], projecting an ever-growing availability of this compound from natural resources. In particular, as new processes were developed to store/generate H_2 using bio-derived FA (Figure 12.2) as an energy carrier [24–26], FA has attracted a tremendous recent upsurge of interest in the context of the future hydrogen economy. In this regard, biogenic FA offers a mild and sustainable way to produce H_2, particularly given its fivefold higher energy density per volume than lithium-ion batteries [26]. FA can also be considered as a liquid surrogate of syngas [27], given that it can be readily converted into CO and water upon thermal decomposition at temperatures above 373K.

Controlled deoxygenation represents a major challenge when dealing with the direct thermocatalytic transformation of biomass derivatives. In this respect, hydrodeoxygenation (HDO) has been identified as an important enabling step in upgrading bio-based feedstocks into fuels and valuable chemicals [28–32]. Most HDO processes employ H_2 as the hydrogen donor [33–36], owing to its wide availability and easy activation on many catalytic metals. However, currently available H_2 is still mainly produced from fossil fuels, which in turn makes the process dependent on fossil carbon. Moreover, the handling of high-pressure H_2 gas incurs significant safety concerns, and thus hefty infrastructure costs on the industrial scale, which in turn poses an economic barrier for developing a sustainable biomass-upgrading economy, especially in the initial stages. By offering a competitive and complementary HDO strategy, the use of safe and inexpensive FA instead of H_2 is particularly appealing in this regard [37–39]. An additional advantage lies in the moderate

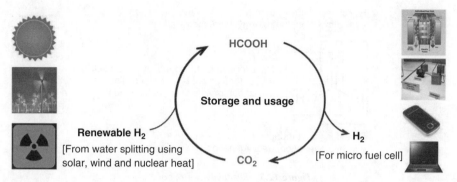

Figure 12.2 *Carbon-neutral H$_2$ store using biorenewable formic acid as an energy carrier.* (See color plate section for the color representation of this figure.)

hydrogenating capability of FA, that can deliver a higher degree of selectivity control in hydrogenation and/or hydrogenolysis, when partially reduced molecules are targeted.

In this chapter, recent progress in the development of new processes for the production of value-added chemicals from bio-derived platform molecules or model compounds, using FA as the essential feedstock sources, is outlined. First, the general properties of FA and the implications for green synthesis are discussed. This analysis helps in understanding strategies for the production of tailor-made chemicals via the FA-mediated transformation of bio-based feedstocks. The main adopted strategy involves the upgrading of FA-mediated platform molecules through controlled deoxygenation reactions to obtain new bio-based compounds in terms of value chains of renewable chemical production. These upgrading reactions generally require reductive conversions through the FA-based HDO of reactive functional groups, such as C=O, acid, and C–OH groups, in which the in-situ generation of H$_2$ gas by efficient and selective catalytic FA decomposition is essential. The concept is also introduced by tandem catalytic upgrading in the presence of FA to achieve cost-competitive and more compact bio-refining strategies, in which the multiple role of FA is highlighted.

12.2 General Properties of FA and Implications for Green Synthesis

In order to facilitate the discussion of some of the general issues involved in the FA-mediated conversion of high-oxygen-content feeds, a general overview of the chemo-catalytic aspects of FA will first be provided. FA has the potential to be considered a valuable platform chemical for next-generation biorefinery because of its particular chemistry (Figure 12.3), in that it has two highly reactive functional groups that enable many synthetic transformations, including esterification [40], dehydration [41], formylation [42,43], addition to olefins [44], and cyclization of toluene diamines [45]. The issues of special interest in developing new processes to enable more efficient biomass upgrading are that any sequences of relevant reactions, such as tandem reduction-formylation [42,43], addition-amination [46] and cascade esterification-dehydroxylation [47] can lead to the direct 'one-pot' construction of diverse, interesting, and complex structures from readily available starting materials. A key prerequisite for realizing such advanced transformation concepts, however, lies in the development of new, versatile catalytic approaches that can allow flexible and controlled FA activation.

Reactions at formyl moiety
e.g. formylation, hydroformylation, methylation

Reactions at carboxyl moiety (as an acid)
e.g. esterification, addition to olefins
cyclization of aldazines, dehydration

Reactions as a reductant
e.g. hydrogenation, deoxygenation

Figure 12.3   *Reactivity portrait of FA.*

Regarding the use of FA as an alternative hydrogen source for selective deoxygenation, it is important to recall that FA can deliver hydrogen to an organic substrate in a process termed catalytic transfer hydrogenation (CTH). Although CTH was introduced more than a century ago by the seminal report of Pd black-catalyzed disproportionation of methyl terephthalate by Knoevenagel [48], it has been largely eclipsed by the success of H_2-based HDO processes until the past few decades. The intense quest for a viable CTH has contributed to a range of heterogeneous, homogeneous – and also some examples of immobilized homogeneous – catalysts capable of addressing the drawbacks of low reaction rates and yields that have long plagued CTH in its early years [49–55]. In this regard, FA-based CTH has proven particularly effective in reducing functional groups by incorporating hydrogen to either unsaturated bonds (hydrogenation), such as C=C [49,50], C≡C [51,52], C=O [53–55], N=O [56–58], N=N [59], and C≡N [58], or single bonds leading to bond cleavage (hydrogenolysis), such as C–O [60–63], C–N [52], C–S [64], and C–X (halogen) [65], although in some cases the involved mechanism remained to be further elucidated.

From the perspective of more efficient resource utilization, there is an additional incentive to make full use of FA as a feedstock source for renewable chemical synthesis. As mentioned earlier with respect to the FA chemistry discussed above, any strategies that can address this issue will be advantageous not only for FA utilization but also for the development of new enabling catalytic technologies based on renewable resources. For instance, starting with FA and a number of structurally diverse alkenes, Porcheddu *et al.* [66] recently showed the potential benefit of utilizing FA as a C_1 building block for the synthesis of a library of alcohols via a catalyzed oxo-synthesis, under green experimental conditions. Along this line, it was recently shown that a facile atom- and step-efficient transformation of nitro compounds can be realized in modular fashion over a single heterogeneous gold-based catalyst [43], simply by controlling the stoichiometry of the employed FA. In this regard, tandem catalytic transformation using FA may bring about new opportunities for the construction of molecular complexity from bio-based feedstocks.

12.3   Transformation of Bio-Based Platform Chemicals

12.3.1   Reductive Transformation Using FA as a Hydrogen Source

One essential component identified for the next-generation biorefineries is γ-valerolactone (GVL), which has been hailed as a sustainable liquid for energy and carbon-based chemicals

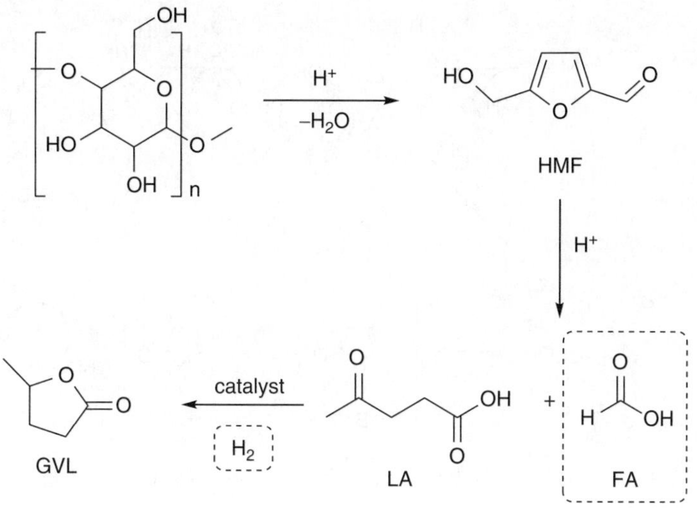

Figure 12.4 *Catalytic conversion of carbohydrate biomass into GVL [68].*

[67]. Targeted GVL production can be achieved by the hydrogenation of LA, a versatile and viable platform molecule that has already been produced efficiently from lignocellulosic biomass on a pilot-plant scale. Given the fact that FA is invariably coproduced in equimolar amounts along with LA during the biomass dehydration process, the most ideal strategy for GVL production is the straightforward use of bio-derived FA as an expedient and convenient source of hydrogen. One of the most beneficial aspects of FA-mediated LA reduction is that it can inherently eliminate the need for an external H_2 supply and avoid the costly purification of LA (Figure 12.4) [68]. However, the implementation of such a truly convenient and green approach for producing GVL remains challenging, due largely to a lack of readily accessible, applicable and reusable solid catalysts that are sufficiently active and selective for H_2 production from the continuous decomposition of large volumes of aqueous FA under typical hydrothermal processing conditions.

Several relevant studies dealing with the use of homogeneous Ru-based catalysts for the direct synthesis of GVL from LA/FA feeds have been reported [69–71]. Indeed, an exceedingly efficient heterogeneous catalytic system has been discovered for the direct conversion of aqueous bio-derived equimolar mixtures of LA/FA streams into GVL. The process is catalyzed by zirconia-supported small Au nanoparticles (Au/ZrO_2) [68], in which the high catalytic performance of supported Au NPs for the in-situ generation of H_2 gas via selective FA decomposition under mild aqueous conditions is essential. Recently, a noble-metal-free production of GVL from FA/LA streams based on earth-abundant copper-based catalysts (Cu/ZrO_2) under relatively mild aqueous conditions (160–200 °C) was developed [72]. Although being over 130-fold less active in terms of the mass-specific reaction rates ($mmol_{GVL}$ g_{metal}^{-1} h^{-1}) based on total metal atoms as compared to that of the corresponding Au/ZrO_2 catalytic system under identical reaction conditions, the use of earth-abundant copper may be justified on its cost (less than US$ 0.008069 g^{-1}) and availability.

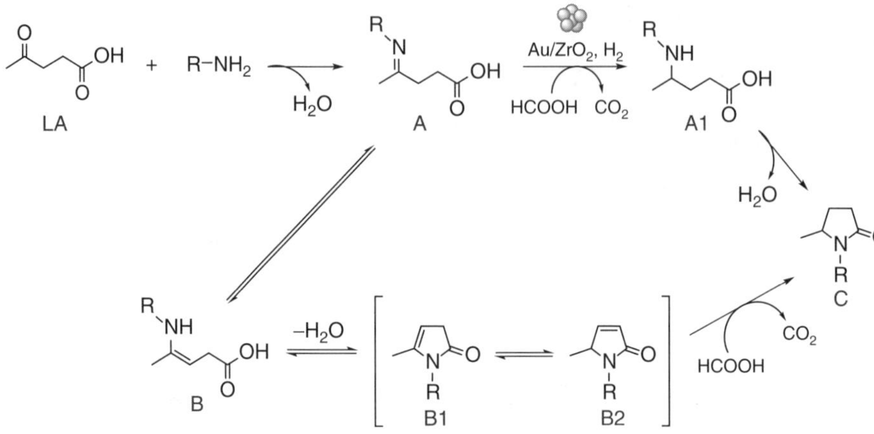

Figure 12.5   *Mechanistic proposal for reductive amination of LA [68].*

Similarly, targeted pyrrolidones production can also be achieved by the FA-mediated reductive amination (RA) of LA. Pyrrolidones are high value-added products possessing a wide variety of applications as solvents, surfactants, chelating agents, agrochemical components, aerosol formulations, transdermal patches. They may also serve as important intermediates in the synthesis of pharmaceutical and agricultural bioactive compounds [73]. Recently, a one-pot synthesis of pyrrolidones was established on the basis of the above-mentioned Au/ZrO$_2$–FA-mediated RA methodology [68], which enables the conversion of a 1:1:1 neat mixture of LA, FA, and amine into 5-methyl-2-pyrrolidones in high yield (88–97% with different amines at 130 °C). The mechanism proposed for this transformation (see Figure 12.5) highlights the key dehydrogenative FA activation of the in-situ H$_2$ generation step that is required to enable the desired transformation. Following the publication of this information, examples of LA-mediated RA production of pyrrolidones by several homogeneous and heterogeneous catalytic systems (Ru, Ir) were reported, although they suffered from limited scope or harsh conditions [74,75].

1,6-Hexanediol (HDO) is used extensively in the production of polyesters for polyurethane elastomers, coatings, adhesives, and polymeric plasticizers. The major route for its large-scale production involves the hydrogenation of adipic acid or its esters [76]. In this context, the direct synthesis of HDO from HMF is particularly attractive because HMF is one of the most promising bio-based platform chemicals. Converting HMF to HDO, however, can be challenging because HMF can undergo several reduction pathways which compete with each other; hence, the chemoselectivities of these reactions must be strictly controlled to avoid the formation of intractable mixtures of polyhydroxyl compounds. Recently, Ebitani *et al.* achieved a surprisingly high yield of HDO (ca. 43%) over a reusable Pd/zirconium phosphate (ZrP) catalyst at 413K in the presence of FA as hydrogen source [77]. The authors suggested that the acidity of the surface of the ZrP support, for ring cleavage, the presence of a transition metal (i.e., Pd) for FA dissociation, and thereby the utilization of in-situ-generated H$_2$ for hydrogenation, are the important aspects governing chemoselective HDO formation from HMF (Figure 12.6).

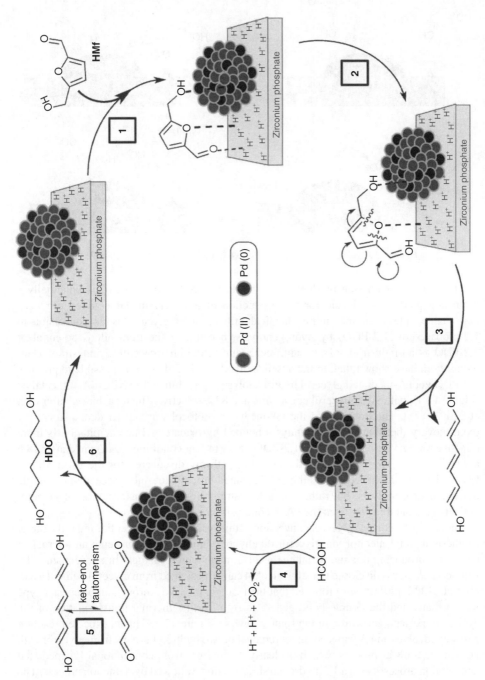

Figure 12.6 *Proposed mechanism of Pd/ZrP-catalyzed hydrogenolysis of HMF to HDO in the presence of FA [77].*

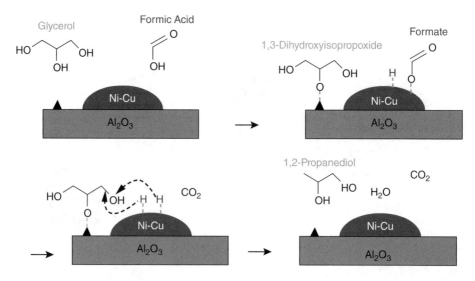

Figure 12.7   *Glycerol hydrogenolysis by FA over Ni–Cu/Al₂O₃ catalyst [78].*

Glycerol, the main side product of the biodiesel production process, is potentially a promising platform molecule for the production of a variety of value-added products. One viable option for the future chemical utilization of glycerol is its conversion to 1,2-propanediol (1,2-PDO) by hydrogenolysis, because of the tremendous potential of 1,2-PDO as an industrial solvent, antifreeze, and in the pharmaceutical industries. Gandarias *et al.* have shown that, in the absence of externally fed H_2, it is possible to produce 1,2-PDO via FA-mediated glycerol hydrogenolysis over a bimetallic Ni-Cu/Al₂O₃ catalyst (Figure 12.7), with 90% glycerol conversion and 82% selectivity being achieved after 24 h at 220 °C [78]. The potential of the FA-mediated protocol was further demonstrated by the discovery that the Cu/ZrO₂ catalyst obtained by oxalate–gel coprecipitation is more effective for converting glycerol to 1,2-PDO under milder conditions (94% yield after 18 h at 200 °C) [79]. Consistent with the essential pathway discovered for the FA-mediated LA-to-GVL, as described above, this FA-mediated process does not proceed via a simple transfer hydrogenolysis, but rather via a less straightforward transformation of glycerol with H_2 generated *in situ* from the FA decomposition.

Another FA-mediated glycerol conversion pathway that deserves mentioning is the direct synthesis of allyl alcohol via glycerol didehydroxylation, which presents an alternative choice for producing this industrial chemical from a green feedstock. In a recent study by Ellman *et al.*, it was disclosed that allyl alcohol can be produced from glycerol without a catalyst at 240 °C [80]. It was proposed that the reaction proceeds through an orthoester-type intermediate, and the double bond generates from further thermal transformation of the cyclic orthoesters, according to the route shown in Figure 12.8. This unexpected reaction pathway involves an FA-induced direct removal of two hydroxyl groups from glycerol with no hydride transfer process, which is clearly a departure from conventional FA-mediated dehydration processes via a 1,2-hydrogen shift reaction, followed by reduction process. The feasibility of this FA-mediated glycerol conversion technique was further underscored by a continuous operation that resulted in a significantly improved productivity of allyl alcohol

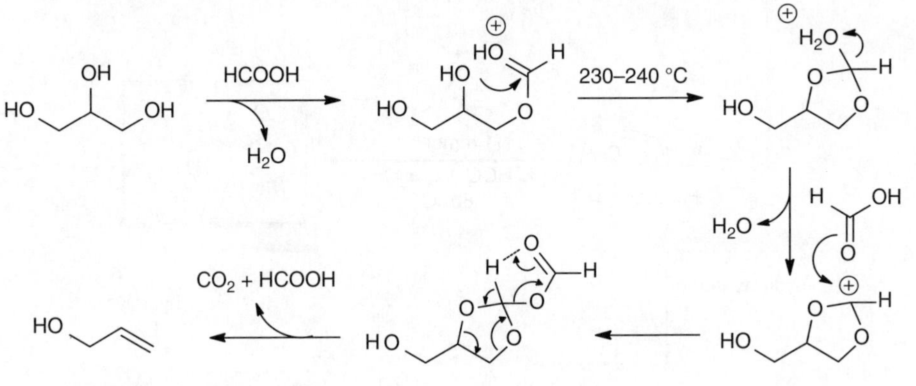

Figure 12.8 *The mechanism proposed for FA-mediated didehydroxylation [80].*

[81], which is obviously more desirable for practical as well as economic reasons. A similar dehydroxylation reaction to synthesize benzoic acid from glucose-derived quinic acid was also reported by Ellman *et al.* [47]. Such one-pot, one-reagent deoxygenation of biogenic polyhydroxy compounds with FA constituted a valuable alternative for the manufacture of reduced-oxygen content products in biorefinery.

12.3.2   Tandem Transformation Using FA as a Versatile Reagent

Due to the potential of FA to serve as an acid catalyst, a solvent, a source of hydrogen and a deoxygenation (DO) agent, several multi-step, one-pot tandem transformations using FA as a key reagent were explored. Rauchfuss and coworkers have demonstrated a facile FA-mediated synthesis of liquid fuel 2,5-dimethylfuran (DMF) from fructose [38], wherein HMF produced by FA-catalyzed fructose dehydration is first reduced to 2,5-bis(hydroxymethyl)-furan (BHMF) with FA in the presence of Pd/C, followed by the targeted generation of DMF via FA-mediated DO of the diformate ester of BHMF as a result of H_2SO_4-assisted BHMF esterification with FA (Figure 12.9). In this case,

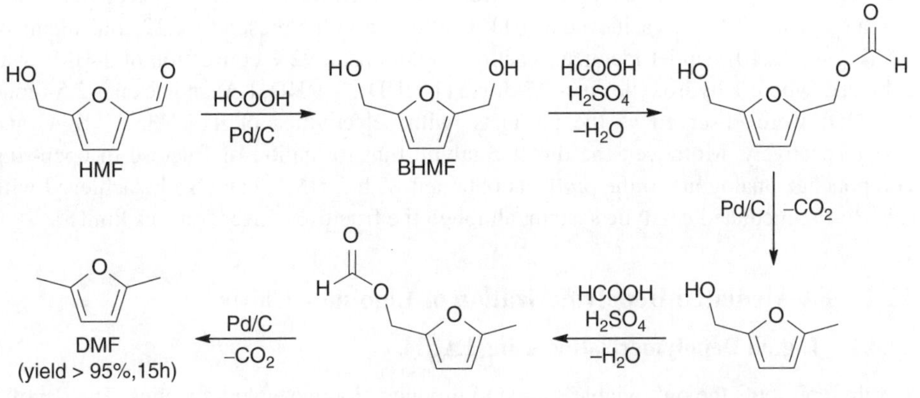

Figure 12.9 *Pathway for DMF from HMF using FA as a reagent [38].*

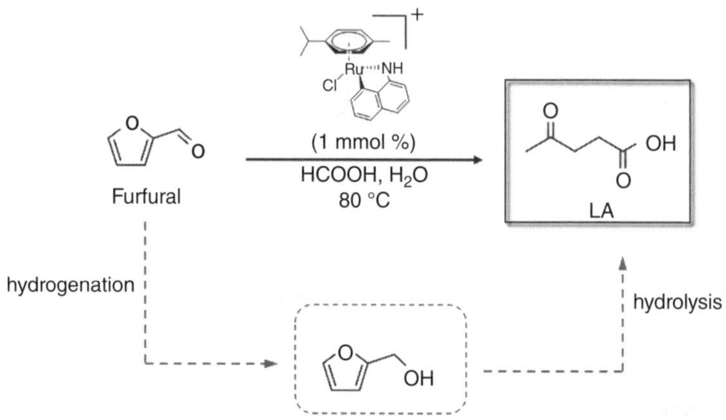

Figure 12.10   *Ru-catalyzed transformation of furfural to LA with FA [83].*

FA provides a milder pathway for DMF synthesis and achieves an excellent yield (>95%). The same group also described the production of hybrid fuels from sugars, further highlighting the potential utility of FA as both solvent and catalyst for the reactions of fructose, glucose, and cellulose [82]. As such, 5-mesitylmenthyfurfural in moderate yields of 20–70% can be obtained via a simple one-pot process comprising sugar dehydration followed by a subsequent Friedel– Crafts arylation of HMF under mild conditions.

Ketoacids and diketones obtained from the catalytic hydrogenation and hydrolytic ring-opening of furans have been extensively used as platform chemicals for the production of several important fine chemicals (e.g., esters, alcohols, lactones, amines, cycloketones), or biofuel components such as blenders for gasoline. Most studies on the transformation of furans have been performed using mineral acids with H_2 gas where a high pressure and a high temperature are necessary. Recently, Singh *et al.* reported a tandem transformation of furans to ketoacids and diketones in the presence of FA with Ru-based homogeneous catalysts [83]. In this case, the significant dual role of FA serving as a hydrogen donor as well as tuning the acidity, facilitated the crucial transfer hydrogenation step followed by the ring-opening reaction. Complete conversion of furfural with >99% selectivity to LA (isolated yield 42%) was achieved at 100 °C after 8 h in the presence of 12 equivalents of FA (Figure 12.10). Under identical reaction conditions, a 92% conversion of 5-HMF was achieved, where 1-hydroxyhexane-2,5-dione (1-HHD), 3-HHD, LA, and hexane-2,5-dione (2,5-HD) were observed as the products, with selectivities of 41%, 36%, 18%, and 5%, respectively. Moreover, the direct catalytic transformation of fructose to open-ring components, analogous to the products obtained with 5-HMF, can also be achieved with this Ru-FA-mediated catalytic system, although the fructose conversion was limited.

12.4   FA-Mediated Depolymerization of Lignin or Chitin

12.4.1   Lignin Depolymerization using FA

Lignin represents the only viable source to produce biorenewable aromatics. The depolymerization of lignin biomass into tailor-made fuels and chemicals is an integral part of

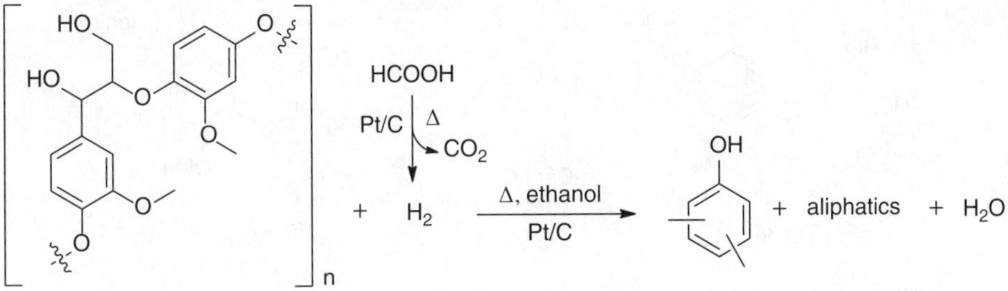

Figure 12.11   *Lignin depolymerization and HDO in the presence of FA [37].*

modern biorefinery. The development of selective catalytic processes capable of dehydration and HDO with a preserved aromatic nature represents a promising, yet challenging, approach. In this respect, biogenic FA has been identified as a reagent with unique favorable properties as opposed to the H_2 gas in facilitating an efficient disassembly of lignin [37,84–86]. For example, switchgrass lignin can be directly converted into phenolic monomers with yields of up to 50% by combining the depolymerization and HDO reactions in a single step [37]. As shown in Figure 12.11, the conversion process involves the thermal treatment of lignin in ethanol or water at 300–350 °C, with FA as the active hydrogen donor. The use of a typical HDO catalyst, Pt/C, is found to be essential to enable FA decomposition that results in the generation of H_2 gas, which combines with oxygen from the methoxy groups of lignin to form water. Since both depolymerization and HDO occur simultaneously, such a splitting reaction can result in monomers with low oxygen contents in a single step.

One challenge that remains for deconstructing lignin is to remove the need for highly energy-intensive conditions. Of special importance in this respect is the FA-mediated catalysis. Notably, Samec *et al.* found that a simple combination of FA and heterogeneous palladium catalyst supported on activated charcoal (Pd/C) showed great promise for producing aromatic compounds from lignin under very mild conditions [87]. The method was notable as a mild and robust procedure in which the catalysis can be performed at 80 °C in air, without using specialized equipment (Figure 12.12). This heterogeneous FA-mediated procedure was found to be particular effective for C–O bond cleavage of various lignin models containing the β-O-4'-linkage, the most abundant structural motif in lignin. Kinetic deuterium-labeling experiments supported an initial Pd-catalyzed alcohol dehydrogenation mechanism involving a palladium-formato complex as the key reaction intermediate. Degradation experiments with more challenging pine-derived native lignin, however, afforded only a moderate shift toward lower-molecular-weight fragments.

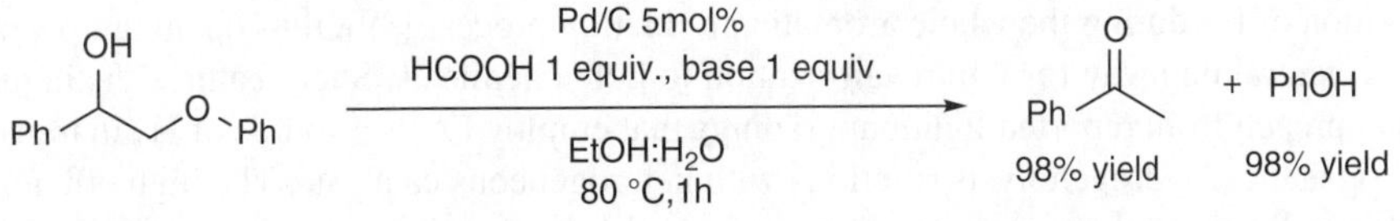

Figure 12.12   *FA-mediated cleavage of the β-O-4'-ether bond of model lignin compounds [87].*

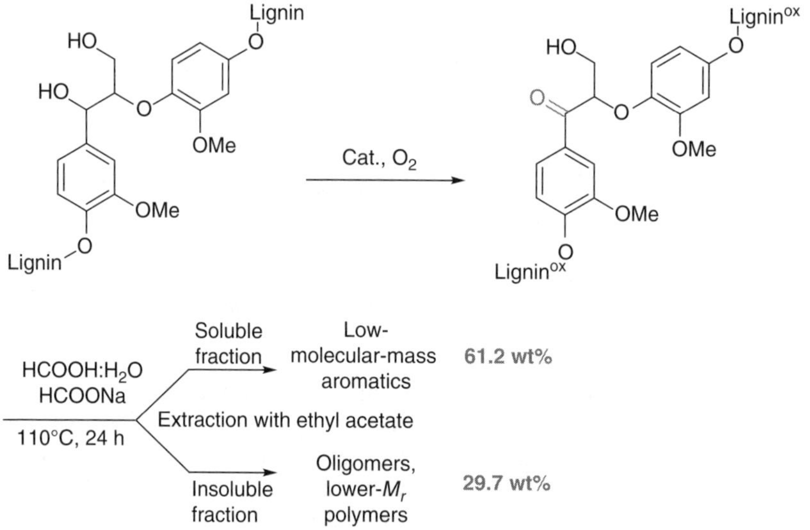

Figure 12.13   *Depolymerization of aspen lignin with FA [39].*

As a further demonstration of the feasibility of achieving facile C–O cleavage for extensive lignin depolymerization, FA chemistry has shown particular promise in meeting the ultimate goal of producing a significant fraction of higher-value mono- aromatic streams for further upgrading (Figure 12.13) [39]. As opposed to known procedures, this FA-mediated depolymerization process is notable for its mild conditions, with relatively low temperatures (110 °C) and low pressures, as well as the lack of any need for expensive metal catalysts. The overall yield (>60 wt%) of structurally identified, monomeric aromatics obtained is the highest reported so far for lignin depolymerization. Key to this process is to subject oxidized lignin to FA processing, in which the crucial formylation of the oxidized β-O-4'-linkage enabling a facile C–O cleavage involving subsequent FA elimination from the formyl intermediates is essential. Mechanistically, the beneficial effect of lignin oxidation may be attributed to the ability of the benzylic carbonyl group to polarize the C–H bond and lower the barrier for the rate-limiting E2 elimination reaction.

These observations highlight the importance of developing chemical conversion technologies for S-, G-, and H-derived aromatics, and suggest that plants containing lignin with a high β-O-4' content (up to 85% has been observed) could be particularly appealing feedstocks for biomass valorization. It should be noted that the depolymerization methods applied so far in lignin conversion suffer from harsh reaction conditions and yield complex mixtures of products, from which the isolation of pure products is an issue. Another crucial aspect that should be mentioned for this particular system is that there is no net consumption of FA during the whole aromatic production process [39], thus opening up a new redox-neutral pathway for lignin valorization to bulk chemicals. Such features distinguish this approach from reported lignin upgradings that employ FA as a source of H_2 in transfer hydrogenation/hydrogenolysis reactions with heterogeneous catalysts. The high efficiency of this redox-neutral process, operating even under mild ambient conditions, exemplifies once more the unique opportunities offered by FA-mediated biorefining.

12.4.2 Chitin Depolymerization using FA

Chitin, which exists mainly in the shells of insects and crustaceans, is one of the major biomass resources on Earth but has been far less studied than lignocellulosic biomass [88]. Chitin is constructed from *N*-acetylglucosamine (NAG) units with β-1,4-glycosidic linkages (Figure 12.14). The biologically fixed nitrogen in the structure of chitin means that it occupies a unique position as a starting material for N-containing chemicals, provided that simple, innovative and effective transformations can be established. One of the simplest processes to depolymerize chitin into soluble small molecules is liquefaction, which has been previously studied in ethylene glycol (EG) with sulfuric acid as catalyst. Unfortunately, the susceptibility of the amide group in the presence of a strong acid at elevated temperatures imposes additional obstacles in chitin liquefaction. As opposed to commonly used mineral acid catalysts, FA is a promising alternative agent for chitin liquefaction as it is much weaker in acidity than sulfuric acid and therefore would minimize undesired side reactions. Moreover, it can be easily recycled by distillation due to its low boiling point.

Very recently, Yan *et al.* reported a simple and convenient protocol for chitin liquefaction using neat FA under mild conditions [89]. An up to 60% combined yield of a series of structurally identified monomeric products was obtained after a FA-mediated processing of ball-milled chitin at 100 °C for 12 h. This simple system could further generate a single compound, 5-(formyloxymethyl)furfural (FMF; a derivative of the well-known platform chemical HMF), with up to 35% yield after a prolonged reaction time. The stages of the proposed chitin depolymerization pathway are shown in Figure 12.15: (i) the process starts with partial formylation of the hydroxyl groups in the chitin side chains, generating soluble polymeric derivatives; (ii) FA catalyzes polymer chain breakage in a non-conventional, non-hydrolytic pathway, to form dehydrated monomers and oligomers; (iii) as formylation continues, water accumulates in the system, which induces further monomer and oligomer generation via hydrolysis, and the formation of rehydrated products. As such, water is constantly generated via formylation and subsequently consumed in hydrolysis and rehydration, making the process self-sustained in nature. Close to 100% liquefaction was achieved even when the chitin loading was increased to 17 wt% (with respect to FA), indicating the remarkably high capacity of the FA in the functionalization and depolymerization of ball-milled chitin. Markedly, proto-chitin in raw shrimp shells could also be directly liquefied in FA without any pretreatment, thus offering great potential for practical applications.

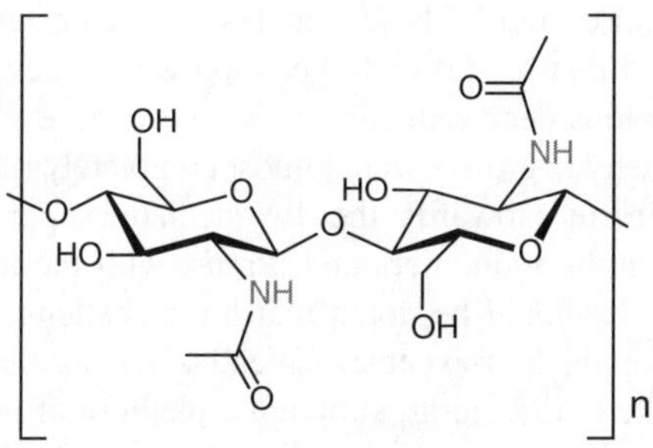

Figure 12.14 *The chemical structure of chitin.*

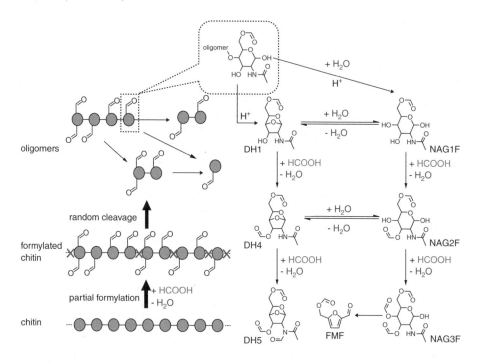

Figure 12.15   *Proposed major reaction pathways for FA-mediated chitin liquefaction [89].*

12.5   Upgrading of Bio-Oil and Related Model Compounds

One prospective method for producing fuels in the future is to produce bio-oil via biomass fast pyrolysis, wherein upgrading is desirable to remove the oxygen and in this way make it resemble crude oil. As is the case with many of the above-mentioned conversions of various forms of bio-derived feedstocks, it can also be envisioned that the FA-mediated reductive strategy could afford an efficient and cost-effective protocol for bio-oil upgrading under HDO conditions, in view of the central importance of FA as an alternative hydrogen source in implementing biorefinery concepts. Xiong *et al.* showed recently that for the processing of crude bio-oil in methanol with liquid FA, the properties of the liquid product can be significantly improved (Figure 12.16) [90]. Among various catalysts investigated, Ru appeared to be more effective than Pd- and Ni-based catalysts. Furthermore, the reaction routes of main components showed that partial hydrogenation was achieved without obvious coke or tar formation. No obvious deep reduction was seen to take place, but the compounds containing alkenyl or aldehyde groups were almost completely reduced.

Efforts were also directed towards the FA-mediated upgrading of bio-oil model compounds, such as phenolic monomers and furans, with the aim of understanding the underlying mechanisms. It should be noted that huge challenges and also opportunities are presented for developing less-expensive metal catalysts that are sufficiently stable and active for FA-mediated biorefining, although a plethora of precious metals – notably palladium and gold – have been shown to be particular active for FA dehydrogenation. In a case study focusing on the HDO of phenol and furfural via in-situ H_2 generated from FA,

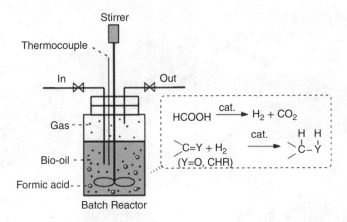

Figure 12.16   *Bio-oil hydroprocessing using FA as an in-situ hydrogen source [90].*

Zhang *et al.* reported that commercial Pd/C and Ni-based catalysts can be applied to give the corresponding hydrogenated products [91]. Their findings also pointed to a strong dependency on the type of catalyst used. Taking the case of FA-mediated phenol deoxygenation as an example, very poor results have been identified for Pd/C as compared to the Ni-based catalyst, despite its notably higher capability to promote selective FA dehydrogenation, mostly due to a lack of sufficient catalytic hydrogenation activity under the described conditions.

Most recently, the use of vanillin (4-hydroxy-3-methoxybenzaldehyde) as a model compound in FA-mediated biorefining has been studied by Xiao and coworkers [92]. Benefiting from the dual presence of Pd/TiO$_2$ sites for FA dehydrogenation and Pd/N-C sites for vanillin hydrogenation, a bifunctional Pd/TiO$_2$@N-C catalyst prepared by supporting Pd NPs onto a composite support comprising TiO$_2$ and nitrogen-modified porous carbon can achieve a full hydrogenation of vanillin into 2-methoxy-4-methyl-phenol (MMP) as sole product at 150 °C. A more efficient dual-functional Pd$_{50}$Ag$_{50}$/Fe$_3$O$_4$/N-rGO catalyst was reported by Singh *et al.* for the sequential tandem FA decomposition–hydrogenation of vanillin to obtain MMP under milder conditions (130 °C) [93]. In this case, the PdAg alloy surface is the main active site, and Fe$_3$O$_4$ particles promote the reaction by chemisorbing FA and transferring the resulting formate and hydrogen to the PdAg surface for hydrogenation process (Figure 12.17). Such synergy of individual catalytic phases might open an alternative route for designing and developing catalysts for future bio-oil upgrading.

12.6   FA as the Direct Feedstock for Bulk Chemical Synthesis

Beyond serving as a versatile feedstock for biomass upgrading, wherein FA is the nexus between many bioconversion pathways, FA is also considered a potential feedstock for direct use and process, for example, as an alternative raw material to produce methanol, which is currently produced from fossil fuels, such as natural gas, coal and oil products (e.g., heavy refinery residues, naphtha) [94–97]. Methanol is not only one of the most versatile and popular chemical commodities in the world, but is also considered as the key to weaning the world off oil in the future [98–101]. As a result of both the high demand for new reactions that utilize renewable resources, and the importance of methanol as the

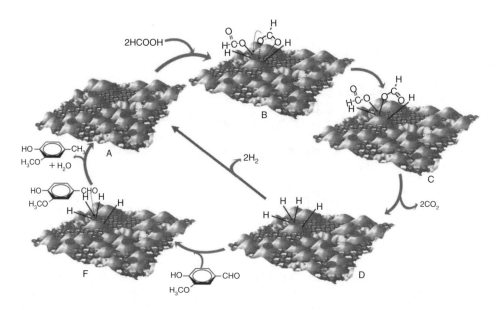

Figure 12.17 *Proposed pathways for tandem hydrodeoxygenation of vanillin with FA [93].*

basis of C_1 chemistry, it is highly desirable that a clean and sustainable methanol synthesis can make full use of renewable resources. In this context, the direct synthesis of methanol from biorenewable FA represents an attractive alternative to large-volume organic chemicals and renewable fuels. At this juncture, it may be recalled that Sabatier, in early studies, had already disclosed that some dimethoxymethane was produced upon the thermolysis of FA over ThO_2, thereby providing indirect evidence for a methanol-producing pathway.

Notwithstanding this, surprisingly few studies have been directed to obtain methanol directly from FA, presumably due to perceived problems associated with the unproductive decomposition of FA to produce CO_2/H_2 or CO/H_2O. In 2013, a pioneering publication by Miller and coworkers showed the potential for homogeneous catalysts to be employed for producing methanol directly via catalytic disproportionation of FA [102]. The use of a well-defined organoiridium catalyst such as [Cp*Ir -(bpy)(H$_2$O)][OTf]$_2$ (1, bpy = 2,2′-bipyridine, OTf = trifluoromethanesulfonate) was shown to be essential in promoting the FA-to-MeOH conversion at 80 °C. Albeit still very limited in the overall methanol yield (<2%), a notable feature of this reaction was that it could occur smoothly in an acidic aqueous solution (3 M, pH 1.4), without the need for any organic solvent or externally added hydrogen. Mechanistically, the reaction was proposed to proceed by hydrogen transfer from FA to the Ir catalyst, generating cationic Ir–H complexes. The subsequent reduction of protonated FA via the intermediacy of formaldehyde then occurred to yield methanol (Figure 12.18).

These studies were followed by a report from Cantat *et al.* which expanded on the use of Ru-based complexes for targeted methanol production via FA disproportionation [103]. In this latter case it was noted that, by using Ru(II) complexed by external phosphine ligands, it is possible to 'switch off' the competitive dehydrogenation of FA to H_2 and CO_2 (Figure 12.19). Interestingly, under relatively mild conditions (150 °C) in the presence of

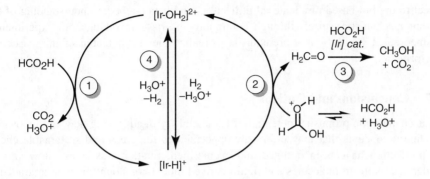

Figure 12.18   *Proposed pathways for the Ir-catalyzed disproportionation of FA to methanol [102].*

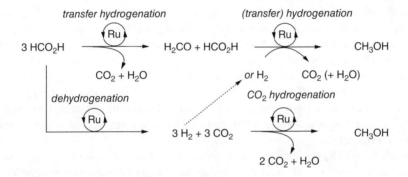

Figure 12.19   *Proposed pathways for the Ru-catalyzed disproportionation of FA to methanol [103].*

0.6 mol% [Ru(COD)(methylallyl)$_2$]+triphos and 1.5 mol% methanesulfonic acid (MSA), FA can be fully decomposed within 1 h, leading to the formation of MeOH in 50.2% yield. Regarding the essential factors that may play a role in facilitating the desired reaction pathway, one aspect that deserves special mention is that acid promoters such as MSA could significantly boost the catalytic efficiency of [Ru(COD)(methylallyl)$_2$]+triphos in FA disproportionation. In line with the detailed mechanistic studies investigating the methylation of amines with CO_2/H_2 by the groups of Beller [104], the crucial role of MSA can be attributed to its effectiveness in enabling an extremely facile formation of reactive Ru–H species.

The field of homogeneous metal-catalyzed disproportionation of FA to produce methanol, including its extension for the production of value-added chemicals from renewable sources, has continued to grow. While ongoing efforts are aimed at optimizing the process parameters necessary for further enhancing MeOH productivity, it can be anticipated that the successful design of an Earth-abundant metal catalyst, preferably in its heterogeneous form, with superior selectivity may ultimately render such FA-based C_1 interconversion processes a feasible option for renewable MeOH production. The results of a very recent study showed that, in the presence of dialkylboranes, FA can be converted to formaldehyde and methanol derivatives without a need for an external reductant [105]. However, the atom economy of these metal-free transformations is not maximized with respect to the bio-based raw material utilization, and hence a better understanding of the relevant mechanistic aspects through continuous efforts – involving both experimental and theoretical studies – is a prerequisite to realize the full potential of this particular transformation.

12.7   Conclusions and Outlook

In this chapter, the promising potential of FA chemistry in advancing new and innovative biorefinery concepts that may offer new opportunities for green and sustainable chemical production have been outlined and exemplified. First, it was shown how various bio-derived platform molecules and lignin-derived model compounds can be upgraded to a range of value-added compounds via FA-mediated HDO processing capable of offering efficient alternatives to conventional H_2-based processes. Some examples of these platform molecules include LA, glycerol, FAL, and HMF. In the presence of FA, these molecules can be transformed through controlled deoxygenation reactions to obtain tailor-made products and monofunctional intermediates suitable for further upgrading to produce renewable fuels and chemicals. It was indicated, with various representative examples, that the employment of safe and convenient FA as a hydrogen source can introduce alternative pathways other than those based on H_2 gas, which adds new space in regulating the product distribution. Therefore, when formulating a new processing strategy, it is important to consider what useful functional molecules can be obtained, how, and what type of hydrogen sources can be employed for controlled removal of oxygen from the bio-based feedstocks.

Throughout this chapter, examples were presented on how the feasibility of a new tandem catalytic process can be explored, by taking full advantage of the unique reaction chemistry of FA. As shown for the FA-mediated conversion of bio-derived furans to valuable ketoacids and diketones, it is important to determine the conditions for which the desired reaction is kinetically favorable. In addition, it is necessary to address selectivity issues for the desired product by identifying series or parallel competitive reactions. In order to maximize the yield of the desired products, key factors that control the selectivity should be identified.

Finally, a better understanding of the following aspects through continuous efforts is needed to realize the full potential of the FA-mediated bio-processing: (i) to identify the key factors determining the ability of a suitable catalytic material to activate FA in the presence of co-feeds; (ii) to identify the synergy between metal and acid–base sites; (iii) to determine the molecular mechanism(s) of the complex reactions that take place; and (iv) to determine the mechanism(s) through which solvents interact with catalytic sites and affect reaction rates.

References

1. Vispute, T.P., Zhang, H., Sanna, A., Xiao, R., Huber, G.W. (2010) Renewable chemical commodity feedstocks from integrated catalytic processing of pyrolysis oils. *Science*, **330** (6008), 1222–1227.
2. Olcay, H., Subrahmanyam, A.V., Xing, R., Lajoie, J., Dumesic, J.A., Huber, G.H. (2013) Production of renewable petroleum refinery diesel and jet fuel feedstocks from hemicellulose sugar streams. *Energy Environ. Sci.*, **6** (1), 205–216.

3. Li, G., Li, N., Li, S., Wang, A., Cong, Y., Wang, X., Zhang, T. (2013) Synthesis of renewable diesel with hydroxyacetone and 2-methyl-furan. *Chem. Commun.*, **49** (51), 5727–5729.

4. Xu, Z., Yan, P., Liu, K., Wan, L., Xu, W., Li, H., Liu, X., Zhang, Z.C. (2016) Synthesis of bis(hydroxylmethylfurfuryl)amine monomers from 5-hydroxy- methylfurfural. *ChemSusChem*, **9** (11), 1255–1258.

5. Chieffi, G., Braun, M., Esposito, D. (2015) Continuous reductive amination of biomass-derived molecules over carbonized filter paper-supported FeNi alloy. *ChemSusChem*, **8** (21), 3590–3594.

6. Gelmini, A., Albonetti, S., Cavani, F., Cesari, C., Lolli, A., Zanotti, V., Mazzoni, R. (2016) Oxidant free one-pot transformation of bio-based 2,5-bis-hydroxy- methylfuran into α-6-hydroxy-6-methyl-4-enyl-2H-pyran-3-one in water. *Appl. Catal. B*, **180**, 38–43.*

7. Li, G., Li, N., Wang, X., Sheng, X., Li, S., Wang, A., Cong, Y., Wang, X., Zhang, T. (2014) Synthesis of diesel or jet fuel range cycloalkanes with 2-methylfuran and cyclopentanone from lignocellulose. *Energy Fuels*, **28** (8), 5112–5118.

8. Li, G., Li, N., Yang, J., Li, L., Wang, A., Wang, X., Cong, Y., Zhang, T. (2014) Synthesis of renewable diesel range alkanes by hydrodeoxygenation of furans over Ni/Hβ under mild conditions. *Green. Chem.*, **16** (2), 594–599.

9. Jia, X., Ma, J., Wang, M., Ma, H., Chen, C., Xu, J. (2015) Catalytic conversion of 5-hydroxymethylfurfural into 2,5-furandiamidine dihydrochloride. *Green Chem.*, **18** (4), 974–978.

10. Müller, C., Diehl, V., Lichtenthaler, F.W. (1998) Building blocks from sugars. Part 23. Hydrophilic 3-pyridinols from fructose and isomaltulose. *Tetrahedron*, **54** (36), 10703–10712.

11. Lichtenthaler, F.W., Brust, A., Cuny, E. (2001) Sugar-derived building blocks. Part 26. Hydrophilic pyrroles, pyridazines and diazepinones from D-fructose and isomaltulose. *Green Chem.*, **3** (5), 201–209.

12. Zakrzewska, M.E., Bogel-Łukasik, E., Bogel-Łukasik, R. (2011) Ionic liquid-mediated formation of 5-hydroxymethylfurfural – a promising biomass-derived building block. *Chem. Rev.*, **42** (20), 397–417.

13. Corma, A., Iborra, S., Velty, A. (2007) Chemical routes for the transformation of biomass into chemicals. *Chem. Rev.*, **107** (6), 2411–2502.

14. Zhao, H., Holladay, J.E., Brown, H., Zhang, Z.C. (2007) Metal chlorides in ionic liquid solvents convert sugars to 5-hydroxymethylfurfural. *Science*, **316** (5831), 1597–1600.

15. Yong, G., Zhang, Y., Ying, J. Y. (2008) Efficient catalytic system for the selective production of 5-hydroxymethylfurfural from glucose and fructose. *Angew. Chem. Int. Ed.*, **47** (48), 9345–9348.

16. Karinen, R., Vilonen, K., Niemelä, M. (2011) Biorefining: heterogeneously catalyzed reactions of carbohydrates for the production of furfural and hydroxymethylfurfural. *ChemSusChem*, **4** (8), 1002–1016.

17. Grasemann, M., Laurenczy, G. (2012) Formic acid as a hydrogen source – recent developments and future trends. *Energy Environ. Sci.*, **5** (5), 8171–8181.

18. Reutemann, W., Kieczka, H. (2002) *Ullmann's Encyclopedia of Industrial Chemistry*, Wiley-VCH.

19. Calvert, J.G., Derwent, R.G., Orlando, J.J., Tyndall, G.S., Wallington, T.J. (2008) *Mechanisms of Atmospheric Oxidation of the Alkanes*, Oxford University Press.

20. Shreiber, E.H., Roberts, G.W. (2000) Methanol dehydrogenation in a slurry reactor: evaluation of copper chromite and iron/titanium catalysts. *Appl. Catal. B. Environ.*, **26** (2), 119–129.

21. Guerreiro, E.D., Gorriz, O.F., Larsen, G., Arrúa, L.A. (2000) Cu/SiO$_2$ catalysts for methanol to methyl formate dehydrogenation: A comparative study using different preparation techniques. *Appl. Catal A. Gen.*, **204** (1), 33–48.

22. Di Girolamo, M., Lami, M., Marchionna, M., Sanfilippo, D., Andreoni, M., Galletti, A.M., Sbrana, R.G. (1996) Methanol carbonylation to methyl formate catalyzed by strongly basic resins. *Catal. Lett.*, **38** (1), 127–131.

23. Kamm, B., Gruber, P.R., Kamm, M. (2006) *Biorefineries-Industrial Processes and Products: Status Quo and Future Directions*, Wiley-VCH.

24. Grasemann, M., Laurenczy, G. (2012) Formic acid as a hydrogen source – recent developments and future trends. *Energy Environ. Sci.*, **5** (5), 8171–8181.

25. Bi, Q.Y., Lin, J.D., Liu, Y.M., Du, X.L., He, H.Y., Cao, Y. (2014) An aqueous rechargeable formate-based hydrogen battery driven by heterogeneous Pd catalysis. *Angew. Chem. Int. Ed.*, **53** (49), 13583–13587.

26. Bi, Q.Y., Du, X.L., Liu, Y.M., Cao, Y., He, H.Y., Fan, K.N. (2012) Efficient subnanometric gold-catalyzed hydrogen generation via formic acid decomposition under ambient conditions. *J. Am. Chem. Soc.*, **134** (21), 8926–8933.

27. Konishi, H., Manabe, K. (2014) Formic acid derivatives as practical carbon monoxide surrogates for metal-catalyzed carbonylation reactions. *Synlett*, **45** (49), 1971–1986.

28. Shiramizu, M., Toste, F.D. (2012) Deoxygenation of biomass-derived feedstocks: oxorhenium-catalyzed deoxydehydration of sugars and sugar alcohols. *Angew. Chem. Int. Ed.*, **124** (32), 8082–8086.

29. Sanna, A., Andrésen, J.M. (2012) Bio-oil deoxygenation by catalytic pyrolysis: new catalysts for the conversion of biomass into densified and deoxygenated bio-oil. *ChemSusChem*, **5** (10), 1944–1957.

30. Dutta, S. (2012) Deoxygenation of biomass-derived feedstocks: hurdles and opportunities. *ChemSusChem* **2012**, **5** (11), 2125–2127.

31. Shi, D., Vohs, J.M. (2015) Deoxygenation of biomass-derived oxygenates: reaction of furfural on Zn-modified Pt (111). *ACS Catal.*, **5** (4), 2177–2183.

32. Nakagawa, Y., Tamura, M., Tomishige, K. (2013) Catalytic reduction of biomass-derived furanic compounds with hydrogen. *ACS Catal.*, **3** (12), 2655–2668.

33. Ota, N., Tamura, M., Nakagawa, Y., Okumura, K., Tomishige, K. (2015) Hydrodeoxygenation of vicinal OH groups over heterogeneous rhenium catalyst promoted by palladium and ceria support. *Angew. Chem. Int. Ed.*, **54** (6), 1897–1900.

34. Amada, Y., Ota, N., Tamura, M., Nakagawa, Y., Tomishige, K. (2014) Selective hydrodeoxygenation of cyclic vicinal diols to cyclic alcohols over tungsten oxide-palladium catalysts. *ChemSusChem*, **7** (8), 2185–2192.

35. Raju, S., Jastrzebski, J.T.B.H., Lutz, M., Klein Gebbink R.J.M. (2013) Catalytic deoxydehydration of diols to olefins by using a bulky cyclopentadiene-based trioxorhenium catalyst. *ChemSusChem*, **6** (9), 1673–1680.

36. Mascal, M., Dutta, S., Gandarias, I. (2014) Hydrodeoxygenation of the angelica lactone dimer, a cellulose-based feedstock: simple, high-yield synthesis of branched C7–C10 gasoline-like hydrocarbons. *Angew. Chem. Int. Ed.*, **53** (7), 1854–1857.

37. Xu, W., Miller, S.J., Agrawal, P.K., Jones, C.W. (2012) Depolymerization and hydrodeoxygenation of switchgrass lignin with formic acid. *ChemSusChem*, **5** (4), 667–675.

38. Thananatthanachon, T., Rauchfuss, T.B. (2010) Efficient production of the liquid fuel 2,5-dimethylfuran from fructose using formic acid as a reagent. *Angew. Chem. Int. Ed.*, **49** (37), 6616–6618.

39. Rahimi, A., Ulbrich, A., Coon, J.J., Stahl, S.S. (2014) Formic-acid-induced depolymerization of oxidized lignin to aromatics. *Nature*, **515** (7526), 249–252.

40. Indu, B., Ernst, W.R., Gelbaum, L.T. (1993) Methanol-formic acid esterification equilibrium in sulfuric acid solutions: influence of sodium salts. *Ind. Eng. Chem. Res.*, **32** (5), 981–985.

41. Jiang, N., Huang, R., Qi, W., Su, R., He, Z. (2012) Effect of formic acid on conversion of fructose to 5-hydroxymethylfurfural in aqueous/butanol media *Bioenerg. Res.*, **5** (2), 380–386.

42. Tao, L., Zhang, Q., Li, S.S., Liu, X., Liu, Y.M., Cao, Y. (2015) Heterogeneous gold-catalyzed selective reductive transformation of quinolines with formic acid. *Adv. Synth. Catal.*, **357** (4), 753–760

43. Yu, L., Zhang, Q., Li, S.S., Huang, J., Liu, Y.M., He, H.Y., Cao, Y. (2015) Gold-catalyzed reductive transformation of nitro compounds using formic acid: mild, efficient, and versatile. *ChemSusChem*, **8** (18), 3029–3035.

44. Wang, Y., Ren, W., Shi, Y. (2015) An atom-economic approach to carboxylic acids via Pd-catalyzed direct addition of formic acid to olefins with acetic anhydride as a co-catalyst. *Org. Biomol. Chem.*, **13** (31), 8416–8419.

45. Das, V.K., Devi, R.R., Raul, P.K., Thakur, A.J. (2012) Nano rod-shaped and reusable basic Al_2O_3 catalyst for N-formylation of amines under solvent-free conditions: A novel, practical and convenient 'NOSE' approach. *Green Chem.*, **14** (3), 847–854.

46. Zhang, Q., Li, S.S., Zhu, M.M., Liu, Y.M., He, H.Y., Cao, Y. (2016) Direct reductive amination of aldehydes with nitroarenes using bio-renewable formic acid as a hydrogen source. *Green Chem.*, **18** (8), 2507–2513.

47. Arceo, E., Ellman, J.A., Bergman, R.G. (2010) A direct, biomass-based synthesis of benzoic acid: formic acid-mediated deoxygenation of the glucose-derived materials quinic acid and shikimic acid. *ChemSusChem*, **3** (7), 811–813.

48. Brieger, G., Nestrick, T.J. (1974) Catalytic transfer hydrogenation. *Chem. Rev.*, **74** (5), 567–580.

49. Brieger, G., Nestrick, T.J., Fu, T.H. (1979) Catalytic transfer reduction: scope and utility. *J. Org. Chem.*, **44** (11), 1876–1878.

50. Cortese, N.A., Heck, R.F. (1978) Palladium-catalyzed reductions of α, β-unsaturated carbonyl compounds, conjugated dienes, and acetylenes with trialkylammonium formates. *J. Org. Chem.*, **43** (20), 3985–3987.

51. Johnstone, R.A.W., Wilby, A.H. (1981) Metal-assisted reactions – Part 101: Rapid, stereoselective and specific catalytic transfer reduction of alkynes to cis-alkenes. *Tetrahedron*, **37** (21), 3667–3670.

52. Weir, J.R., Patel, B.A., Heck, R.F. (1980) Palladium-catalyzed triethylammonium formate reductions. 4. Reduction of acetylenes to cis-monoenes and hydrogenolysis of tertiary allylic amines. *J. Org. Chem.*, **45** (24), 4926–4931.

53. Aramendía, M.A., Borau, V., Jiménez, C., Marinas, J.M., Ruiz, J.R., Urbano, F.J. (2001) Liquid-phase heterogeneous catalytic transfer hydrogenation of citral on basic catalysts. *J. Mol. Catal. A. Chem.*, **171** (1), 153–158.

54. Jae, J., Mahmoud, E., Lobo, R.F., Vlachos, D.G. (2014) Cascade of liquid-phase catalytic transfer hydrogenation and etherification of 5-hydroxymethylfurfural to potential biodiesel components over Lewis acid zeolites. *ChemCatChem*, **6** (2), 508–513.

55. Panagiotopoulou, P., Vlachos, D.G. (2014) Liquid phase catalytic transfer hydrogenation of furfural over a Ru/C catalyst. *Appl. Catal. A. Gen.*, **480**, 17–24.

56. Entwistle, I.D., Jonstone, R.A.W., Povall, T.J. (1975) Selective rapid transfer-hydrogenation of aromatic nitro-compounds. *J. Chem. Soc., Perkin Trans.*, **1** (13), 1300–1301.

57. Gowda, D.C., Mahesh, B. (2000) Catalytic transfer hydrogenation of aromatic nitro compounds by employing ammonium formate and 5% platinum on carbon. *Synth. Commun.* **30** (20), 3639–3644.

58. Gowda, S., Gowda, D.C. (2002) Application of hydrazinium monoformate as new hydrogen donor with Raney nickel: a facile reduction of nitro and nitrile moieties. *Tetrahedron*, **58** (11), 2211–2213.

59. Mohapatra, S.K., Sonavane, S.U., Jayaram, R.V., Selvam, P. (2002) Regio- and chemoselective catalytic transfer hydrogenation of aromatic nitro and carbonyl as well as reductive cleavage of azo compounds over novel mesoporous NiMCM-41 molecular sieves. *Org. Lett.*, **4** (24), 4297–4300.

60. Gandarias, I., Requies, J., Arias, P.L., Armbruster, U., Martin, A. (2012) Liquid-phase glycerol hydrogenolysis by formic acid over Ni-Cu/Al_2O_3 catalysts. *J. Catal.*, **290**, 79–89.

61. Liu, X., Lu, G., Guo, Y., Guo, Y., Wang, Y., Wang, X. (2016) Catalytic transfer hydrogenolysis of 2-phenyl-2-propanol over palladium supported on activated carbon. *J. Mol. Catal. A. Chem.*, **252** (1), 176–180.

62. Scholz, D., Aellig, C., Hermans, I. (2014) Catalytic transfer hydrogenation/ hydrogenolysis for reductive upgrading of furfural and 5-(hydroxymethyl) furfural. *ChemSusChem*, **7** (1), 268–275.

63. Gandarias, I., Arias, P.L., Fernández, S.G., Requies, J., El Doukkali, M., Güemez, M.B. (2012) Hydrogenolysis through catalytic transfer hydrogenation: Glycerol conversion to 1, 2-propanediol. *Catal. Today*, **195** (1), 22–31.

64. Kibby, C.L., Swift, H.E. (1976) Study of catalysts for cyclohexane-thiophene hydrogen transfer reactions. *J. Catal.*, **45** (2), 231–241.

65. Cortese, N.A., Heck, R.F. (1977) Palladium catalyzed reductions of halo-and nitroaromatic compounds with triethylammonium formate. *J. Org. Chem.*, **42** (22), 3491–3494.

66. Savourey, S., Lefévre, G., Berthet, J.C., Cantat, T. (2014) Catalytic methylation of aromatic amines with formic acid as the unique carbon and hydrogen source. *Chem. Commun.*, **50** (90), 14033–14036.

67. Huber, G.W., Corma, A. (2007) Synergies between bio- and oil refineries for the production of fuels from biomass. *Angew. Chem. Int. Ed.*, **46** (38), 7184–7201.

68. Du, X.L., He, L., Zhao, S., Liu, Y.M., Cao, Y., He, H.Y., Fan K.N. (2011) Hydrogen-Independent reductive transformation of carbohydrate biomass into γ-valerolactone and pyrrolidone derivatives with supported gold catalysts. *Angew. Chem. Int. Ed.*, **50** (34), 7815–7819.

69. Deng, L., Li, J., Lai, D.M., Fu, Y., Guo, Q.X. (2009) Catalytic conversion of biomass-derived carbohydrates into γ-valerolactone without using an external H_2 supply. *Angew. Chem. Int. Ed.*, **48** (35), 6529–6532.

70. Deng, L., Zhao, Y., Li, J., Fu, Y., Liao, B., Guo, Q.X. (2010) Conversion of levulinic acid and formic acid into γ-valerolactone over heterogeneous catalysts. *ChemSusChem*, **3** (10), 1172–1175.

71. Braden, D.J., Henao, C.A., Heltzel, J., Maravelias, C.C., Dumesic, J.A. (2011) Production of liquid hydrocarbon fuels by catalytic conversion of biomass-derived levulinic acid. *Green Chem.*, **13** (7), 1755–1765.

72. Yuan, J., Li, S.S., Yu, L., Liu, Y.M., Cao, Y., He, H.Y., Fan, K.N. (2013) Copper-based catalysts for the efficient conversion of carbohydrate biomass into γ-valerolactone in the absence of externally added hydrogen. *Energy Environ. Sci.*, **6** (11), 3308–3313.

73. Manzer, L.E. (2005) Production of 5-methyl-N-aryl-2-pyrrolidone and 5-methyl-N-cycloalkyl-2-pyrrolidone by reductive amination of levulinic acid with aryl amines, *US Patent* **6**, 743–819.

74. Huang, Y.B., Dai, J.J., Deng, X.J., Qu, Y.C., Guo, Q.X., Fu, Y. (2011) Ruthenium-catalyzed conversion of levulinic acid to pyrrolidines by reductive amination. *ChemSusChem*, **4** (11), 1578–1581.

75. Wei, Y., Wang, C., Jiang, X., Xue, D., Li, J., Xiao, J. (2013) Highly efficient transformation of levulinic acid into pyrrolidinones by iridium-catalysed transfer hydrogenation. *Chem. Commun.*, **49** (47), 5408–5410.

76. Fischer, R.H., Pinkos, R., Stein, F. (2002) Methods for Producing 1,6-Hexanediol and 6-Hydroxycaproic Acid or Their Esters, US Patent 6,426,438.

77. Tuteja, J., Choudhary, H., Nishimura, S., Ebitani, K. (2014) Direct synthesis of 1,6-hexanediol from HMF over a heterogeneous Pd/ZrP catalyst using formic acid as hydrogen source. *ChemSusChem*, **7** (1), 96–100.

78. Gandarias, I., Requies, J., Arias, P.L., Armbruster, U., Martin, A. (2012) Liquid-phase glycerol hydrogenolysis by formic acid over Ni-Cu/Al$_2$O$_3$ catalysts. *J. Catal.*, **290**, 79–89.

79. Yuan, J., Li, S.S., Yu, L., Liu, Y.M., Cao, Y. (2013) Efficient catalytic hydrogenolysis of glycerol using formic acid as hydrogen source. *Chin. J. Catal.*, **34** (11), 2066–2074.

80. Arceo, E., Marsden, P., Bergman, R.G., Ellman, J.A. (2009) An efficient didehydroxylation method for the biomass-derived polyols glycerol and erythritol. Mechanistic studies of a formic acid-mediated deoxygenation. *Chem. Commun.*, (23), 3357–3359.

81. Li, X., Zhang, Y. (2016) Highly efficient process for the conversion of glycerol to acrylic acid via gas phase catalytic oxidation of an allyl alcohol intermediate. *ACS Catal.*, **6** (1), 143–150.

82. Zhou, X., Rauchfuss, T.B. (2013) Production of hybrid diesel fuel precursors from carbohydrates and petrochemicals using formic acid as a reactive solvent. *ChemSusChem*, **6** (2), 383–388.

83. Dwivedi, A.D., Gupta, K., Tyagi, D., Rai, R.K., Mobin, S.M., Singh, S.K. (2015) Ruthenium and formic acid based tandem catalytic transformation of bioderived furans to levulinic acid and diketones in water. *ChemCatChem*, **7** (24), 4050–4058.

84. Huang, S., Mahmood, N., Tymchyshyn, M., Yuan, Z., Xu, C. (2014) Reductive de-polymerization of Kraft lignin for chemicals and fuels using formic acid as an in-situ hydrogen source. *Bioresource Technol.*, **171**, 95–102.

85. Toledano, A., Serrano, L., Balu, A.M., Luque, R., Pineda, A., Labidi, J. (2013) Fractionation of organosolv lignin from olive tree clippings and its valorization to simple phenolic compounds. *ChemSusChem*, **6** (3), 529–536.

86. Forchheim, D., Gasson, J.R., Hornung, U., Kruse, A., Barth, T. (2012) Modeling the lignin degradation kinetics in a ethanol/formic acid solvolysis approach. Part 2. validation and transfer to variable conditions *Ind. Eng. Chem. Res.*, **51** (46), 15053–15063.

87. Galkin, M.V., Sawadjoon, S., Rohde, V., Dawange, M., Samec, J.S.M. (2014) Mild heterogeneous palladium-catalyzed cleavage of β-O-4'-ether linkages of lignin model compounds and native lignin in air. *ChemCatChem*, **6** (1), 179–184.

88. Kim, S.K. (2010) *Chitin, Chitosan, Oligosaccharides and Their Derivatives: Biological Activities and Applications*, Taylor & Francis.

89. Zhang, J., Yan, N. (2016) Formic acid-mediated liquefaction of chitin. *Green Chem.*, **18** (18), 5050–5058.

90. Xiong, W.M., Fu, Y., Zeng, F.X., Guo, Q.X. (2011) An in situ reduction approach for bio-oil hydroprocessing. *Fuel Process Technol.*, **92** (8), 1599–1605.

91. Tan, Z.C., Xu, X.M., Liu, Y.G., Zhang, C.S., Zhai, Y.P., Peng, L., Zhang, R.Q. (2014) Upgrading bio-oil model compounds phenol and furfural with in situ generated hydrogen *Environ. Prog. Sustain. Energy*, **33** (3), 751–755.

92. Wang, L., Zhang, B., Meng, X., Su, D.S., Xiao, F.S. (2014) Hydrogenation of biofuels with formic acid over a palladium-based ternary catalyst with two types of active sites. *ChemSusChem*, **7** (6), 1537–1541.

93. Singh, A.K., Jang, S., Kim, J.Y., Sharma, S., Basavaraju, K.C., Kim, M.G., Kim, K.R., Lee, J.S., Lee, H.H., Kim, D.P. (2015) One-pot defunctionalization of lignin-derived compounds by dual-functional Pd$_{50}$Ag$_{50}$/Fe$_3$O$_4$/N-rGO catalyst. *ACS Catal.*, **5** (11), 6964–6972.

94. Arutyunov, V. (2014) *Direct Methane to Methanol: Foundations and Prospects of the Process*, Elsevier.

95. Hammond, C., Jenkins, R.L., Dimitratos, N., Lopez-Sanchez, J.A., Ab Rahim, M.H., Forde, M.M., Thetford, A., Murphy, D.M., Hagen, H., Stangland, E.E., Moulijn, J.M., Taylor, S.H., Willock, D.J., Hutchings, G.J. (2012) Catalytic and mechanistic insights of the low-temperature selective oxidation of methane over Cu-promoted Fe-ZSM-5. *Chem. Eur. J.*, **18** (49), 15735–15745.

96. Reubroycharoen, P., Yamagami, T., Vitidsant, T., Yoneyama, Y., Ito, M., Tsubaki, N. (2003) Continuous low-temperature methanol synthesis from syngas using alcohol promoters. *Energy Fuels*, **17** (4), 817–821.

97. Supp, E. (1990) *How to Produce Methanol from Coal*, Springer.

98. Kamarudin, S.K., Achmad, F., Daud, W.R.W. (2009) Overview on the application of direct methanol fuel cell (DMFC) for portable electronic devices. *Int. J. Hydrogen Energy*, **34** (16), 6902–6916.

99. Wasmus, S., Küver, A. (1999) Methanol oxidation and direct methanol fuel cells: A selective review. *J. Electroanal. Chem.*, **461** (1-2), 14–31.

100. Kakati, N., Maiti, J., Lee, S.H., Jee, S.H., Viswanathan, B., Yoon, Y.S. (2014) Anode catalysts for direct methanol fuel cells in acidic media: Do we have any alternative for Pt or Pt-Ru? *Chem. Rev.*, **114** (24), 12397–12429.

101. Zhao, X., Yin, M., Ma, L., Liang, L., Liu, C., Liao, J., Lu, T., Xing, W. (2011) Recent advances in catalysts for direct methanol fuel cells. *Energy Environ. Sci.*, **4** (8), 2736–2753.

102. Gladiali, S., Alberico, E. (2006) Asymmetric transfer hydrogenation: Chiral ligands and applications. *Chem. Soc. Rev.*, **35** (3), 226–236.

103. Talwar, D., Wu, X., Saidi, O., Salguero, N.P., Xiao, J. (2014) Versatile iridicycle catalysts for highly efficient and chemoselective transfer hydrogenation of carbonyl compounds in water. *Chem. Eur. J.*, **20** (40), 12835–12842.

104. Li, Y., Sorribes, I., Yan, T., Junge, K., Beller, M. (2013) Selective methylation of amines with carbon dioxide and H$_2$. *Angew. Chem. Int. Ed.*, **52** (46), 12156–12160.

105. Chauvier, C., Thuéry, P., Cantat, T. (2016) Metal-free disproportionation of formic acid mediated by organoboranes. *Chem. Sci.*, **7** (9), 5680–5685.

Index